Using History to Teach Mathematics

An International Perspective

Cover art: Frontispiece of *Introductio in Analysin Infinitorum,* Leonhard Euler, 1748.

ISBN: 0-88385-163-6
Library of Congress Catalog Card Number 00-103314

Printed in the United States of America

Current Printing
10 9 8 7 6 5 4 3 2

Using History to Teach Mathematics

An International Perspective

Victor J. Katz
Editor

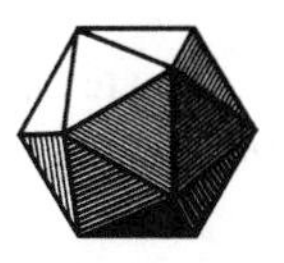

Published and Distributed by The Mathematical Association of America

The MAA Notes Series, started in 1982, addresses a broad range of topics and themes of interest to all who are involved with undergraduate mathematics. The volumes in this series are readable, informative, and useful, and help the mathematical community keep up with developments of importance to mathematics.

MAA Notes

11. Keys to Improved Instruction by Teaching Assistants and Part-Time Instructors, *Committee on Teaching Assistants and Part-Time Instructors, Bettye Anne Case,* Editor.
13. Reshaping College Mathematics, *Committee on the Undergraduate Program in Mathematics, Lynn A. Steen,* Editor.
14. Mathematical Writing, by *Donald E. Knuth, Tracy Larrabee, and Paul M. Roberts.*
16. Using Writing to Teach Mathematics, *Andrew Sterrett,* Editor.
17. Priming the Calculus Pump: Innovations and Resources, *Committee on Calculus Reform and the First Two Years,* a subcomittee of the Committee on the Undergraduate Program in Mathematics, *Thomas W. Tucker,* Editor.
18. Models for Undergraduate Research in Mathematics, *Lester Senechal,* Editor.
19. Visualization in Teaching and Learning Mathematics, *Committee on Computers in Mathematics Education, Steve Cunningham and Walter S. Zimmermann,* Editors.
20. The Laboratory Approach to Teaching Calculus, *L. Carl Leinbach et al.,* Editors.
21. Perspectives on Contemporary Statistics, *David C. Hoaglin and David S. Moore,* Editors.
22. Heeding the Call for Change: Suggestions for Curricular Action, *Lynn A. Steen,* Editor.
24. Symbolic Computation in Undergraduate Mathematics Education, *Zaven A. Karian,* Editor.
25. The Concept of Function: Aspects of Epistemology and Pedagogy, *Guershon Harel and Ed Dubinsky,* Editors.
26. Statistics for the Twenty-First Century, *Florence and Sheldon Gordon,* Editors.
27. Resources for Calculus Collection, Volume 1: Learning by Discovery: A Lab Manual for Calculus, *Anita E. Solow,* Editor.
28. Resources for Calculus Collection, Volume 2: Calculus Problems for a New Century, *Robert Fraga,* Editor.
29. Resources for Calculus Collection, Volume 3: Applications of Calculus, *Philip Straffin,* Editor.
30. Resources for Calculus Collection, Volume 4: Problems for Student Investigation, *Michael B. Jackson and John R. Ramsay,* Editors.
31. Resources for Calculus Collection, Volume 5: Readings for Calculus, *Underwood Dudley,* Editor.
32. Essays in Humanistic Mathematics, *Alvin White,* Editor.
33. Research Issues in Undergraduate Mathematics Learning: Preliminary Analyses and Results, *James J. Kaput and Ed Dubinsky,* Editors.
34. In Eves' Circles, *Joby Milo Anthony,* Editor.
35. You're the Professor, What Next? Ideas and Resources for Preparing College Teachers, *The Committee on Preparation for College Teaching, Bettye Anne Case,* Editor.
36. Preparing for a New Calculus: Conference Proceedings, *Anita E. Solow,* Editor.
37. A Practical Guide to Cooperative Learning in Collegiate Mathematics, *Nancy L. Hagelgans, Barbara E. Reynolds, SDS, Keith Schwingendorf, Draga Vidakovic, Ed Dubinsky, Mazen Shahin, G. Joseph Wimbish, Jr.*
38. Models That Work: Case Studies in Effective Undergraduate Mathematics Programs, *Alan C. Tucker,* Editor.
39. Calculus: The Dynamics of Change, *CUPM Subcommittee on Calculus Reform and the First Two Years, A. Wayne Roberts,* Editor.
40. Vita Mathematica: Historical Research and Integration with Teaching, *Ronald Calinger,* Editor.
41. Geometry Turned On: Dynamic Software in Learning, Teaching, and Research, *James R. King and Doris Schattschneider,* Editors.

42. Resources for Teaching Linear Algebra, *David Carlson, Charles R. Johnson, David C. Lay, A. Duane Porter, Ann E. Watkins, William Watkins,* Editors.
43. Student Assessment in Calculus: A Report of the NSF Working Group on Assessment in Calculus, *Alan Schoenfeld,* Editor.
44. Readings in Cooperative Learning for Undergraduate Mathematics, *Ed Dubinsky, David Mathews, and Barbara E. Reynolds,* Editors.
45. Confronting the Core Curriculum: Considering Change in the Undergraduate Mathematics Major, *John A. Dossey,* Editor.
46. Women in Mathematics: Scaling the Heights, *Deborah Nolan,* Editor.
47. Exemplary Programs in Introductory College Mathematics: Innovative Programs Using Technology, *Susan Lenker,* Editor.
48. Writing in the Teaching and Learning of Mathematics, *John Meier and Thomas Rishel.*
49. Assessment Practices in Undergraduate Mathematics, *Bonnie Gold,* Editor.
50. Revolutions in Differential Equations: Exploring ODEs with Modern Technology, *Michael J. Kallaher,* Editor.
51. Using History to Teach Mathematics: An International Perspective, *Victor J. Katz,* Editor.

MAA Service Center
P. O. Box 91112
Washington, DC 20090-1112
800-331-1622 fax: 301-206-9789

Preface

This book continues a long tradition of relating the history of mathematics to the teaching of mathematics. Throughout the twentieth century, mathematics educators at various levels have argued that the history of mathematics is a marvelous resource for motivating and exciting students studying mathematics. During this time, there has been a constant stream of articles showing in detail how to use history in teaching. And during the final decades of the twentieth century, there has in fact been an increasing use of history in the teaching of mathematics at all levels. In 1972, the International Commission on Mathematics Instruction (ICMI) approved the founding of an affiliated study group, called the International Study Group on the Relations Between History and Pedagogy of Mathematics (HPM). This group now has regular quadrennial meetings in connection with the International Congress on Mathematical Education (ICME), as well as other meetings, both in the U.S. and abroad. Because of the work of HPM, ICMI authorized an ICMI Study on the Role of the History of Mathematics in the Teaching of Mathematics. This study has resulted in a new volume in the ICMI Study Series which is appearing in August, 2000. Among other indications of increased interest in the use of history in the teaching of mathematics, we note that the National Science Foundation has funded several grant projects dealing with the use of the history of mathematics at both the undergraduate and the secondary level. These projects will result in several publications demonstrating to teachers how they can use history in the mathematics classroom.

The current volume brings together articles in the history of mathematics and its use in teaching, all based on papers presented at ICME in Seville, Spain and at the Quadrennial Meeting of HPM in Braga, Portugal, both in the summer of 1996. The latter meeting brought together close to five hundred teachers from around the world interested in using history with most having some experience in doing so. There was a great sharing of ideas, as usually occurs during an international meeting, and all who participated felt that they could go back home and convince their colleagues that the use of history was an idea whose time had come. The articles in this volume present but some of the fruits of those meetings. They vary in technical and/or teaching level and also in the level of generality. Some of the articles deal very specifically with how one can use history in the teaching of a particular topic, such as quadratic equations or the rank of a matrix. Others are more general and present an overview of reasons why history is useful in the mathematics classroom. And some articles deal primarily with the history of a particular area of mathematics, but including suggestions as to how one can incorporate that history in teaching those ideas.

Part One of this volume contains three articles dealing in general terms with the use of history in teaching. Man-Keung Siu gives us the ABCDs of using history - four general categories of how history has and can be used in the classroom. Frank Swetz shows us how the pedagogy of early mathematical texts can help in today's classroom. And Anne Michel-Pajus provides an example of the use of history in university classrooms in France.

In Part Two, four authors discuss historical ideas as they can influence the pedagogy in our classrooms. Lucia Grugnetti gives us some general examples, concentrating in particular on the work of Cavalieri. Wann-Sheng Horng compares the work of Euclid and Liu Hui on several geometric topics, drawing some interesting conclusions as to how the ideas of both can be used. Fulvia Furinghetti resurrects an old Italian journal for high school teachers and demonstrates how the articles in it can still be useful in today's teaching. And Frank Swetz shares some interesting historical problems which can be valuable additions to the repertory of any practicing teacher.

Part Three contains five articles dealing very specifically with the use of history in the teaching of a particular topic. Luis Radford and Georges Guerette give us details on how to use the geometric methods of the Babylonians and the Islamic mathematicians in teaching the quadratic formula. Janet Barnett illustrates how the historical resolution of several important anomalies in mathematics can be used to teach the relevant topic today. Evelyne Barbin takes us a tour of the teaching of geometry, showing how what is "obvious" has changed over the centuries. Jean-Luc Dorier gives us some results of a research study in the teaching of linear algebra, while describing how the use of some of Euler's original ideas clarify concepts with which today's students have difficulty. And Constantinos Tzanakis relates the teaching of physics to certain advanced mathematical topics through attention to their mutual history.

In Part Four we see how the history of mathematics has been successfully used in teacher training. Ian Isaacs, V. Mohan Ram, and Ann Richards describe a historically based course for preservice primary teachers in Australia, while Greisy Winicki demonstrates the use of the rule of false position to improve understanding of linear equations with prospective secondary teachers in Israel. Maxim Bruckheimer and Abraham Arcavi then describe a long-running project in Israel on developing teaching materials in several areas using original sources and guided readings.

The fifth and final part of the volume is devoted to articles on the history of mathematics, virtually all of which contain numerous suggestions as to how this history is applicable to the teaching of mathematics. Eleanor Robson presents the background to Mesopotamian mathematics, an area often missing from standard history books, while Man-Keung Siu provides a brief overview of ancient Chinese mathematics. George Heine takes us back to medieval Islam to look at the question of the value of mathematics. The Islamic answers are compared with modern ones. We then move to Renaissance Italy, where Uwe Gellert looks at the construction of a cathedral and shows how it reflects the mathematical knowledge of the day and the gradual shift to humanism. Luis Moreno and Guillermina Waldegg describe the history of the idea of "number" through the ages, an idea which must be communicated to students at various levels. Robin Wilson gives us a brief, though comprehensive, look at the history of combinatorics and provides a number of historical problems for students to solve. Torkil Heiede then presents the history of non-Euclidean geometry and tries to answer the question as to why a strong knowledge of this history is necessary for today's secondary school geometry teachers. Livia Giacardi's article is a complement to this one, presenting a more detailed history of Beltrami's contribution to the understanding of non-Euclidean geometry, while Gavin Hitchcock describes the world of Augustus DeMorgan through a scene from a mathematical play. The volume concludes with two preliminary studies of the history of mathematics in areas of the world still little studied. Jaime Carvalho, Antonio Duarte and Joao Queiro discuss the history of mathematics in Portugal, while Ubiratan D'Ambrosio provides the same service for South and Central America. In both cases we see how the study of mathematics was strongly impacted by the social and political currents of the day, even up through the twentieth century.

I wish to thank not only the contributors to this volume, but also the many others who refereed the papers and gave valuable advice both to the authors and to me. These referees include Tom Archibald, Marcia Ascher, Janet Barnett, Al Buccino, Ronald Calinger, Louis Charbonneau, Dan Curtin, Florence Fasanelli, John Fauvel, Alejandro Garciadiego, Fernando Gouvêa, Judith Grabiner, Charles Jones, Herb Kasube, Israel Kleiner, Stacy Langton, Karen Michalowicz, Daniel Otero, David Pengelley, Kim Plofker, Eleanor Robson, Amy Rocha, Ed Sandifer, Man-Keung Siu, Frank Swetz, Jim Tattersall, Erica Voolich, and Robin Wilson. Thanks are also due to the members of the MAA Notes Committee, including Dan Curtin, Phil Straffin, Tina Straley, Anita Solow, and Barbara Reynolds for their help in getting the book approved and for their detailed editorial work as well. Finally, I wish to thank the staff at the MAA, including Elaine Pedreira and Beverly Ruedi, for all their excellent, efficient work in producing the book.

Victor J. Katz
Washington, DC
May, 2000

Dedicated to Eduardo Veloso,
for all his magnificent work in
organizing the Braga meeting of HPM.

Contents

Part V: The History of Mathematics

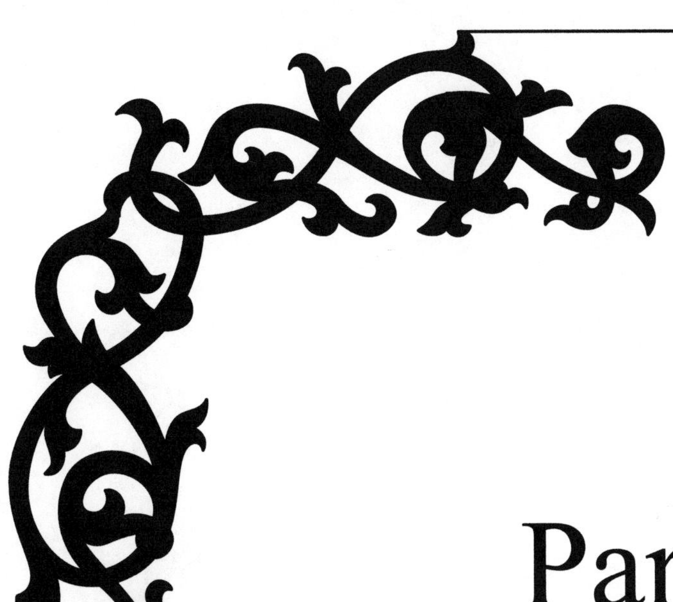

Part I

General Ideas on the Use of History in Teaching

The ABCD of Using History of Mathematics in the (Undergraduate) Classroom

Siu Man-Keung
University of Hong Kong

"The history of science is science itself."
—Johann Wolfgang von Goethe, *Theory of Colour (1808).*

Introduction

Mathematics is a human endeavour which has spanned over four thousand years; it is part of our cultural heritage; it is a very useful, beautiful and prosperous subject. In his Presidential Address delivered to the British Association for the Advancement of Science in 1897 Andrew Russ Forsyth (1858–1942) said, "Mathematics is one of the oldest of sciences; it is also one of the most active; for its strength is the vigour of perpetual youth." [1, Chapter VII] This quotation hints at a peculiar feature of mathematics, which other sciences do not seem to possess, or at least not to the same extent, viz. the past, the present and the future of the subject are intimately interrelated, making mathematics a cumulative science with its past forever assimilated in its present and future [2, 3]. No wonder in another Presidential Address to the British Association for the Advancement of Science in 1890, J.W.L. Glaisher (1848–1928) said that "no subject loses more than mathematics by any attempt to dissociate it from its history" [1, Chapter VI]. The great French mathematician Henri Poincaré (1854–1912) even said, "If we wish to forsee the future of mathematics, our proper course is to study the history and present condition of the science."

For many years now various authors in different parts of the world have written on the important role played by the history of mathematics in mathematics education. A good summary of some reasons for using the history of mathematics in teaching mathematics and of some ways in carrying it out can be found in [4, pp. 4–5]. Perhaps it can be added that not only does the appropriate use of the history of mathematics help in teaching the subject, but that in this age of "mathematics for all", history of mathematics is all the more important as an integral part of the subject to afford perspective and to present a fuller picture of what mathematics is to the public community.

Be that as it may, enough has been said on a propagandistic level. Some enthusiasts have already channeled their effort into actual implementation, resulting in a corpus of interesting material published in recent years in the form of books or collections of papers [5, 6, 7, 8, 9, 10, 11, 12, 13, 14]. With this in mind I wish to share with readers some of my experience in integrating the history of mathematics into the day-to-day teaching in the (undergraduate) classroom. Do not be misled by the title into thinking that this article is a guide to the use of the history of mathematics in the classroom! The letters A, B, C, D refer to four categories, or levels, of the use of the history of mathematics in the classroom: A for anecdotes, B for broad outline, C for content, and D for development of mathematical ideas. Except for the last category, which describes a course by itself, the first three categories represent three aspects of the use of the history of mathematics. Following a good practice in teaching, I shall illustrate each category with examples taken from actual classroom experience instead of just explaining what each category means in words. Even though such examples are admittedly piecemeal, I hope readers can still get an impression of how the four categories contribute to impart a sense of history in the study of mathematics in a varied and multifarious way.

A for Anecdotes

Everybody agrees that anecdotes about mathematics and mathematicians can contribute to the teaching of the subject in various ways. In the preface to his book [15], Howard Eves sums it up beautifully, "These stories and anecdotes have proved very useful in the classroom—as little interest-rousing atoms, to add spice and a touch of entertainment, to introduce a human element, to inspire the student, to instill respect and admiration for the great creators, to yank back flagging interest, to forge some links of cultural history, or to underline some concept or idea." (For more anecdotes readers can consult two more books of a similar title by the same author [16, 17].)

When we make use of anecdotes we usually brush aside the problem of authenticity. It may be strange to watch mathematicians, who at other times pride themselves upon their insistence on preciseness, repeat without hesitation apocryphal anecdotes without bothering one bit about their authenticity. However, if we realize that these are to be regarded as anecdotes rather than as history, and if we pay more attention to their value as a catalyst, then it presents no more problem than when we make use of a heuristic argument to explain a theorem. Besides, though many anecdotes have been embroidered over the years, many of them are based on some kind of real occurrence. Of course, an ideal situation is an authentic as well as amusing or instructive anecdote. Failing that we still find it helpful to have a good anecdote which carries a message.

There are plenty of examples of anecdotes which serve to achieve the aims set out in Eves' preface. I will give only two examples. The first example illustrates the function mentioned last in Eves' list—to underline some concept or idea. The second example, besides introducing a human element, illustrates that mathematics is not an isolated intellectual activity.

The first example is an anecdote about the German mathematician Hermann Amandus Schwarz (1843–1921), reported by Hans Freudenthal [18]. Schwarz, who was noted for his preciseness, would start an oral examination at the University of Berlin as follows.

Schwarz: Tell me the general equation of fifth degree.
Student: $x^5 + bx^4 + cx^3 + dx^2 + ex + f = 0$.
Schwarz: Wrong!
Student: ...where e is not the base of the natural logarithms.
Schwarz: Wrong!
Student: ...where e is *not necessarily* the base of the natural logarithms.

This anecdote, whether it is true, semitrue or even false, makes for a perfect appetizer to the main course of the general equation of degree m. It drives the point home as to how *special* a general equation is! I have made use of this anecdote several times in a (second) course on abstract algebra, and each time students love it. After listening to it, they appreciate much better the definition of a general equation of degree m to be given subsequently.

The second example is a real historical document, a letter dated March 6, 1832 from Carl Friedrich Gauss (1777–1855) to his friend Farkas Bolyai (1775–1856), seven weeks after receipt of the amazing work on non-Euclidean geometry by the latter's son, János Bolyai (1802–1860). We can imagine the dismay (but not without a trace of delight!) of the proud father when he read the letter which said, "If I commenced by saying that I am unable to praise this work [by János], you would certainly be surprised for a moment. But I cannot say otherwise. To praise it, would be to praise myself. Indeed the whole contents of the work, the path taken by your son, the results to which he is led, coincide almost entirely with my meditations, which have occupied my mind partly for the last thirty or thirty-five years. ...of which up till now I have put little on paper; my intention was not to let it be published during my lifetime. ...On the other hand it was my idea to write down all this later so that at least it should not perish with me. It is therefore a pleasant surprise for me that I am spared this trouble, and I am very glad that it is just the son of my old friend, who takes the precedence of me in such a remarkable manner." [19, p.100] From this passage we can unfold an interesting discussion on the interaction between philosophy and mathematics, and realize better how mathematics forms a "subculture" within a broader culture.

B for Broad Outline

It is helpful to give an overview of a topic or even of the whole course at the beginning, or to give a review at the end. That can provide motivation and perspective so that students know what they are heading for or what they have covered, and how that relates to knowledge previously gained. In either case we can look for ideas in the history of the subject, even though in some cases the actual path taken in history was much too tortuous to be recounted to pedagogical advantage.

One good example which permeates different levels in the study of mathematics is the concept of a function. (See [20] for a discussion of an attempt to incorporate this mathematical-historical vein into the teaching of mathematics at various levels, from secondary school to university.) Let me give a more "localized" (to just one subject) example here, that of the differential geometry of surfaces.

With the invention of calculus came its application in the study of plane curves and later space curves. One crucial description is captured in the notion of curvature with its several different but equivalent definitions. If we view the curvature κ as the rate of change of a turning tangent, then it is not surprising that, as proved by Abraham Gotthelf Kästner (1719–1800) in 1761,

$$\oint_C \kappa ds = 2\pi$$

for a simple closed curve C on the plane. In the 18th century, knowledge about space curves allowed mathematicians to study a surface S in space, notably its curvature, through the investigation of intersecting curves on S by planes through the normal at a point. Leonhard Euler (1707–1783) introduced the notion of principal curvatures κ_1, κ_2, which are the maximum and minimum values of the curvatures of sectional curves so obtained on a pair of mutually orthogonal planes. The product $\kappa = \kappa_1\kappa_2$ turns out to be of significance and is known as the Gaussian curvature, which can also be described through the "Gauss map", which measures how fast the surface bends away from the tangent plane by measuring the "dispersion of directions" of unit normal vectors at all points in a neighbourhood. Calculation of these quantities involves the use of coordinates, i.e., the surface S is regarded as something sitting in three-dimensional Euclidean space. For this reason we say that such quantities are extrinsically determined.

Mathematicians would like to talk about the intrinsic geometry of a surface, i.e., describe the surface as someone living on the surface without having to leave the surface and look at it from above or below! In a famous memoir of 1827 titled "Disquisitiones generales circa superficies curvas" (General investigation of curved surfaces) Gauss initiated this approach. (Remember that Gauss did a lot of survey work and mapmaking, and in those days one surveyed the terrain on the ground, not from the air!) The crucial notion is that of a geodesic, the line of shortest distance on the surface between two given points. Two surfaces which are applicable the one to the other by bending but without stretching, so that the distance between two given points remains the same, will have the same geometry. For instance, the geometry on a cylindrical surface will be the same as that on a plane surface, but will be different from that on a spherical surface. However, both the plane surface and the spherical surface enjoy a common property, viz. a small piece cut on each will be applicable to any other part on that same surface; in other words, they are both surfaces of constant curvature. Indeed, Gauss proved in his memoir that the (Gaussian) curvature κ is an intrinsic property, a result so remarkable that he named it "Theorema Egregium". He further showed that for a triangle $\triangle$ on S whose sides are geodesics,

$$\iint_\triangle \kappa dS = 2\pi - (\text{sum of exterior angles}).$$

(More generally, for a simple closed curve C on a surface S, the analogue to Kästner's result is

$$\oint_C \kappa_g ds = 2\pi - \iint_R \kappa dS,$$

where κ_g is the so-called geodesic curvature and R is the region on the surface bounded by C. For a geodesic triangle $\triangle$, the first integral becomes the sum of exterior angles.) This important result was later generalized by Pierre-Ossian Bonnet (1819–1892) in 1848 and by other mathematicians still later into the deep Gauss-Bonnet Theorem, which relates the topology of a surface to the integral of its curvature. (For the continued development initiated by the famous 1854 *Habilitationsvortrag* of Georg Friedrich Bernhard Riemann (1826–1866), readers can consult [21, Chapters 11–15].)

C for Content

In [22] David Rowe points out that a major challenge facing the history of mathematics as a discipline will be to establish a constructive dialogue between the "cultural historians" (those who approach mathematics as historians of science, ideas, and institutions) and the "mathematical historians" (those who study the history of mathematics primarily from the standpoint of modern mathematicians). In this connection one should also consult [23, 24, 25] to savour the different views held by some mathematicians and some historians of mathematics. I learn and benefit from both groups in my capacity as a teacher and student of mathematics, for I agree with what Charles Henry Edwards, Jr. says in the preface to his book [10], "Although the study of the history of mathematics has an intrinsic appeal of its own, its chief *raison d'être* is surely the illumination of mathematics itself ... to promote a more mature appreciation of [theories]." In this section I will give four examples borrowed from pages in the history of mathematics with an eye to the enhancement of understanding of the mathematics. This is a particularly pertinent function of the history of mathematics for a mathematics teacher's day-to-day work.

The first example has appeared in [26], which is in turn gleaned from [27, Appendix I]. (It also appears as one example in [28].) In 1678 Gottfried Wilhelm Leibniz (1646–1716) announced a "law of continuity" which said that if a variable at all stages enjoyed a certain property, then its limit would enjoy the same property. Up to the early 19th century mathematicians still held this tenet, so that Augustin-Louis Cauchy (1789–1857) might have been guided by it to arrive at the following result in 1821: If $\{f_n\}$ is a sequence of continuous functions with limit f, i.e., $\lim_{n\to\infty} f_n(x) = f(x)$, then f is a continuous function. Whenever I teach a calculus class I present Cauchy's "proof" to the class as follows. For sufficiently large n, $|f_n(x) - f(x)| < \varepsilon$. For sufficiently large n, $|f_n(x+h) - f(x+h)| < \varepsilon$. Choose a specific n so that both inequalities hold, then

$$|f_n(x) - f(x)| + |f_n(x+h) - f(x+h)| < 2\varepsilon \ .$$

For this chosen f_n, we have $|f_n(x+h) - f_n(x)| < \varepsilon$ for sufficiently small $|h|$, since f_n is continuous at x. Hence, for sufficiently small $|h|$, we have

$$\begin{aligned} |f(x+h) - f(x)| &\le |f(x+h) - f_n(x+h)| \\ &\quad + |f_n(x+h) - f_n(x)| \\ &\quad + |f_n(x) - f(x)| \\ &< 3\varepsilon. \end{aligned}$$

With ε being arbitrary to begin with, this says that f is continuous at x. (A picture will make the argument even more convincing!)

While many students are still nodding their heads, I tell them that Jean Baptiste Joseph Fourier (1768–1830) at about the same time showed that certain very discontinuous functions could be represented as limits of trigonometric polynomials! In hindsight we see that the work of Fourier provided counter-examples to the "theorem" of Cauchy. But at the time it was not regarded in this light. Actually, when the Norwegian mathematician Niels Henrik Abel (1802–1829) offered in his memoir of 1826 the example

$$\sin\phi - \sin 2\phi/2 + \sin 3\phi/3 - \cdots,$$

he remarked that "it seems to me that there are some exceptions to Cauchy's theorem" and asked instead what "the safe domain of Cauchy's theorem" should be [27, Appendix I]. Abel resolved the puzzle by restricting attention to the study of power series, but in so doing, missed an opportunity to investigate the way infinite series (of functions) converge.

I ask the class to wrestle with the "proof" of Cauchy and see what is amiss. If they cannot spot it, I tell them not to feel bad since Cauchy could not spot it either, and it was left to Philipp Ludwig von Seidel (1821–1896) to find out the mistake twenty-six years later! Rectification of the proof led later to the new notion of uniform convergence explicitly explained by Karl Theodor Wilhelm Weierstrass (1815–1897). With this historical overture we pass naturally on to a discussion of the mathematics of uniform convergence. (See [5, Chapter 5; 12, Chapter III.4] for an enlightening discussion of the mathematics.)

The second example has been used several times in an advanced elective course on algebra. It started with an announced "proof" of Fermat's Last Theorem by Gabriel Lamé (1795–1870) in the meeting of the Paris Academy on March 1, 1847. The key step lies in the factorization

$$z^p = x^p + y^p = (x+y)(x+y\zeta)\cdots(x+y\zeta^{p-1})$$

in the ring of cyclotomic integers $\mathbb{Z}[\zeta]$ (modern terminology), where ζ is a primitive pth root of unity. For an interesting account of the pursuit of this question in subsequent meetings of the Paris Academy, readers can consult [29, Chapter 4]. The account includes the deposit of "secret packets" with the Academy by Cauchy and Lamé—an institution of the Academy which allowed members to go on record as having been in possession of certain ideas at a certain time without revealing the content, in case a priority dispute developed later. The packets remained secret and the matter was put to rest when Joseph Liouville (1809–1882) read a letter from his friend Ernst Eduard Kummer (1810–1893) in the meeting of the Paris Academy on May 24, 1847. In the letter Kummer pointed out that the "proof" broke down owing to failure of unique factorization in $\mathbb{Z}[\zeta]$ in general. He even included a copy of his memoir, published three years earlier, in which he demonstrated that unique factorization failed for $p = 23$. He went on to say that he could save unique factorization by introducing a new kind of complex number he christened "ideal complex numbers". With suitably chosen illustrative examples to supplement the story, this is a natural point to launch into a detailed discussion on the unique factorization of ideals in a Dedekind domain.

The third example is also on algebra. It concerns the basic result known as the Chinese Remainder Theorem. I will skip both the statement of the result in the language of abstract algebra and the origin of the result found in Problem 26 of Chapter 3 of *Sunzi Suanjing* (Master Sun's Mathematical Manual, c. 4th century), which can be found in most textbooks, such as [30]. I will also skip the application of this type of problem, viz. $x \equiv a_i \bmod m_i$, $i \in \{1, 2, \ldots, N\}$, in ancient Chinese calendrical reckoning. I will only highlight what I would do next after going through the two aforesaid issues with the class. I discuss with them an algorithmic method devised by Qin Jiushao (1202–1261), known as the "Dayan art of searching for unity" and explained in his book *Shushu Jiuzhang* (Mathematical Treatise in Nine Sections) of 1247. It is instructive to see how to find a set of "magic numbers" from which a general solution can be built by linear combination. It suffices to solve separately single linear congruence equations of the form $kb \equiv 1 \bmod m$, by putting $m = m_i$ and $b = (m_1 \cdots m_N)/m_i$. The key point in Qin's method is to find a sequence of ordered pairs (k_i, r_i) such that $k_i b \equiv (-1)^i r_i \bmod m$ and the r_i's are strictly decreasing. At some point $r_s = 1$ but $r_{s-1} > 1$. If s is even, then $k = k_s$ will be a solution. If s is odd, then $k = (r_{s-1} - 1)k_s + k_{s-1}$ will be a solution. This sequence of ordered pairs can be found by using "reciprocal subtraction" (known as the Euclidean algorithm in the West), viz. $r_{i-1} = r_i q_{i+1} + r_{i+1}$ with $r_{i+1} < r_i$, and $k_{i+1} = k_i q_{i+1} + k_{i-1}$. If one looks into the calculation actually performed at the time, one will find that the method is even more streamlined and convenient. Consecutive pairs of numbers are put at the four corners of a counting board, starting with

$$\boxed{\begin{array}{cc} 1 & b \\ 0 & m \end{array}}, \quad \text{ending in} \quad \boxed{\begin{array}{cc} k & 1 \\ * & * \end{array}}.$$

The intermediate processes are shown below:

$$\boxed{\begin{array}{cc} k_i & r_i \\ k_{i-1} & r_{i-1} \end{array}} \longrightarrow \boxed{\begin{array}{cc} k_i & r_i \\ k_{i-1} & r_{i+1} \end{array}}$$
$$\longrightarrow \boxed{\begin{array}{cc} k_i & r_i \\ k_{i+1} & r_{i+1} \end{array}} \quad \text{if } i \text{ is even },$$

or

$$\boxed{\begin{array}{cc} k_{i-1} & r_{i-1} \\ k_i & r_i \end{array}} \longrightarrow \boxed{\begin{array}{cc} k_{i-1} & r_{i+1} \\ k_i & r_i \end{array}}$$
$$\longrightarrow \boxed{\begin{array}{cc} k_{i+1} & r_{i+1} \\ k_i & r_i \end{array}} \quad \text{if } i \text{ is odd }.$$

The procedure is stopped when the upper right corner becomes a 1, hence the name "searching for unity". Students will be amazed by noting how the procedure outlined in *Shushu Jiuzhang* can be phrased word for word as a computer program!

The fourth example is about the Cayley-Hamilton theorem taught in a linear algebra course, viz. $\chi(A) = 0$ where $\chi(X)$ is the characteristic polynomial of the $n \times n$ matrix A. First I show students a letter dated November 19, 1857 from Arthur Cayley (1821–1895) to James Joseph Sylvester (1814–1897) [31, pp. 213–214]. The letter illustrated the theorem by exhibiting a concrete 2×2 case. This is particularly pertinent for an average student, who at this stage may even be confused by the mere statement of the result, not to mention the explanation of why it is true. Hence I emphasize the point in class by repeating the words made by Cayley himself (in 1858), "The determinant, having for its matrix a given matrix less the same matrix *considered as a single quantity* (italics mine) involving the matrix unity, is equal to zero." To drive the point home I continue to produce a "joke-proof": set $X = A$ in the expression $\det(A - XI)$, hence $\det(A - AI) = \det 0 = 0$, which means A satisfies the characteristic polynomial $\chi(X) = \det(A - XI)$. Students are requested to find out why this is not a valid proof. I will give a valid proof in the next lecture, and for the more mathematically oriented students I may further explain how to turn the "joke-proof" into a rigorous proof by regarding both $\chi(X)$ and $A - XI$ as polynomials over the ring of $n \times n$ matrices (or more precisely, as polynomials over the *commutative* subring generated by A). My experience tells me that students are stimulated into discussion by the letter of Cayley, perhaps because they see from it that mathematicians do not work alone but talk shop with each other and engage in social interaction. Students are particularly "sympathetic" to the statement made by Cayley in his 1858 memoir after demonstrating the theorem for a 2×2 matrix: "I have not thought it necessary to undertake the labour of a formal proof of the theorem in the general case of a matrix of any degree." [32, p.624] As teachers we know how best to handle this sentiment!

D for Development of Mathematical Ideas

"Development of Mathematical Ideas" is the title of a course I have been teaching at my university since 1976. As an elective course for upper-level mathematics students (with an occasional few in other majors) with a moderate class size of around twenty, the course does not have a fixed syllabus nor a fixed format in teaching and assessment, thus allowing me to try out freely new approaches and new teaching material from year to year. Some past experience has been reported in [33, 34].

For the academic year 1995–96 I built the course around the anthology "Classics of Mathematics" edited by Ronald Calinger [35], which became more readily available as a textbook through its re-publication in 1995. The idea is to let students read some selected primary source material and to "learn from the masters". The year-long course was roughly divided into five sections: (1) Euclid's *Elements*, (2) Mathematical Thinking, (3) From Pythagoras, Eudoxus, . . . (Incommensurable Magnitudes) to Dedekind, Cantor, . . . (Real Numbers), (4) Non-Euclidean Geometry, (5) Gödel's Incompleteness Theorem. Passages in [35] were fitted into these five sections. Besides the primary source material, some of the general historical accounts (named "Introduction" of each chapter) make for useful assigned reading, to be supplemented by a general text such as [32]. Lectures were devoted to a more in-depth discussion, with more emphasis on the mathematics. I needed to add some extra source material from time to time, especially material on ancient Chinese mathematics. For instance, in the part on mathematical thinking I tried to let students experience, through the writings of mathematicians such as Liu Hui (c.250), Yang Hui (c.1250), Leonhard Euler (1707–1783), Julius Wilhelm Richard Dedekind (1831–1916), Henri Poincaré (1854–1912), and George Pólya (1887–1985), how working mathematicians go about their jobs. Students would learn that the logical and axiomatic approach exemplified in Euclid's *Elements* is not the only way. The textbook by Calinger [35], with its extensive bibliography, also provides useful support for the project work (in groups of two), which consists of an oral presentation and a written report on a topic of the students' choice. The course itself is in fact the presentation of my project work!

Conclusion

Using history of mathematics in the classroom does not necessarily make students obtain higher scores in the subject overnight, but it can make learning mathematics a meaningful and lively experience, so that (hopefully) learning will come easier and will go deeper. The awareness of this evolutionary aspect of mathematics can make a teacher more patient, less dogmatic, more humane, less pedantic. It will urge a teacher to become more reflective, more eager to learn and to teach with an intellectual commitment. I can attest to the benefits brought by the use of history of mathematics through my personal experience. The study of history of mathematics, though it does not make me a better mathematician, does make me a happier man who is ready to appreciate the multi-dimensional splendour of the discipline and its relationship to other cultural endeavours. It does enhance the joy derived from my job as a mathematics teacher when I try to share this kind of feeling with my class. I attempt to sow the seeds of appreciation of mathematics as a cultural endeavour in them. It is difficult to tell when these seeds will blossom forth, or whether they ever will. But the seeds are there, and I am content. I like the view proclaimed by the noted historian of science George Sarton (1884–1956), who said, "The study of the history of mathematics will not make better mathematicians but gentler ones; it will enrich their minds, mellow their hearts, and bring out their finer qualities." [36, p. 28]

References

[1] R.E. Moritz, *Memorabilia Mathematica: The Philomath's Quotation Book*, Mathematical Association of America, Washington, D.C., 1993; original edition in 1914; reprinted by Dover, New York, 1958.

[2] F.K. Siu, M.K. Siu, History of mathematics and its relation to mathematical education, *Int. J. Math. Educ. Sci. Technol.* 10 (1979), pp. 561–567.

[3] M.K. Siu, Implications and inspirations from the history of mathematics (in Chinese), *Dousou Bimonthly* 17 (1976), pp. 46–53.

[4] J. Fauvel, Using history in mathematics education, *For the Learning of Mathematics* 11(2) (1991), pp. 3–6.

[5] D. Bressoud, *A Radical Approach to Real Analysis*, Mathematical Association of America, Washington, D.C., 1994.

[6] R. Calinger (ed.), *Vita Mathematica: Historical Research and Integration with Teaching*, Mathematical Association of America, Washington, D.C., 1996.

[7] Commission Inter-IREM Epistémologie et Histoire des Mathématiques, *Histoires de Problèmes, Histoire des Mathématiques*, IREM de Lyon, 1992.

[8] Commission Inter-IREM Epistémologie et Histoire des Mathématiques, *History and Epistemology in Mathematics Education: First European Summer University Proceedings*, IREM de Montpellier, 1995.

[9] Commission Inter-IREM Epistémologie et Histoire des Mathématiques, *Contribution à une Approche Historique de l'Enseignement des Mathématiques: Actes de la 6ème Université d'Eté Interdisciplinaire sur l'Histoire des Mathématiques*, IREM de Besancon, 1996.

[10] C.H. Edwards, Jr., *The Historical Development of the Calculus*, Springer-Verlag, New York, 1979; 3rd printing, 1994.

[11] J. Fauvel (ed.), *History in the Mathematics Classroom: the IREM Papers*, Mathematical Association, London, 1990.

[12] E. Hairer, G. Wanner, *Analysis By Its History*, Springer-Verlag, New York, 1996.

[13] S. Nobre (ed.), *Proceedings of the Meeting of the International Study Group on Relations Between History and Pedagogy of Mathematics*, Blumenau, 1995.

[14] F. Swetz *et al* (ed.), *Learn From the Masters*, Mathematical Association of America, Washington D.C., 1995.

[15] H.W. Eves, *In Mathematical Circles: A Selection of Mathematical Stories and Anecdotes*, Quadrants I, II, III, IV, Prindle, Weber & Schmidt, Boston, 1969.

[16] H.W. Eves, *Mathematical Circles Revisited: A Second Collection of Mathematical Stories and Anecdotes*, Prindle, Weber & Schmidt, Boston, 1971.

[17] H.W. Eves, *Return to Mathematical Circles: A Fifth Collection of Mathematical Stories and Anecdotes*, PWS-Kent Publishing, Boston, 1988.

[18] H. Freudenthal, A bit of gossip: Schwarz, *Math. Intelligencer* 7(1) (1985), p.79.

[19] R. Bonola, *Non-Euclidean Geometry: A Critical and Historical Study of Its Development*, first English translation by H.S. Carslaw, Open Court Publishing Company, 1912; reprinted by Dover, New York, 1955.

[20] M.K. Siu, Concept of function—Its history and teaching, in [14], pp. 105–121.

[21] J. Gray, *Ideas of Space: Euclidean, Non-euclidean and Relativistic*, Clarendon Press, Oxford, 1989.

[22] D.E. Rowe, New trends and old images in the history of mathematics, in [6], pp. 3–16.

[23] J. Dauben, Mathematics: An historian's perspective, *Philosophy and the History of Science* 2 (1993), pp. 1–21; also in *The Intersection of History and Mathematics*, C. Sasaki *et al* (eds.), Birkhäuser, Basel, 1994, pp. 1–13.

[24] I. Grattan-Guinness, A residual category: Some reflections on the history of mathematics and its status, *Math. Intelligencer* 15(4) (1993), pp. 4–6.

[25] A. Weil, History of mathematics, why and how, *Proc. Intern. Cong. Math. at Helsinki*, 1978, pp. 227–236.

[26] M.K. Siu, History of [(Mathematics)] Teachers, Mathematical Tall Timbers, *Supp. Math. Log* 11 (1985); French translation in *Bull. de l'Association des Prof. de Math.* n° 354 (1985), pp. 309–319.

[27] I. Lakatos, *Proofs and Refutations: The Logic of Mathematical Discovery*, Cambridge University Press, Cambridge, 1976.

[28] V.F. Rickey, My favorite ways of using history in teaching calculus, in [14], pp. 123–134.

[29] H.M. Edwards, *Fermat's Last Theorem: A Genetic Introduction to Algebraic Number Theory*, Springer-Verlag, New York, 1977.

[30] Y. Li, S.R. Du, *Chinese Mathematics: A Concise History* (translated by J.N. Crossley and A.W.C. Lun from the original Chinese edition, Zhonghua, Beijing, 1963), Clarendon Press, Oxford, 1987.

[31] T. Crilly, Cayley's anticipation of a generalized Cayley-Hamilton theorem, *Historia Mathematica* 5 (1978), pp. 211–219.

[32] V.J. Katz, *A History of Mathematics: An Introduction*, Harper Collins, New York, 1993.

[33] K.T. Leung, M.K. Siu, A course on development of mathematical ideas (in Chinese), *Dousou Bimonthly* 41 (1980), pp. 38–44; also in *Retrospect and Outlook on Mathematics Education in Hong Kong* (in Chinese), M.K. Siu (ed.), Hong Kong University Press, Hong Kong, 1995, pp. 57–68.

[34] M.K. Siu, Mathematical thinking and history of mathematics, in [14], pp. 279–282.

[35] R. Calinger (ed.), *Classics of Mathematics*, original edition by Moore Publishing Company, 1982; Prentice-Hall, Englewood Cliffs, 1995.

[36] G. Sarton, *The Study of the History of Mathematics and the Study of the History of Science*, reprinted by Dover, New York, 1957; originally published in 1936.

This article appeared in Volume 1 (1998) of the *Bulletin of the Hong Kong Mathematical Society* and is reprinted by permission of the copyright holder, Gordon and Breach Publishers, Lausanne, Switzerland.

Mathematical Pedagogy: An Historical Perspective

Frank Swetz
The Pennsylvania State University

Introduction

Old mathematical texts can tell us many things. Certainly they provide information on the development of mathematical knowledge and procedures, the uses of mathematics, and the types of problems that were important to our forebearers. They provide insights into the culture and times within which they were written and give us hints as to the forces that shaped and controlled mathematical concerns. But if we look beyond the mathematics itself, and attempt to discern the author's intentions, "What is he attempting to teach?", "How is he doing it?", a perspective of early mathematical pedagogy emerges.

An examination and analysis of didactical trends in historical material can take place along several lines:

1. The organization of material; the sequential ordering of topics and specific problems.
2. The use of an instructional discourse and techniques of motivation contained within the discourse.
3. A use of visual aids; diagrams, illustrations and colors, to assist in the grasping of concepts on the part of the learner.
4. The employment of tactile aids, either directly or by reference, to clarify a mathematical concept.

It is impossible, in a limited discussion of this nature, to consider all of these aspects in some historical depth, but I would like to survey a few examples of pedagogic practices evident in old texts. Hopefully, other researchers will pursue more detailed investigations.

The Organization and Format of Mathematical Presentations

The teaching of mathematics has a structure that proceeds from the simple to the complex, from the concrete to the abstract. For example, contemporary school children are introduced to the natural numbers before they encounter the concept of integers; simple fractions before rational numbers; geometric proofs involving triangles before those considering circles and circular relations and so on. It appears that authors of mathematical texts have always followed such a scheme. British Museum cuneiform tablet 15285 from the Old Babylonian period (1800–1600 BCE) contains a series of geometrical diagrams. Each diagram presents a problem to its viewer. It is believed that this tablet originally contained over 40 systematically arranged exercise problems; however, only 30 are wholly or partially preserved.[1] These problems are reminiscent of present-day geo- or peg-board exercises. Early Assyriologists who studied the tablet and its contents were confused as to its purpose. Initially it was described as a surveyors manual; however, later interpreters determined that it is a series of exercises for mathematical scribes.[2] Each problem involves a square whose side measures 1 ÚS,[3] and each square is partitioned into smaller regions by the use of straight lines and circular arcs. The partitioning is accomplished by the use of equal and symmetric divisions of the area. Accompanying text enumerates the resulting regions and refers to them by name. Students are requested "to put down" or "draw" and "to touch" the regions in question. Whether this latter direction refers to the physical sense of touch, urging a multi-sensual approach to problem solving, or merely indicates an intellectual "touch" i.e., think about, is open to speculation.

Much of the Babylonian shape-designating terminology is readily translatable into familiar figures; thus the tablet's author speaks of squares, rectangles and circles but other terms describe "double bows", "ox eyes" and "deep-going boats" and require the use of a modern reader's imagination. One particular class of regions known to the Babylonians as *abusamikku* is especially interesting (Figure 1e). A modern viewer might describe them as concave equilateral triangles or squares bounded by the tangents of three or four congruent circles. Finding the area of such regions makes its historical debut as a problem in BM 15285. Later consideration of these concave regions would appear in the works of Heron of Alexandria (c. 75 CE) and the Indian mathematician Mahavira (c. 850). The tablet's student user is required to find the area of a specific region within each square (my black shadings in Figure 1). Geometric intuition and problem solving skills are challenged. The problems are sequenced from the simple to the complex and a student-learner must work his way through prerequisite problems before the later, more complex, problems can be solved. It appears evident that the author of the tablet sequenced these problems in a pedagogically purposeful manner.

It is also evident in the Egyptian *Rhind Papyrus* of 1650 BCE that a series of problems has been arranged in a controlled order to facilitate learning. Problems 41–60 of the Chace translation concern geometry.[4] Computations of volume are required (problems 41–46); area calculations follow (problems 48–55); finally, problems 56–60 require the application of triangle knowledge in work with pyramids. Just as children interact with three-dimensional solids before they appreciate the geometrical properties of plane shapes, the author of the Rhind problems has his scribes consider simple problems of volume before they attempt more intricate calculations involving triangles and applications of triangles.

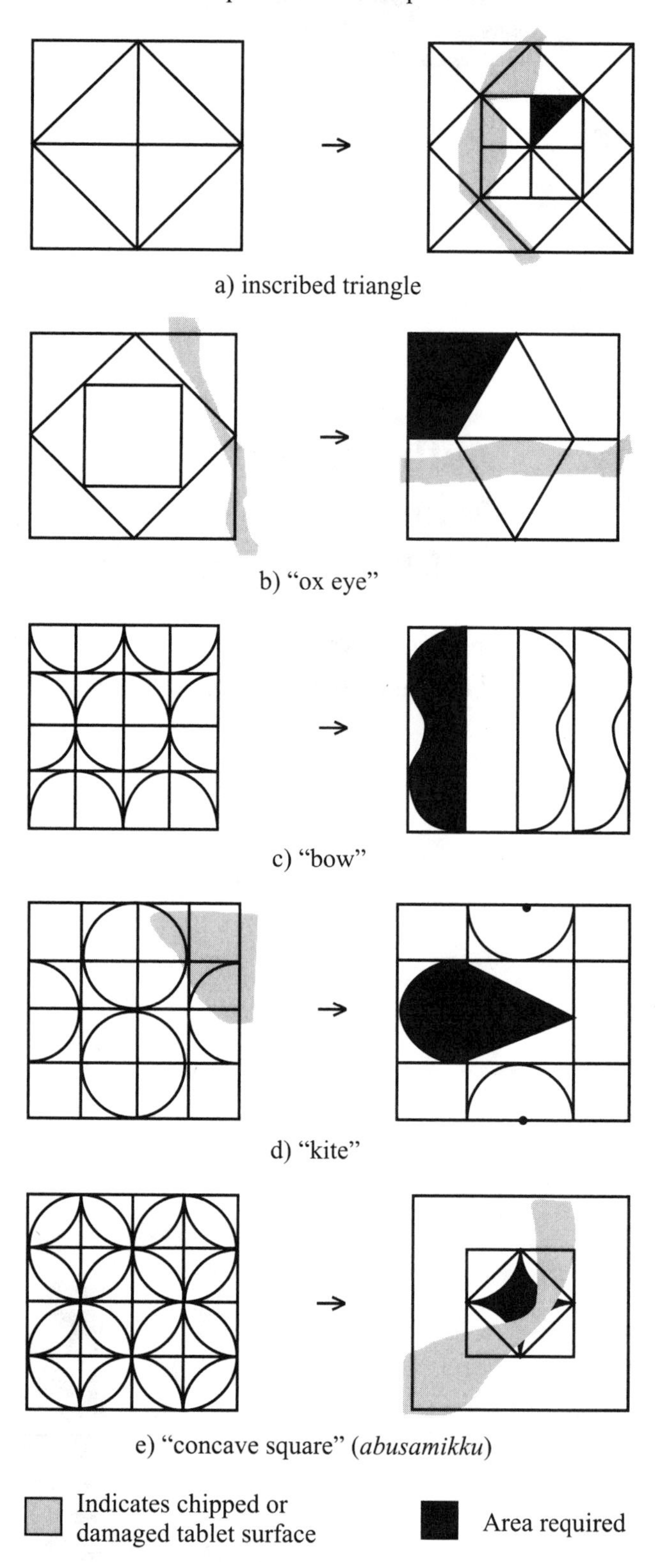

FIGURE 1
Old Babylonian Geometry Problems

Perhaps the most comprehensive collection of problems from the ancient world, in terms of both scope and mathematical content, is the *Jiuzhang suanshu* [*Nine Chapters of the Mathematical Art*] (c. 100 CE) from Han China. The *Jiuzhang* is comprised of 246 problems divided into nine chapters according to their methods and applications. In each chapter, the sequencing of problems carefully progresses from the basic, demonstrating the principles to be learned or techniques to be mastered, to the theoretical and complex, where problem solving strategies are sharpened. A pedagogical analysis has already been undertaken on the contents of the ninth chapter concerning right triangles.[5] Let us briefly examine the organization of the eighth chapter, entitled *fang cheng* [square tabulation], which teaches methods of solution for systems of simultaneous equations. The "square tabulation" method involves the use of algorithmic computing rod techniques and parallels what today is known as the method of "Gaussian Elimination". Of the eighteen problems of this chapter, eight (problems 2, 4–6, 7, 9–11) involved two equations in two unknowns, six (problems 1, 3, 8, 12, 15, 16) concern three equations in three unknowns, and two (14, 17) four equations in four unknowns. Problem 13 involves five equations in six unknowns—an indeterminate situation,[6] and the last problem (18) has five equations in five unknowns. Note how the complexity of the situation is gradually increased throughout the sequence until eventually the student is introduced to an indeterminate situation.

Use of Visual Aids: A Chinese Example

The mathematical classics of ancient China contain many pedagogical features that are recently being recognized. Commentaries on the *Zhoubi suanjing* [*Mathematical classic of the Zhou gnomon*] (c. 100 BCE) contain one of the first documented proofs of the "Pythagorean theorem". It's *xian thu* diagram employs the use of 3-4-5 right triangles, a superimposed grid network, and colors (red and yellow) to assist in its dissection proof strategy.

Liu Hui (c. 263), one of the great mathematical commentators and mathematicians of old China, urged his readers to make diagrams on paper and to cut and rearrange the pieces in order to justify mathematical statements. For the Chinese, paper cutting and folding was a readily accepted

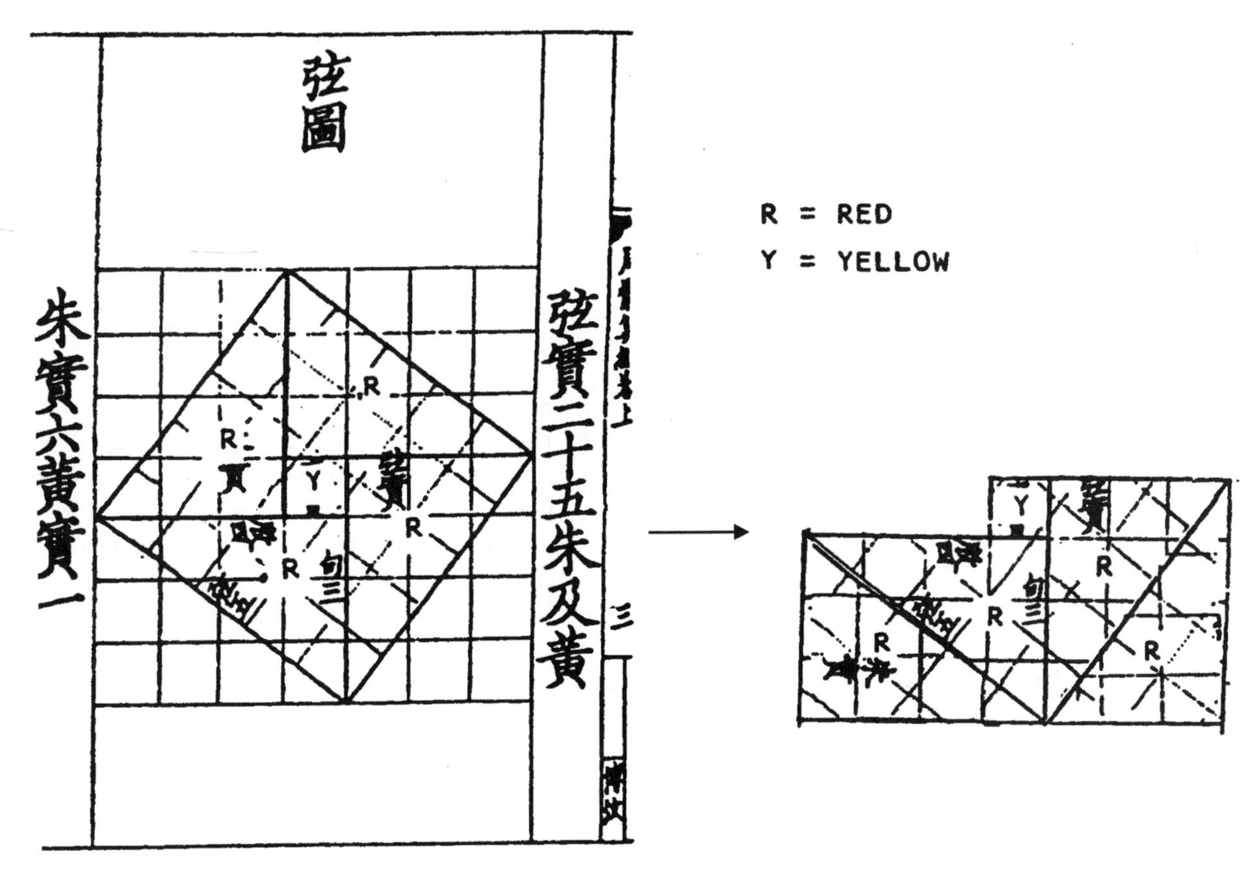

FIGURE 2

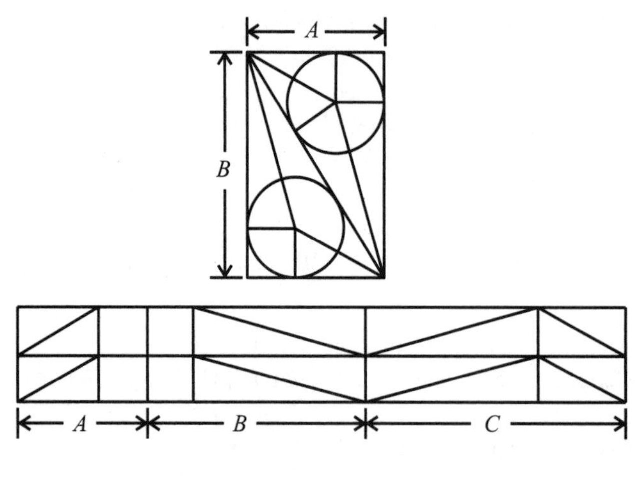

FIGURE 3

method of mathematical demonstration.[7] A further illustration of this technique is provided by Liu's algebraic-geometric justification of a solution formula for the sixteenth problem of the *Jiuzhang*'s ninth chapter, where the reader is asked to find the diameter of the largest circle that could be inscribed in a right triangle of given dimensions. Using modern notation, let the length of the legs of the triangle be given by A and B, the hypotenuse by C and the unknown diameter of the circle by D. Liu conceived of the right triangle as being one half of a rectangle of area AB. He used two such rectangles partitioned into sets of congruent right triangles and squares, then cut and rearranged the pieces to obtain a visual statement of the relationships of the unknowns. Liu's diagrams demonstrated $D = 2AB/(A + B + C)$, the correct result.

The manipulation of plane rectilinear shapes to confirm algebraic relations conforms to an ancient Chinese methodology of the "out-in complementary principle."[8] This form of visual mathematical thinking is also employed in early Babylonian and Greek works. The puzzle-game of Tangrams, which is operationally similar, originated in China. An interesting question arises, "Were there pedagogical designs in the conception and use of Chinese tangrams?"

Visual Aids and the Printed Image

When the first printed edition of Euclid's *Elements* appeared in Europe in 1480, it attracted much attention not only for its contents but also for its visual impact, particularly its prolific use of diagrams and illustrations.[9] With the advent of printing, books moved from the realm of being passive repositories of information to becoming vehicles for active learning. In the transition to the geometry book, the geometric diagram became firmly implanted as a pedagogical tool. As the printer's art developed so too did the complexity and scope of illustrations, diagrams and collections of mathematical exercises for the readers themselves to solve.

Almost a century later (1570) when the first English language version of Euclid was published in London, it was obvious that the book was composed with the learning needs of the reader firmly in mind.[10] In his preface to the reader, Henry Billingsley, its author, comments on the feature:

> Whereunto I have added easy and plain declaration and examples by figures, or definitions. In which book also you shall in due place find manifold additions, Scholia, Annotations, and Inventions: which I have gathered out of many of the most famous and chief mathematicians, both of old time, and in our age: as by diligent reading it in course, you shall well perceive. The fruit and gain which I require for these my pains and travail, shall be nothing else, but only that you gentle reader, will gratefully accept the same: and that you may thereby receive some profit and moreover to excite and stir up other learned, to do the like, and to take pains in that behalf.[11]

The volume abounds with diagrams and explanatory notes and includes folded paper pop-up figures. These three dimensional solids allow the reader to interact with the polyhedra discussed. Thomas Heath in his review of the book noted:

> The print and appearance of the book are worthy of its contents; and, in order that it may be understood how no pains were spared to represent everything in the clearest and most perfect form, I need only mention that the figures of the proposition in Book XI are nearly all duplicated, one being the figure of Euclid, the other an arrangement of pieces of paper (triangular, rectangular, etc.) pasted at the edges on to the page of the book so that the pieces can be turned up and made to show the real form of the solid figures represented.[12]

A copy of Billingsley's geometry housed at the Princeton University Library still contains 38 operational pop-up models. Originally, the volume contained more; how many more is not clear. Since Billingsley makes no special mention of these models in his preface, it can be assumed that such mathematical teaching aids were known and used in sixteenth century England.[13]

Colors were also effectively used as a visual and discriminatory aid in later British geometry texts as explained in the title of Oliver Byrne's 1847 book, *The First Six Books of the elements of Euclid in which Coloured Diagrams and Symbols are Used Instead of letters for the Greater Ease of the Learner.* In his text Byrne employed the colors of red, yellow, blue and black. Using these colors, he visually coded the geometric elements under consideration and placed the colors systematically throughout his discussion and proofs. For example, when he states the Pythagorean theorem, a right triangle is depicted with a red hypotenuse and blue and yellow legs. When he mentions the hypotenuse, a red line is shown; similarly, blue and yellow lines in the text represent legs of the triangle. An accompanying illustration then shows the squares constructed on the sides of the triangle: a red square, a blue square and a yellow square. Byrne's book has been described as "one of the oddest and most beautiful of the whole [nineteenth] century."[14] Its use of color in teaching mathematics was revolutionary.

Mathematical authors of the early European Renaissance were truly imaginative in their use of picturesque schemes to import to their readers the techniques of algorithmic computation employing "Hindu-Arabic" numerals. Pacioli (1494) offered his audience eight different schematic techniques to obtain the product of two multi-digit numbers.[15] Multiplication of two two-digit numbers could easily be accomplished *per crocetta* or "by the cross." The computation to multiply 32 by 57 was done mentally and goes as follows: $2 \times 7 = 14$; write down the four retain the one; "by the cross" $(3\times 7)+(5\times 2)+1 = 31+1 = 32$; write the two retain the three; $(5 \times 3) + 3 = 15 + 3 = 18$, write it down. Thus, the product is 1824.

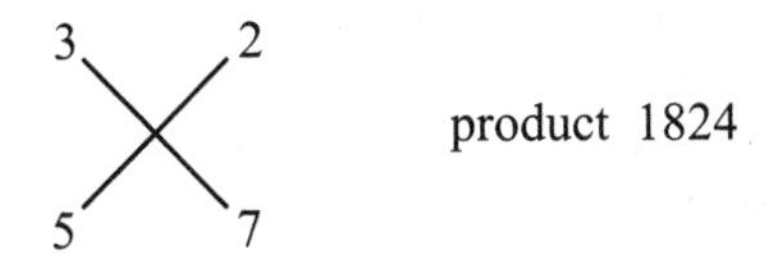

An increase in the number of digits and the resulting increase of partial products often caused place value confusion. To help remedy this situation, numerical configurations were devised to assist in ordering the partial products. These configurations were associated with common objects. In a sense they became visual algorithms. Now a problem solver confronting a higher order multiplication could obtain a correct product with the aid of "the little castle," *per castellucio,* "by the chalice," *per coppa* or "by the bell," *per campana.*[16] The following example of finding a product of two three-digit numbers illustrates the benefits of employing a visual algorithm. (To understand the processes better, the reader should attempt to reconstruct the partial products for the original information given):

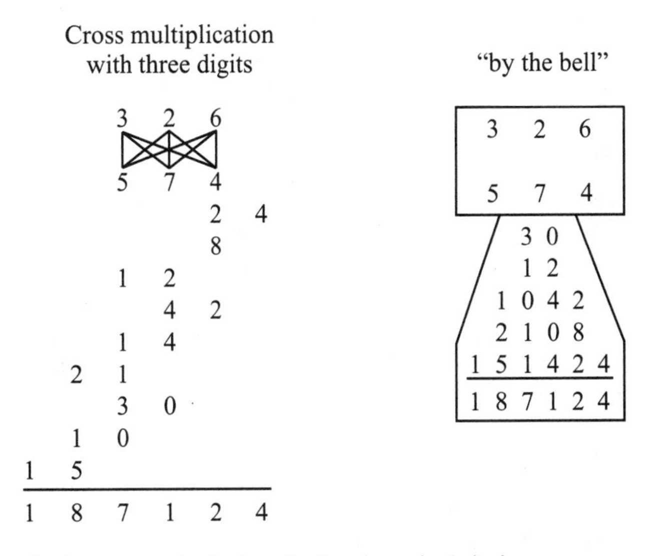

Such mnemonic devices had pedagogical designs.

Conclusion

Mathematical pedagogy, that is a conscious, organized approach to imparting mathematical processes and concepts to a learner, has a long and multifaceted history. It appears that from earliest times mathematical teacher-authors were devising techniques to facilitate the understanding of their discipline—to make mathematics learning easier. By employing diagrams, using color and tactile and visual aids, they incorporated the learner's senses of sight and touch into the processes of understanding and increased the receptive dimensions of learning. Concrete operational teaching is not a product of the twentieth century! Also obvious in old mathematical texts is a purposeful and sequential ordering of topics and problems allowing the students to construct their own edifice of understanding. As heirs and perpetuators of this history, we should be both mindful and proud of the traditions of associating good pedagogy with mathematics learning and teaching.

Endnotes

[1] The author is indebted to Jöran Friberg of the University of Gothenburg for sharing his impressions of this tablet and its content with me. For more specific information on this tablet see, H.W. Saggs, "A Babylonian Geometric Text," *Revue d'assyriologic et d'archeologic orientali* (1960) 54: 131–145.

[2] See C.J. Gadd, "Forms and colours" and R. Caratini "Quadrature du Cercle et Quadratures des Iunules en Msopotamie", *Revue*

d' assyriologic et d' archeologic orientate (1992) 19:149–159; (1957) 51: 11–20 respectively.

[3] In the Babylonian metrology of the period: 1 ÚS = 60 gar, 1 gar = 12 cubits, where a cubit in contemporary measure is 46–50 cm.

[4] A.B. Chace, *The Rhind Mathematical Papyrus,* Reston, VA: National Council of Teachers of Mathematics, 1967 (reprint of 1927–1929 edition).

[5] See F.J. Swetz, "Right Triangle Concepts in Ancient China", *History of Science* (1993) 31: 421–439. A complete translation of the ninth chapter is available in F.J. Swetz and T.I. Kao, *Was Pythagoras Chinese? An Examination of Right Triangle Theory in Ancient China,* University Park, PA: Pennsylvania State University Press, 1977.

[6] "There is a common well belonging to five families; [if we take] 2 lengths of rope of family X, the remaining part equals 1 length of rope of family Y; the remaining part from 3 ropes of Y equals 1 rope of Z; the remaining part from 4 ropes of Z equals 1 rope of V; the remaining part from 5 ropes of V equals 1 rope of U; the remaining part from 6 ropes of U equals 1 rope of X. In all instances if one gets the missing length of rope, the combined lengths will reach [the water]. Find the depth of the well and the lengths of the ropes."

If W is allowed to be the depth of the well, then the situation is:

$$2X + Y = W,$$

$$3Y + Z = W,$$

$$4Z + V = W,$$

$$5V + U = W,$$

$$6U + X = W.$$

[7] M.K. Siu, "Proof and Pedagogy in Ancient China—Examples from Lui Hui's commentary on Jiuzhang Suanshu", *Educational Studies in Mathematics* (1993) 24: 345–357.

[8] The Chinese "out-in complementary principle" applies to situations involving rectangles and depends on the fact that the complements of rectangles about a diagonal of a given rectangle are equal in area. This concept was formalized by Euclid as Proposition 43 in Book I of his *Elements.* See discussion in Wu Wenchun, "The Out-In Complementary Principle" in *Ancient Chinese Technology and Science,* Beijing: Foreign Languages Press, 1983, pp. 66–89.

[9] The Johannes Campanus translation *Preclarissimus liber elementorum Euclides* Venice: Erhard Ratdolt, 1482. For discussion of such texts, see: Charles Thomas-Stanford, *Early Editions of Euclid's Elements,* London: The Bibliographic Society, 1926.

[10] Henry M Billingsley, *The Elements of Geometrie of the Most Ancient Philosopher Euclide of Megara,* London: John Day, 1570. See R.C. Archibald, "The first Translation of Euclid's Elements into English and its Source," *The American Mathematical Monthly* (1950) 57: 443–452.

[11] Billingsley, *op cit,* p. 2 of "Translator to Reader".

[12] Thomas L. Health, *The Thirteen Books of Euclid's Elements Translated from the Text of Heiberg,* New York: Dover Publications, 1956, p. 110 (reprint of 1926 edition).

[13] Billingsley's pop-up solids appear to be an historical first in book publishing. Mechanical devices, that is, movable parts incorporated into illustrations, appeared in texts as early as 1345. Fourteenth century anatomical books contained illustrations with layers of superimposed plates that when lifted revealed interior parts of the body and movable wheels or "volvelles" appeared in fortune-telling books and in material on secret codes. Historians of book design have ignored Billingsley's innovation. Peter Haining in his *Movable Books: An Illustrated History,* London: New English Library, 1979 claims there were no movable books before 1700; see also, Edwina Evers, "A Historical Survey of Movable Books," *AB Bookman's Weekly* (August, 1985) 76: 1204–1205.

[14] Ruari McLean, *Victorian Book Design and Colour Printing,* New York: 1963, p. 51.

[15] The techniques Pacioli discusses are:

1. *Per scachieri,* Venetian for tesselated.
2. *Castellucio,* Florentine for "little castle"
3. *A Traveletta* or *per colona,* by the table or column.
4. *Per quadrilatero,* by the quadrilateral.
5. *Per crocetta* or *casella,* by the cross or pigeonhole.
6. *Per gelosia* or *graticola,* method of cells.
7. *Per repiego,* method of decomposition of factors.
8. *A scapezza,* distributing, separating, and multiplying by the parts.

For further information on these techniques see: Frank Swetz, *Capitalism and Arithmetic: The New Math of the 15th Century,* La Salle, IL: Open Court, 1987.

[16] For information on these various visual algorithms see D.E. Smith, *History of Mathematics,* New York: Dover Publications, 1958, 2: 101–128 (reprint of 1923 edition).

On the Benefits of Introducing Undergraduates to the History of Mathematics—A French Perspective

Anne Michel-Pajus
*IREM PARIS VII**

Some aspects of the reform of the preparatory levels

France has a centralized education system that officially prescribes the various courses of instruction that students follow. In the system I teach in, some 50,000 students undergo their first two years of tertiary education. At the end of this time, the great majority of them take competitive examinations to gain entry to engineering schools; some others enter the Ecoles Normales Supérieures in order to become researchers and teachers at either tertiary or secondary level.

In mathematics, the same teacher teaches one class 16–20 hours a week. The basic class of about 45 students either works as a whole group or is divided into subgroups according to the activity. For two to five hours a week, the students work on exercises in groups of ten to twenty, with or without a computer. One hour every two weeks they work in groups of three for oral questions.

This system has just undergone an important reform, affecting both its structure and its programs of instruction. This reform aims to reduce the importance of mathematics itself by relating it more closely to physics and engineering science. It also aims to develop a spirit of initiative in the students.

The mathematics syllabus is in two parts, with accompanying comments. The first sets out general educational aims while the other deals with the topics of linear algebra, calculus, and geometry. The students must know how to use both calculators and programs that perform symbolic manipulations (*Maple*, in my institution). There are differences in content according to the major followed (mathematics and physics, physics and engineering sciences, physics and chemistry, physics and technology) but the educational objectives are the same. In sum, mathematical education must "simultaneously develop intuition, imagination, reasoning and rigor."

A short addition concerns the history of mathematics: "It is important that the cultural content of mathematics should not be simply sacrificed to its technical aspects. In particular, historical texts and references allow the analysis of the interaction between mathematical problems and the construction of concepts, and bring to the fore the central role played by scientific questioning in the theoretical development of mathematics. Moreover, they show that the sciences, and mathematics in particular, are in perpetual evolution and that dogmatism is not advisable."

Another innovation is the introduction of project work. The history of mathematics is also mentioned in relation to this: "The study of a subject brings an increasing depth of theoretical understanding together with experimental aspects and applications as well as the application of computing methods. It may include an historical dimension." In the first year, students choose their project freely. In their second year, they must fix their area and subject in a very wide framework (for example, dynamical systems in mathematics, measurement in physics). At their final assessment, each student has to present a page long summary and to speak for twenty minutes about their work before two examiners.

The evaluation deals mainly with the communication abilities of the students, but for the teachers who are constrained by a heavy, prescriptive program, this project, which takes up two hours a week, allows the freedom for them to introduce the history of mathematics.

These reforms began in 1995, and so I have only a limited experience to draw upon in writing this article. However, I have found working in the area of the history of mathematics most enriching, and if my readers can glean some ideas from it, its purpose will have been achieved.

Texts available in France

The first difficulty is to find documents students can work on. One of our basic principles for students undertaking such a project is that any historical study must involve at least one extract from an original text. But since it is usually difficult to read them, secondary studies need also to be made available. The majority of useful documents in France have been published by the Commission Inter-IREM d'Histoire et Epistémologie. IREM stands for the Institut de Recherche sur l'Enseignement des Mathématiques (Institute for Research in the Teaching of Mathematics). In France these institutes bring both secondary and tertiary teachers together. Those teachers working on the application of the history of mathematics in the regional IREMs come together nationally three times a year. These meetings comprise lectures and workshops and allow experiences to be exchanged. Proceedings and many other documents are published. Among these are the "problem-documents" in which explanations and questions are linked to selected passages of an historical text.[1] The collection provides homework for students in various classes. An example of such a document is attached (Appendix 1). A selection has been similarly translated and published by The British Mathematical Association.[2] This network is also committed to publishing texts—sources which are difficult to find in French in any other way. As a result I have participated in the edition of *A History of Algorithms* which systematically presents the historical texts, translated if necessary, situated in their cultural context and accompanied by mathematical explanations. The majority of the texts to which I refer here are drawn from this work.[3]

Why choose approximating algorithms?

Apart from the fact that my work in this area has allowed me to collect a large number of documents, this choice presents many advantages.

- the same topic allows one to cross centuries and civilizations, and to meet many problems which are still open. For example, students are always fascinated by research into the decimal expansion of π.
- The question under consideration is clear: it is easy for students to compare the efficiency of procedures, which they can even test using a computer, and thus see the usefulness of theoretical concepts whose depth and generality they otherwise find difficult to evaluate for the lack of a theoretical overview. For example, iterative methods of solution of linear systems involve the topology of spaces of matrices.[4] The interdisciplinary aspect of this theme is worth noting, as the first studies arose from astronomy and surveying.
- Students can see how technical necessities drive the construction of new concepts and sometimes inspire them. Consequently, there is no boundary between pure and applied mathematics—the noble mathematics of mathematicians and that of engineers.
- Being able to write an algorithm in a computer language, such as *Maple* or *Mathematica,* involves the effort of freeing oneself from the restrictions of one specific language and from a particular set of ideas. This is similar to the effort needed to understand a historical text.
- There are, of course, other advantages tied up with course work, such as the connection with the theme of dynamical systems.

Some examples

1. A Project: The approximation of π. Six students chose to divide this topic up among themselves for their project. To assist them in understanding the sources I furnished them with problem-documents put together by secondary teachers for their students. They obviously obtained other general references concerning the history of mathematics. An example is given as Appendix 1: it is a problem-document relating to a text of Euler, which was originally intended for students in their final year of secondary school.

As part of the task of carrying out this project, I insisted that they state precisely the problems under consideration, the concepts, and the tools in use at the period of the text they were studying by asking them to reflect on the following questions:

- What is the author seeking to determine—a number, an area, a surface? What is he seeking to prove?
- Why do it?
- What tools—notation, theories, theorems—are available to him? What is implicit?
- Does it seem rigorous to you?

The texts which the students chose to present to their class were:

Archimedes[5]: Measurement of a circle (1st century BCE)

Descartes[6]: De la quadrature du cercle (1701) [Concerning the quadrature of the circle]

Leibniz[7]: Lettre à La Roque (circa 1780) [Letter to la Roque]

Euler[8]: Des quantités transcendantes qui naissent du cercle (1748) [Transcendant Quantities arising from the Circle)

Richardson[9]: The deferred approach of the limit (1927)

With reference to the historical aspects:

- students are astonished to see how slowly the idea that π was actually a number arose. Until the seventeenth century π was never written, or even thought of, as a number in a formula. Instead we find a method of calculation, written in the language of the time, which can be used to calculate the area or perimeter of a circle. (This can be seen, not only in the works of Archimedes, but also in the work of the Chinese). Unhappily the text which is available to us is not very enticing, while students are regaled with the geometry of the triangle used by Archimedes, going so far as to find four different proofs of one theorem.
- they are surprised to see in 1927, an almost contemporary mathematician, Richardson, picking up the example of Archimedes in order to illustrate his method of extrapolation. Since this is a little known work, an extract is attached (Appendix 2).
- they have determined that there are several stages in the development of approximation.
  - first stage: the basic idea of an approximate value (interesting for its technical uses).
  - second stage: the determination of an interval of validity. (This is a point that students find difficult to grasp. Generally, even in their exercises they give numerical results which show all the numerals which appear on their calculator display, without troubling themselves with the precision of the procedure they are using, even when the objective of the problem is to evaluate this procedure!) The text of Archimedes illustrates this step in a remarkable manner.
  - third stage: practical methods which provide more and more rapid approximations such as the use of series, continued fractions, and so on. The region of validity is sometimes made explicit by giving an inequality as in the works of Wallis or Huygens, for example, and is sometimes implied, as in the work of Leibniz and Euler who give an alternating series. (Leibniz first established that the partial sums are alternatively above and below the sum to infinity.)
  - concerns as to how to speed up the rate convergence lead to theoretical methods of the acceleration of convergence which are useful in other contexts. One example, which has the advantage of being recent, is found in the article by Richardson.
  - The concepts of infinite series, convergence, and rate of convergence are being built up concurrently with mathematicians carrying out theoretical studies as to the nature of π and its relation to the quadrature of the circle.

2. A computer-related study: Newton's method. This is equally well known under the names of "the tangent method" or "Newton-Raphson" even though neither Newton nor Raphson spoke of tangents or even of geometry in this context. Moreover, a century passed before Lagrange saw that they all came down to the same method. The algorithm put forward by Newton is very clear and easy to put into effect with symbolic manipulation software. (A *Maple* program is appended.). Proposed before the invention of fluxions, it does not use derivatives, but only the idea that some quantities are negligible with respect to others. The principle of a calculus of approximations is not made explicit in the form of a recurrence relation, unlike that which appears in Raphson. However, the writings of Raphson are more difficult to read and a much clearer explanation can be found in Euler. These authors treat particular cases in which a recurrence relation appears, although they do not use the derivative explicitly in their writing.

The task is for students to enter into the logical processes of a seventeenth or eighteenth century mathematician and to translate their methods into their own programming language. I make the further demand that they justify the fact that the two methods lead to the formula with the derivative. This historical aspect reinforces the idea that the derivative allows one to make a first order approximation, or that it allows one to replace a function by its first order expansion—the tangent giving a geometrical illustration of this (cf. Appendix 3).

3. Commentaries associated with the course: The Cauchy-Lipschitz Theorem. I have searched for examples in which *techniques* of approximating solutions have led to the *theoretical* proof of a theorem. The topic of differential equations provides a remarkable example of this, but one which lies outside the technical capabilities of my students. I can only refer to it in the course.

The syllabus involves stating the Cauchy-Lipshitz Theorem on the existence of a unique solution to the problem of Cauchy, without proof. It also includes research on approximating solutions to differential equations by Euler's method. Now the first proof given by Cauchy in a text of 1824, which remained unedited until its recent rediscovery, begins precisely with the approximation furnished by Euler's method applied to the interval $[x_0, X]$ for a given partition. By considering upper bounds on the error, Cauchy showed that, under very general assumptions, the approximate value tends toward a limit which only depends on X, as the steps of the subdivision tend to zero. It then only remained to prove that the function defined in making X vary across a neighborhood of x_0 is, in fact, the solution of the given differential equation. Picard's work, which completes this theory, also refers back to Euler's method, although the object of the successive approximations is no longer the value of a function at a given point (as in Euler) but a function defined on an interval.[10]

Conclusion

I would like to conclude by revisiting the educational aspects of my syllabus. I have tried to "stimulate intuition and imagination" by offering multiple viewpoints of the same mathematical object; "struggle against dogmatism" by demonstrating that the process of mathematical creation is cumulative but not linear, and that each era constructs its methods, concepts and proofs with a rigour appropriate to its intellectual framework; and finally show that there is no hierarchy between "pure" and "applied" mathematics, theoretical concepts and their applications, and that even for future engineers mathematics is not merely a "serviceable discipline" but also a marvelous success of human thought.

I am not able to say to what degree, greater or lesser, I have been able to attain these objectives. I do know that this work has not only given me pleasure, but that it has also given some of my students pleasure as well. And that's quite something.

Endnotes

* I would like to thank Stuart Laird (Rangitoto College, New Zealand) for his translation, for his pertinent comments and for his careful rereading of this article and for his New Zealand warmth so welcome in this wintry season.

[1] Problem documents for 15/16 years old in *Brochure M:A.T.H.* n° 61, IREM PARIS VII, la mesure du cercle, pp. 8–13; in *Brochure M:A.T.H.* n° 79, IREM PARIS VII, trois fiches sur le calcul de pi, pp. 36–45; la mesure du cercle, p. 46; in *History in the mathematics classroom,* The Mathematical Association, 1990, Reading Archimedes' measurement of a Circle, Martine Bühler, pp. 43–58.

[2] *History in the Mathematics Classroom,* edited by John Fauvel, The Mathematical Association, 1990.

[3] *Histoire d'Algorithmes,* J.L. Chabert et al., Belin, Paris, 1994. *History of Algorithms,* J.L. Chabert et al., trans. by Chris Weeks, Springer-Verlag, New York, 1999.

[4] The texts of Gauss, Jacobi, Seidel et Nekrasov are to be found in *History of Algorithms,* chapter 9.

[5] The English translation of T.L. Heath, in *The Works of Archimedes,* Cambridge, Cambridge University Press, 1897, is very far removed from the Greek text, much more so than the French translation of Paul Ver Eecke, *Les oeuvres complètes d'Archimède,* 1921, réed. Blanchard, Paris, 1961, pp. 130–134. English translation and commentaries can be found in *History of Algorithms,* pp. 140–145.

[6] Circuli quadratio, excerpta ex MS.R. Des Cartes, Ed. Amsterdam, 1701, in *Les oeuvres de Descartes,* published by C.A. P. Tannery, t.X, 1908, pp. 304–305; re-edited Vrin, Paris, 1974. Translation and comments J.L. Chabert in *History of Algorithms,* pp. 153–156.

[7] Leibniz, Lettre á La Roque, Mathematische Schriften, t.V, pp. 88–92, Olms, Hildesheim, 1962; *History of Algorithms,* pp. 158–161. M.F. Jozeau, M. Hallez, M. Bühler, Séries et quadratures chez Leibniz, *Histoire d'infini,* IREM de Brest, 1994, pp. 273–297.

[8] Euler, Introduction á l'analyse infinitésimale, Livre I, Ch VIII, Des quantités transcendantes qui naissent du cercle. English translation: J.D. Blanton, *Introduction to the Analysis of the Infinite,* NY, Springer, 1988, 1990.

[9] Lewis Richardson, *Philosophical Transactions of the Royal Society of London,* Series A, Vol 226, 1927, pp. 300–305.

[10] cf. *History of Algorithms,* pp. 378–381.

Appendix 1

Euler and the calculation of π*
A. Preliminary Activities

Define F on $\mathbb{R}$ by $F(t) = \int_0^t \frac{1}{1+x^2}dx$

1. a) Justify the existence and differentiability of F. Calculate $F(t)$ and determine whether F is increasing or decreasing.
2. a) Show that: $\forall t \in \mathbb{R}$, $\frac{1}{1+x^2} = 1 - x^2 + x^4 - x^6 + \frac{x^8}{1+x^2}$

   b) Hence show that $F(t) = P(t) + R(t)$ where P is a polynomial of degree 7 and $R(t) = \int_0^t \frac{x^8}{1+x^2}dx$

   c) Show that $\forall t \in [0,1], 0 \leq R(t) \leq \frac{t^9}{9}$
3. Define G on $I = \left]-\frac{\pi}{2}, \frac{\pi}{2}\right[$ by $G(z) = F(\tan z)$.

   a) Justify that G is differentiable and calculate $G'(z)$ for z in I.

   b) Calculate $G(0)$.

   c) From the preceding deduce that $\forall z \in I$, $G(z) = z$.
4. a) Read lines 1 to 2 of the adjoined text.

   b) What is the relationship between this text and the preceding questions?

   c) Explain the &c (= etcetera) line 2.
5. a) Calculate $F(1)$ and $F(1/\sqrt{3})$.

   b) Read lines 2 to 4 of the text.

   c) What is meant by the word "series"?

B. Approximation of π

1. a) Calculate $P(1)$ (see 2 above) to 2 decimal places. How close an approximation to π does this value give? Explain the &c of line 3.

   b) Read lines 5 to 15 of the text. Make explicit Euler's computation, by calculating each square root to four decimal places. How close an approximation to π does this value give?
2. Improving the approximation

   a) Read lines 16 to 20. Verify Euler's calculations. In particular, prove the formula giving $\tan(a+b)$.

   b) Suppose $\tan a = \frac{1}{2}$. Calculate $\tan b$ (where $a + b = \frac{\pi}{4}$). What are the values of $F(\frac{1}{2})$ and $F(\frac{1}{3})$? Calculate $P(\frac{1}{2})$ and $P(\frac{1}{3})$ to 6 decimal places. Give upper bounds for $R(\frac{1}{2})$ and $R(\frac{1}{3})$.

   c) Read the end of the text. Explain, using the notation of the problem, the equality of line 23. How close an approximation to π do you now have? Compare this with your earlier values.

Euler, Introduction to Infinitesimal Analysis (1748)

Thus set tang $\zeta = t$, so that ζ is the arc whose tangent is t, and which we designate thus: $A\,\text{tang}\,t$, giving $\zeta = A\,\text{tang}\,t$. Knowing the tangent, the corresponding arc will be $\zeta = \frac{t}{1} - \frac{t^3}{3} + \frac{t^5}{5} - \frac{t^7}{7} + \frac{t^9}{9} -$ &c. Then, assuming the tangent t is equal to the radius 1, the arc ζ becomes equal to the arc of 45° or $\frac{\pi}{4}$, and we find $\frac{\pi}{4} = 1 - \frac{1}{3} + \frac{1}{5} - \frac{1}{7} +$ &c; the series which Leibnitz first gave to express the value of the circumference of the circle. But in order to obtain the arc length of a circle quickly, by means of such a series, it is clear that it is necessary to take a sufficiently small fraction for the value of the tangent t. In this way, with the help of this series, it is easy to find the arc length ζ, whose tangent $t = \frac{1}{10}$ as this arc will

---

* M. Bühler, Une approximation de pi, *Mnémosyne* n° 10, IREM PARIS VII, pp. 67–71.

be $\zeta = \frac{1}{10} - \frac{1}{3000} + \frac{1}{500000} - \&c$—a series whose value may be easily found using decimals. But the measure of such an arc tells us nothing about the total length of the circumference as it is not possible to assign a relationship between the arc, with tangent $= t$, and the entire circumference. This is why, in this research, it is necessary to take an arc which is an aliquot part of the circumference and whose tangent is sufficiently small and can be conveniently expressed. To fulfill this aim a 30° arc whose tangent $= 1/\sqrt{3}$ is normally chosen, as tangents of smaller arcs, which have a measurable relationship with the circumference, are too irrational. Thus, as the arc of $30° = 1/\sqrt{3}$, we will have that

$$\frac{\pi}{6} = \frac{1}{\sqrt{3}} - \frac{1}{3.3\sqrt{3}} + \frac{1}{5.3^2\sqrt{3}} - \&c \quad \text{and} \quad \pi = \frac{2\sqrt{3}}{1} - \frac{2\sqrt{3}}{3.3} + \frac{2\sqrt{3}}{5.3^2} - \frac{2\sqrt{3}}{7.3^3} + \&c.$$

And it is by means of this series, and an incredible amount of work, that the goal of finding the value of π which we gave above can be attained.

This calculation is all the more laborious as all the terms are irrational and each term is barely less than a third of the preceding one; but this disadvantage can be remedied as follows: still choose an arc of 45° or $\frac{\pi}{4}$. Although the value of this arc is represented by a barely convergent series $= 1 - \frac{1}{3} + \frac{1}{5} - \frac{1}{7} + \&c$; we retain it nevertheless, and imagine it divided into two arcs a and b, such that $a + b = \frac{\pi}{4} = 45°$. Then since $\operatorname{tang}(a+b) = 1 = \dfrac{\operatorname{tang} a + \operatorname{tang} b}{1 - \operatorname{tang} a \operatorname{tang} b}$ we will have

$$1 - \operatorname{tang} a \operatorname{tang} b = \operatorname{tang} a + \operatorname{tang} b \quad \text{and} \quad \operatorname{tang} b = \frac{1 - \operatorname{tang} a}{1 + \operatorname{tang} a}.$$

Now set $\operatorname{tang} a = \frac{1}{2}$; we will find $\operatorname{tang} b = \frac{1}{3}$; then the two arcs a and b can be expressed by a rational series much more convergent than the preceding, and their sum will give the value of the arc $\frac{\pi}{4}$. Thus

$$\pi = 4\left\{\begin{array}{l} \frac{1}{1.2} - \frac{1}{3.2^3} + \frac{1}{5.2^5} - \frac{1}{7.2^7} + \frac{1}{9.2^9} - \&c \\ \frac{1}{1.3} - \frac{1}{3.3^3} + \frac{1}{5.3^5} - \frac{1}{7.3^7} + \frac{1}{9.3^9} - \&c \end{array}\right\}$$

In this way, by using the series we have previously given, the length of the semi-circumference can be found far more readily than it has been.

Appendix 2

Lewis Richardson, The deferred approach of the limit, *Philosophical Transactions of the Royal Society of London, Series A*, vol 226, 1927, pp. 300–305.

Various problems concerning infinitely many, infinitely small, parts had been solved before the infinitesimal calculus was invented; for example, ARCHIMEDES on the circumference of the circle. The essence of the invention of the calculus appears to be that the passage to the limit was thereby taken at the earliest possible stage, where diverse problems had operations like d/dx in common. Although the infinitesimal calculus has been a splendid success, yet there remain problems in which it is cumbrous or unworkable. When such difficulties are encountered it may be well to return to the manner in which they did things before the calculus was invented, postponing the passage to the limit until after the problem had been solved for a moderate number of moderately small differences.[...]

The h^2-extrapolation was discovered by a hint from theory followed by arithmetical experiments, which gave pleasing results.[...]

Imagine that we are back in the time of ARCHIMEDES.

As a first, obvious very crude, approximation, take the perimeter of an inscribed square $= 4\sqrt{2} = 5.6568$. As a second approximation, take the perimeter of an inscribed hexagon $= 6$ exactly.

The errors of these two estimates should be to one another as $\frac{1}{4^2} : \frac{1}{6^2}$ that is $\frac{9}{4}$, if the error is proportional to the square of the coordinate difference. Thus the extrapolated value is $6 + \frac{4}{5}(6 - 5.6568) = 6.2746$.

The error of the extrapolated value is thus only $1/33$ of the error in the better of the two values from which it was derived; so that extrapolation seems a useful process. To get as good a result from a single inscribed regular polygon it would need to have 35 sides.

Appendix 3: Newton , Euler and Maple

a) extract of "The method of fluxions," *The mathematical papers of Isaac Newton,* vol III, Whiteside ed, Cambridge University Press, 1969, pp. 43–47. [Written between 1664 and 1671, published in 1736, and explained by Wallis in his *Treatise of Algebra* in 1685.]

When, however, affected equations are proposed, the manner in which their roots might be reduced to this sort of series should be more closely explained, the more so since their doctrine, as hitherto expounded by mathematicians in numerical cases, is delivered in a roundabout way (and indeed with the introduction of superfluous operations) and in consequence ought not to be brought in to illustrate the procedure in species. In the first place, then, I will discuss the numerical resolution of affected equations briefly but comprehensively, and subsequently explain the algebraical equivalent in similar fashion.

Let the equation $y^3 - 2y - 5 = 0$ be proposed for solution and let the number 2 be found, one way or another, which differs from the required root by less than its tenth part. I then set $2 + p = y$, and in place of y in the equation I substitute $2+p$. From this there arises the new equation $p^3 + 6p^2 + 10p - 1 = 0$, whose root p is to be sought for addition to the quotient. Specifically, (when $p^3 + 6p^2$ is neglected because of its smallness) we have $10p - 1 = 0$, or $p = 0.1$ narrowly approximates the truth. Accordingly, I write 0.1 in the quotient and, supposing $0.1 + q = p$, I substitute this fictitious value for it as before. There results $q^3 + 6.3q^2 + 11.23q + 0.061 = 1$. And since $11.23q + 0.061 = 0$ closely approaches the truth, in other words very nearly $q = -0.0054$ (by dividing 0.061 by 11.23, that is, until there are obtained as many figures as places which, excluding the bounding ones, lie between the first figures of this quotient and of the principal one—here, for instance, there are two between 2 and 0.005), I write -0.0054 in the lower part of the quotient seeing that it is negative and then, supposing $-0.0054 + r$ equal to q, I substitute this value as previously. And in this way I extend the operation at pleasure after the manner of the diagram appended.

$$\begin{cases} +2\cdot 10000000 \\ -0\cdot 00544852 \end{cases}$$

$2\cdot 09455148$ [$= y$]

$2+p=y.$	y^3	$+8$	$+12p$	$+6p^2$	$+p^3$
	$-2y$	-4	$-2p$		
	-5	-5			
	Total	-1	$+10p$	$+6p^2$	$+p^3$
$0\cdot 1+q=p.$	$+p^3$	$+0\cdot 001$	$+0\cdot 03q$	$+0\cdot 3q^2$	$+q^3$
	$+6p^2$	$+0\cdot 06$	$+1\cdot 2$	$+6$	
	$+10p$	$+1$	$+10$		
	-1	-1			
	Total	$0\cdot 061$	$+11\cdot 23q$	$+6\cdot 3q^2$	$+q^3$
$-0\cdot 0054+r=q.$	$+q^3$	$-0\cdot 000000157464$	$+0\cdot 0008748r$	$-0\cdot 0162r^2$	$+1r^3$
	$+6\cdot 3q^2$	$+0\cdot 000183708$	$-0\cdot 06804$	$+6\cdot 3$	
	$+11\cdot 23q$	$-0\cdot 060642$	$+11\cdot 23$		
	$+0\cdot 061$	$+0\cdot 061$			
	Total	$+0\cdot 0005416$	$+11\cdot 162r$		
$-0\cdot 00004852+s=r.$					

b) extract of Euler, *Elements of Algebra,* Chapter XVI, Of the Resolution of Equations by Approximations, §784–789

784. When the roots of an equation are not rational, and can only be expressed by radical quantities, or when we have not even that resource, as is the case with equations which exceed the fourth degree, we must be satisfied with determining their values by approximation; that is to say, by methods which are continually bringing us nearer to the true value, till at last the error being very small, it may be neglected. Different methods of this kind have been proposed, the chief of which we shall explain.

785. The first method which we shall mention supposes that we have already determined, with tolerable exactness, the value of one root; that we know, for example, that such a value exceeds 4, and that it is less than 5. In this case, if

we suppose this value $= 4+p$, we are certain that p expresses a fraction. Now, as p is a fraction, and consequently less than unity, the square of p, its cube, and, in general, all the higher powers of p, will be much less with respect to unity; and, for this reason, since we require only an approximation, they may be neglected in the calculation. When we have, therefore, nearly determined the fraction p, we shall know more exactly the root $4+p$; from that we proceed to determine a new value still more exact, and continue the same process till we come as near the truth as we desire.*

786. We shall illustrate this method first by an easy example, requiring by approximation the root of the equation $x^2 = 20$.

Here we perceive, that x is greater than 4, and less than 5; making, therefore, $x = 4 + p$, we shall have $x^2 = 16 + 8p + p^2 = 20$; but as p^2 must be very small, we shall neglect it, in order that we may have only the equation $16 + 8p = 20$, or $8p = 4$. This gives $p = \frac{1}{2}$, and $x = 4\frac{1}{2}$, which already approaches nearer the true root. If, therefore, we now suppose $x = 4\frac{1}{2} + p'$; we are sure that p' expresses a fraction much smaller than before, and that we may neglect p'^2 with great propriety. We have, therefore, $x^2 = 20\frac{1}{4} + 9p' = 20$, or $9p' = -\frac{1}{4}$; and consequently, $p' = -\frac{1}{36}$; therefore $x = 4\frac{1}{2} - \frac{1}{36} = 4\frac{17}{36}$.

And if we wished to approximate still nearer to the true value, we must make $x = 4\frac{17}{36} + p''$, and should thus have $x^2 = 20\frac{1}{1296} + 8\frac{34}{36}p'' = 20$; so that $8\frac{34}{36}p'' = -\frac{1}{1296}$, or $322p'' = -\frac{36}{1296} = -\frac{1}{36}$, and

$$p = -\frac{1}{36 \times 322} = -\frac{1}{11592}:$$

therefore $x = 4\frac{17}{36} - \frac{1}{11592} = 4\frac{5473}{11592}$, a value which is so near the truth, that we may consider the error as of no importance.

787. Now, in order to generalise what we have here laid down, let us suppose the given equation to be $x^2 = a$, and that we previously know x to be greater than n, but less than $n+1$. If we now make $x = n + p$, p must be a fraction, and p^2 may be neglected as a very small quantity, so that we shall have $x^2 = n^2 + 2np = a$; or $2np = a - n^2$, and $p = \dfrac{a-n^2}{2n}$; consequently, $x = n + \dfrac{a-n^2}{2n} = \dfrac{n^2+a}{2n}$.

Now, if n approximated towards the true value, this new value $\dfrac{n^2+a}{2n}$ will approximate much nearer; and, by substituting it for n, we shall find the result much nearer the truth; that is, we shall obtain a new value, which may again be substituted, in order to approach still nearer; and the same operation may be continued as long as we please.

For example, let $x^2 = 2$; that is to say, let the square root of 2 be required; and as we already know a value sufficiently near, which is expressed by n, we shall have a still nearer value of the root expressed by $\dfrac{n^2+2}{2n}$. Let, therefore,

1. $n = 1$, and we shall have $x = \frac{3}{2}$,
2. $n = \frac{3}{2}$, and we shall have $x = \frac{17}{12}$,
3. $n = \frac{17}{12}$, and we shall have $x = \frac{577}{408}$.

This last value approaches so near $\sqrt{2}$, that its square $\frac{332929}{166464}$ differs from the number 2 only by the small quantity $\frac{1}{166464}$, by which it exceeds it.

788. We may proceed in the same manner, when it is required to find by approximation cube roots, biquadrate roots, &c.

Let there be given the equation of the third degree, $x^3 = a$; or let it be proposed to find the value of $\sqrt[3]{a}$.

Knowing that it is nearly n, we shall suppose $x = n + p$; neglecting p^2 and p^3, we shall have $x^3 = n + 3n^2p = a$; so that $3n^2p = a - n^3$, and $p = \dfrac{a-n^3}{3n^2}$; whence

$$x = (n+p) = \frac{2n^3+a}{3n^2}.$$

---

* This is the method given by Sir Is. Newton at the beginning of his "Method of Fluxions." When investigated, it is found subject to different imperfections; for which reason we may with advantage substitute the method given by M. de la Grange, in the *Memoirs of Berlin* for 1768 and 1767.—F.T.

This method has since been published by De la Grange, in a separate Treatise, where the subject is discussed in the usual masterly style of this author.

If, therefore, n is nearly $= \sqrt[3]{a}$, the quantity which we have now found will be much nearer it. But for still greater exactness, we may again substitute this new value for n, and so on.

For example, let $x^3 = a = 2$; and let it be required to determine $\sqrt[3]{2}$. Here, if n is nearly the value of the number sought, the formula $\dfrac{2n^3+2}{3n^2}$ will express that number still more nearly; let us therefore make

1. $n = 1$, and we shall have $x = \frac{4}{3}$,
2. $n = \frac{4}{3}$, and we shall have $x = \frac{91}{72}$,
3. $n = \frac{91}{72}$, and we shall have $x = \frac{162130896}{128634294}$.

789. This method of approximation may be employed, with the same success, in finding the roots of all equations.

To show this, suppose we have the general equation of the third degree, $x^3 + ax^2 + bx + c = 0$, in which n is very nearly the value of one of the roots. Let us make $x = n - p$; and, since p will be a fraction, neglecting the powers of this letter, which are higher than the first degree, we shall have $x^2 = n^2 - 2np$, and $x^3 = n^3 - 3n^2p$; whence we have the equation $n^3 - 3n^2p + an^2 - 2anp + bn - bp + c = 0$, or $n^3 + an^2 + bn + c = 3n^2p + 2anp + bp = (3n^2 + 2an + b)p$; so that $p = \dfrac{n^3 + an^2 + bn + c}{3n^2 + 2an + b}$, and $x = n - \left(\dfrac{n^3 + an^2 + bn + c}{3n^2 + 2an + b}\right) = \dfrac{2n^3 + an^2 - c}{3n^2 + 2an + b}$. This value, which is more exact than the first, being substituted for n, will furnish a new value still more accurate.

790. In order to apply this operation to an example, let $x^3 + 2x^2 + 3x - 50 = 0$, in which $a = 2$, $b = 3$, and $c = -50$. If n is supposed to be nearly the value of one of the roots, $x = \dfrac{2n^3 + 2n^2 + 50}{3n^2 + 4n + 3}$ will be a value still nearer the truth.

Now, the assumed value of $x = 3$ not being far from the true one, we shall suppose $n = 3$, which gives us $x = \frac{61}{21}$; and if we were to substitute this new value instead of n, we should find another still more exact.

791. We shall give only the following example, for equations of higher dimensions than the third.

Let $x^5 = 6x + 10$, or $x^5 - 6x - 10 = 0$, where we readily perceive that 1 is too small, and that 2 is too great. Now, if $x = n$ be a value not far from the true one, and we make $x = n + p$, we shall have $x^5 = n^5 + 5n^4p$; and, consequently,

$$n^5 + 5n^4p = 6n + 6p + 10; \quad \text{or} \quad 5n^4p - 6p = 6n + 10 - n^5. \quad \text{And} \quad p(5n^4 - 6) = 6n + 10 - n^5.$$

Wherefore $p = \dfrac{6n + 10 - n^5}{5n^4 - 6}$, and $x\ (= n + p) = \dfrac{4n^5 + 10}{5n^4 - 6}$. If we suppose $n = 1$, we shall have $x = \frac{14}{-1} = -14$; this value is altogether inapplicable, a circumstance which arises from the approximated value of n having been taken much too small. We shall therefore make $n = 2$, and shall thus obtain $x = \frac{138}{74} = \frac{69}{37}$, a value which is much nearer the truth. And if we were now to substitute for n, the fraction $\frac{69}{37}$, we should obtain a still more exact value of the root x.

c) Program *Maple* for Newton's algorithm

```
>P:= -5 -2 x + x^3 ;
>r:= 2 ; h:= r ;
>for i from 1 to 4 do P:= sort(expand(subs(x = x + h, P))):
>    h:= evalf(- x*op(P)[4] / op(P)[3]);
>    r:= evalf(r + h) ; od ;
```

Part II

Historical Ideas and Their Relationship to Pedagogy

The History of Mathematics and its Influence on Pedagogical Problems

Lucia Grugnetti
University of Parma, Italy

Introduction

Using the history of mathematics as an introduction to a critical and cultural study of mathematics is one of the most important challenges for mathematics teachers and for students. There are many possibilities in mathematics education for the use of history, which we discuss in what follows.

If learning is not only an accumulation of items of knowledge, but a set of critical attitudes about knowledge, then the question is not about the quantity of transmitted knowledge, but about its quality. Why did a certain concept arise? Under which historical conditions? The traditional idea that the development of mathematics is purely cumulative is largely out-of-date. Teachers must remain aware of the inherent relativity of knowledge, and of the fact that, in the long run, providing students with an adequate view of how science builds up knowledge is more valuable than the mere acquisition of facts (von Glasersfeld, 1991). Yet there is a growing debate concerning the role of the history of mathematics in mathematics education.

One of the several risks in introducing the history of mathematics in mathematics education is the anachronism which consists in attributing to an author such conscious knowledge as he never possessed. There is a vast difference between recognizing Archimedes as a forerunner of integral and differential calculus, whose influence on the founders of the calculus can hardly be overestimated, and fancying to see in him, as has sometimes been done, an early practitioner of the calculus (Weil, 1978). If the risk of anachronism is a big one for historians, it is not smaller in doing history of mathematics in mathematics education. So, once we introduce at school a mathematician or, in general, a scientist, it is fundamental to analyze the political, social, and economic context in which he lived. In this way it is possible to discover that facts and theories, studied in different disciplines, are concretely related. Moreover, as Pepe (1990) reminds us, the "meeting" between history and didactics of mathematics must be developed taking into account the negative influences that they can have one on the other. A possible negative influence of history on didactics is an increase in interesting and curious references which are, in effect, not essential. On the contrary, the history of mathematics offers us several examples which gain by an interdisciplinary approach (Pepe, 1990) as, for example: the number systems of the ancients; the use by Galileo of both mathematical and experimental methods; and Descartes' use of the analytical method.

The influence of the history of mathematics on pedagogical problems can be seen by various methods. For example,

- By using old problems, students can compare their strategies with the original ones. This is an interesting way for understanding the economy and the effectiveness of our present algebraic process. In observing the historical evolution of a concept, pupils will find that mathematics is not fixed and definitive.
- History for constructing mathematical skills and concepts .
- An historical and epistemological analysis allows teachers to understand why a certain concept is difficult for the student (as, for example, the concept of the function, the concept of the limit, but also fractions, operations with zero, etc) and can help in the didactical approach and development. An epistemological analysis is not easy to do, and some researchers have worked and are working on this aspect.[1]

By using old problems students can compare their strategies with the original ones

An example of the history of mathematics as used in interdisciplinary teaching could come from the Liber Abbaci (1202) by Leonardo Pisano (known as Fibonacci). This book can be used as a source of problems (from the thirteenth century) which will involve different teachers and subjects. For example, one can look at both Italian and Latin and ask what kind of language is that of the *Liber Abbaci*? In history, one can study the development of the Middle Ages in Europe and Islam. In geography, one can consider the differences between Western Europe and the Middle East. And in mathematics, one can ask why Fibonacci used certain strategies for problem solving and can compare those strategies with current ones. The students can, for example, understand the economy and the effectiveness of present algebraic processes compared to the ancient methods. The activity of recognizing and comparing strategies is one of the most important aspects to develop in mathematics learning. Once students become able to compare different strategies (for solving problems, but also for proving theorems), they can begin the process of generalization.

It could be interesting to ask 13-year-old pupils (as also older ones) to translate the following problem from the *Liber Abbaci*:

> In quodam plano sunt due turres, quarum una est alta passibus 30, altera 40, et distant in solo passibus 50; infra quas est fons, ad cuius centrum volitant due aves pari volatu, descendentes pariter ex altitudine ipsarum; queritur distantia centri ab utraque turri.

Anyway, even if pupils are not able to translate the problem, we can ask them to solve it:

> Two towers, the heights of which are 40 paces and 30 paces, are 50 paces apart at their bases. Between the two towers there is a fountain where two birds, flying down from the two towers at the same speed will arrive at the same time. What is the distance of the fountain from the two towers?

Pupils aged 13 and 14 can solve the problem using the Pythagorean theorem and solving an equation, but the real interest of this problem could be an examination and discussion of Fibonacci's strategies (literally translated):

> If the higher tower is at a distance of 10 from the fountain, 10 times 10 is 100, which added to the higher tower times itself is 1600, which gives 1700, we must multiply the remaining distance times itself, which added to the lower tower times itself, i.e., 900, gives 2500. This sum and the previous one differ by 800. We must move the fountain away from the higher tower. For example by 5, i.e., globally by 15, which multiplied by itself is 225, which added to the higher tower times itself gives 1825, which added to the lower tower times itself gives 2125. The two sums differ by 300. Before the difference was 800. So, when we added 5 paces, we reduced the difference of 500. If we multiply by 300 and we divide by 500, we have 3, which added to 15 paces gives 18 which is the distance of the fountain from the higher tower.

It could be interesting to analyze and discuss with pupils this explanation of Fibonacci, in which arithmetic writing of operations is not given and in which the Pythagorean theorem is implicitly used. Pupils have to interpret Fibonacci's sentences by translating them into mathematical symbolism. This activity can be done in small heterogeneous groups. In modern symbolism Fibonacci's procedure can be written as: $10^2+40^2=100+1600=1700$ and $(50-10)^2+30^2=40^2+30^2=1600+900=2500$ (Fibonacci says: "this sum and the previous one differ by 800"). $15^2+40^2=225+1600=1825$ and $35^2+30^2=1225+900=2125$ (Fibonacci says: "the two sums differ by 300"). Fibonacci uses now the diagram

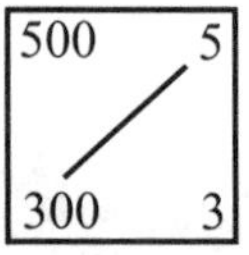

and his last sentence could be written as $(5\times300)/500=3$; $3+15=18$.

The discussion can bring in the method of 'false position', one of the first ways to solve "equations", a method which was used in fact by the Egyptians 4000 years ago. Students can see that it is more 'economical' if they solve this problem using a simple algebraic equation, which Fibonacci could not use. The discussion can now concern the reasons why Fibonacci could not use algebra in our sense. In this way a historical example can give students the opportunity to compare arithmetic and algebraic procedures. However, Fibonacci's problem doesn't finish here! In fact, Fibonacci considers a second strategy to solve it.

After having explained that the triangle agz (the fountain is at z) is isosceles with base ag (with $ae = eg$) by construction, Fibonacci adds: "40 and 30 is 70; the half is 35, in fact the line ef. The lines df and fb are 25 in length, the difference between 35 and the lower tower is 5, which, multiplied by 35 is 175, which divided by the half of the distance between the two towers, in fact 25, gives 7 (the line fz). Therefore dz is 32 and it remains 18 for the line zb."

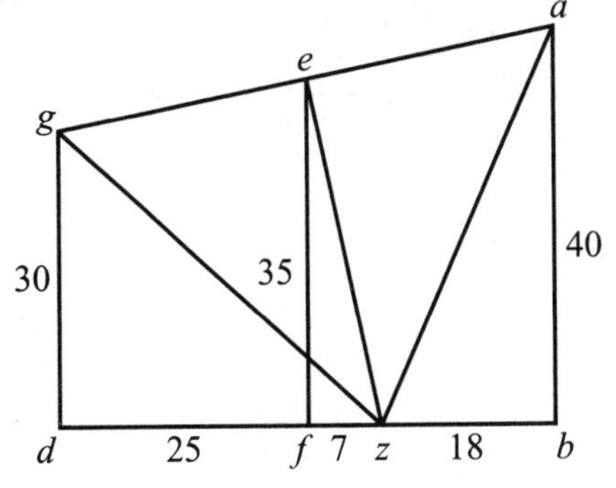

FIGURE 1

It could be interesting to discuss with students Fibonacci's procedure which, as we can see, is based on the similarity of triangles efz and ghe where h is the intersection point of ef and the parallel to df which contains g. Hypotheses of students and the discussion of them is an important element in this activity.

While algebraic procedures in the thirteenth century were not developed, geometrical ones were the same as those our students use nowadays. Here, an important historical chapter could be opened in secondary school.

History for constructing mathematical skills and concepts

The main aim of IREM proposals[2] is that of making teachers and students sensitive to the evolution of mathematics concepts and language. This activity allows, for example, students to work on Euclidean geometry in a critical way.

Concerning three-dimensional geometry and calculus it could be fruitful from a pedagogical point of view to use Cavalieri's theory of indivisibles before introducing modern integration, which soon becomes a technical way for solving exercises. Since the originals by Cavalieri are very difficult, we have to use a "didactical transposition" of that theory.

In the seventeenth century, the systematic use of infinitesimal techniques for area and volume computation was popularized by two influential books written by Bonaventura Cavalieri (1598– 1647)—his *Geometria indivisibilibum* (Geometry of indivisibles) of 1635 and his *Exercitationes geometricae sex* (Six geometrical exercises) of 1647.[3]

Using his method, Cavalieri proceeded by setting up a one-to-one correspondence between the indivisibles of two given geometrical figures. The fundamental idea in the *Geometria indivisibilibum* is that it is possible to compare two continua by comparing their indivisibles. Thus by means of a straight line moving parallel to itself it will be possible to characterize all the lines of the figure, namely the intersections of the moving line with the figure. In a similar way a plane moving parallel to itself will characterize in a solid body all its indivisibles, the intersections of the solid with the moving plane. Together with plane and solid figures we have then other objects, namely all the lines of the plane figures and all the planes of the solid ones, that will be compared with each other and each with the related figure, with the purpose of finding the ratios of the latter.[4]

If corresponding indivisibles of the two given figures had a certain (constant) ratio, Cavalieri concluded that the areas or volumes of the given figures had the same ratio. Typically the area or volume of one of the figures was known in advance, so this gave the other: Cavalieri's principle is stated as: *If two solids have equal altitudes, and if sections made by planes parallel to the bases and at equal distances from them are always in a given ratio, then the volumes of the solids are also in this ratio.* (See Figure 2.)

By using Cavalieri's method it is both possible and interesting to calculate the volume of the solid obtained by intersecting perpendicularly two equal cylinders. As is well known, the volume of such a solid is in general found at university level by the use of double integrals. On the contrary as André Deledicq (1991) shows us, Cavalieri's theory also allows secondary level students to solve this problem, which seems to have been solved by Evangelista Torricelli (1608–1647).

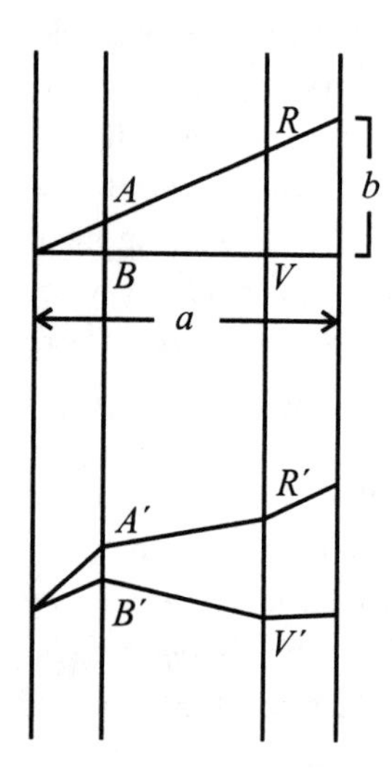

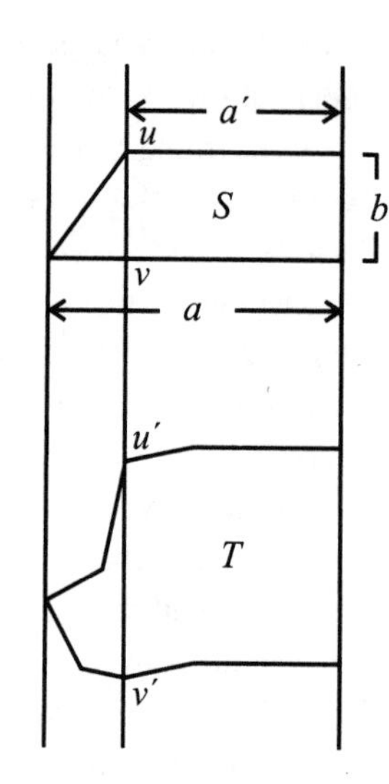

$AB = A'B'$; $RV = R'V'$, etc.
The figures have a ratio 1:1.

$2uv = u'v'$, etc.
$S : T = 1 : 2$
$T = 2S$

$$S = \frac{(a + a')b}{2};\ T = (a + a')b$$

FIGURE 2

We can represent the solid by Figure 3.

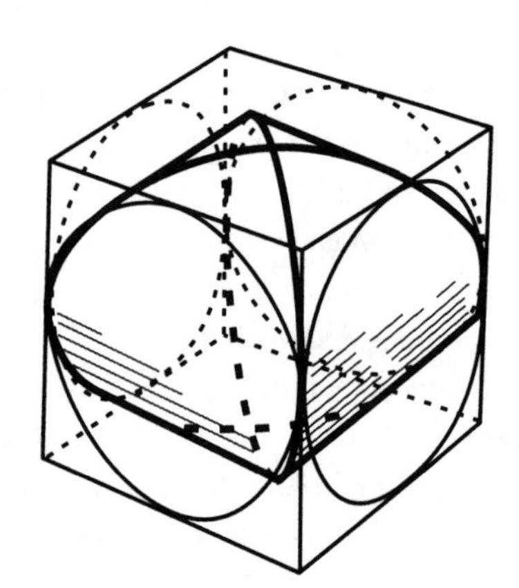

FIGURE 3

At a university level we can find the volume of a part of the solid (the sixteenth part) as in Figure 4.

By Cavalieri's theory:

- in this solid S we can consider a sphere which is tangent to the two cylinders
- if we dissect the sphere by horizontal planes we obtain for each plane a circle
- if we dissect the solid S by horizontal planes we obtain for each plane a square (see Figure 5).

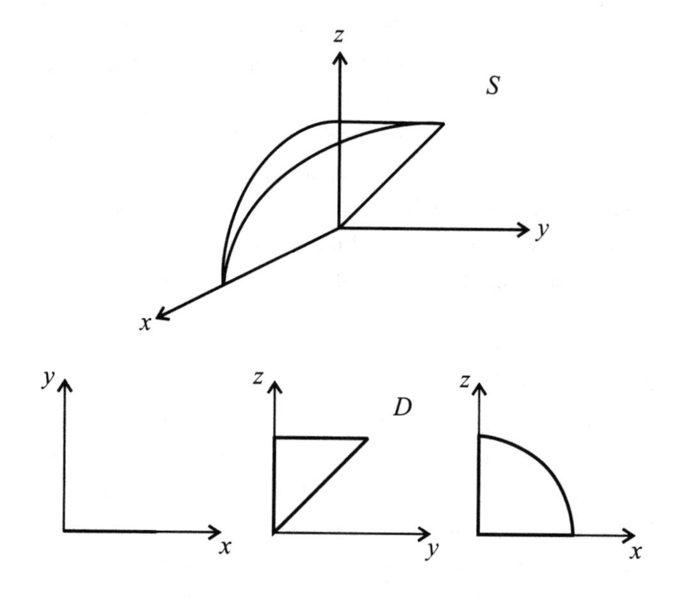

$$\begin{aligned}
V &= 16 \times \iiint_S dx\,dy\,dz \\
&= 16 \times \iint_D dz\,dy \int_0^{\sqrt{r^2-z^2}} dx \\
&= 16 \times \int_0^r \sqrt{r^2 - z^2} dz \int_0^z dy \\
&= 16 \times \int_0^r z\sqrt{r^2 - z^2} dz \\
&= 16 \times \left(-\frac{1}{2}\right) \int_0^r (-2)\ z\ (r^2 - z^2)^{1/2} dz \\
&= -8 \times \frac{2}{3}\left[(r^2 - z^2)^{3/2}\right]_0^r \\
&= -\frac{16}{3}[-r^2]^{3/2} = \frac{16}{3}r^3
\end{aligned}$$

FIGURE 4

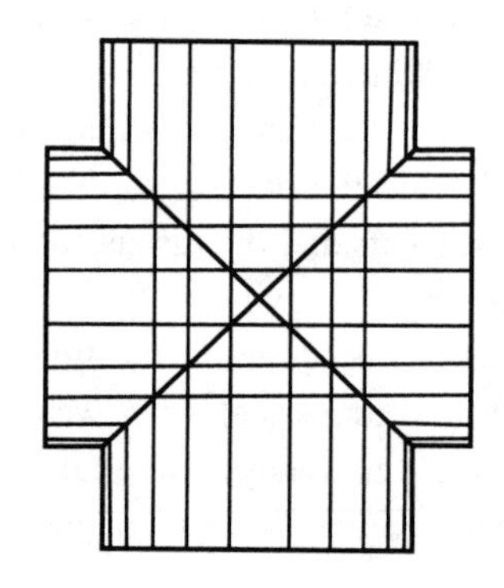

FIGURE 5

In each plane we can see the sections as in Figure 6 where the square is a section of the solid S and the circle is a

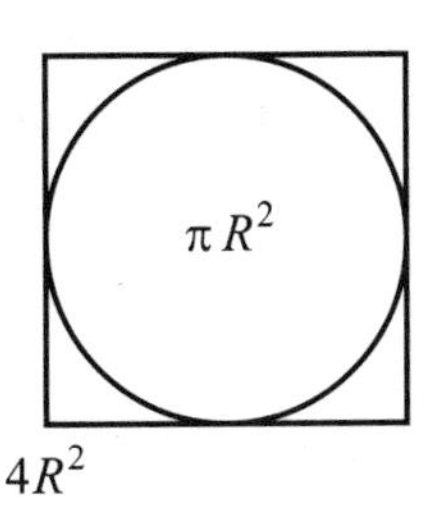

FIGURE 6

section of the sphere. The ratio of the surfaces of the two sections is $4R^2/\pi R^2 = 4/\pi$.

By Cavalieri's principle, the same ratio must exists for the two volumes, so we obtain:

$$\text{Volume } S = (\text{Volume sphere}) \times 4/\pi = (4\pi r^3/3)(4/\pi) = 16r^3/3.$$

Before Torricelli, this solid also interested the artist and mathematician Piero della Francesca (c. 1415–1492) and a half of the solid appears in the famous altar screen in Brera Palace in Milan.

An interesting use, from a didactical point of view, of Cavalieri's theory, can be found also in an Italian textbook (L. Lombardo Radice and L. Mancini Proia, 1979) where it is used to find the area of the ellipse: If we have the circle: $x^2 + y^2 = a^2$ and the ellipse: $\dfrac{x^2}{a^2} + \dfrac{y^2}{b^2} = 1$ and we consider the chords obtained by cutting the circle and the ellipse by straight lines which are parallel to the y-axis (Figure 7), we have:

$$y^2 = a^2 - x^2; \qquad y^2 = \frac{b^2\left(a^2 - x^2\right)}{a^2}$$

The ratio of the two chords is b/a, so we obtain: $E : C = b : a$ (where E and C are respectively the area of the ellipse and of the circle) from which $E = Cb/a$. We know that $C = \pi a^2$; therefore $E = \pi ab$.

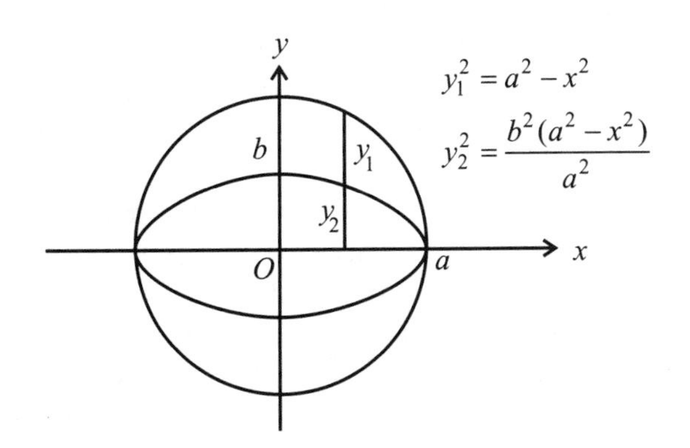

FIGURE 7

This approach allows the teacher to follow with the students the historical process in which the problem of the quadratures (integration) precedes the problem of the tangents (differentiation). Moreover we think that, by this process, students can also appreciate later "modern" integration as need of generalization. At the same time, if we try to follow, in a certain sense, the historical process of calculus and, as in the previous examples, the theory of indivisibles, we have to develop with students a critical analysis concerning that theory. So why was the seventeenth century the moment of the theory of indivisibles? Or why was the calculation of areas and volumes one of the main problems of that century?

The end of the sixteenth century saw the final establishment of geometry as the leading branch of mathematical research, as a result of a process begun half a century earlier with an impressive series of translations, mostly into Latin, of the classical works of the Greek geometers, frequently accompanied by erudite comments (Giusti, 1980). This century saw the wide dissemination and serious study of Euclid's *Elements* and of the works of Archimedes. Taking pride of place in this revival of geometric research were Archimedean themes, in particular the calculus of areas and volumes of geometrical figures and of their centers of mass. This sector expanded very rapidly, and maintained its leadership for most of the following century. The calculation of areas and volumes was one of the main problems of the early years of the seventeenth century, while the method of exhaustion was the only technique handed on by the ancients. This technique was as rigorous as any other, but at the same time was extremely laborious and left little scope for intuition (Giusti, 1980).

Therefore, during the seventeenth century, although Archimedes' accomplishments provided the chief inspiration for the resumption of mathematical progress, the time was ripe for the development of simpler new methods. Methods were sought that could be applied to the investigation of area and volume problems with greater ease than could the method of exhaustion with its tedious double *reductio ad absurdum* proofs. While continuing to regard Archimedean proofs as the ultimate models of rigor and precision, the Renaissance mathematical mind was more interested in quick new results and methods of rapid discovery than in the stringent requirements of rigorous proof. It is the main moment for 'indivisibles' in area and volume computation. Other subjects of present curricula find their "justification" in a historical and epistemological study, as, for example, non-Euclidean geometry introduced in the curriculum of some countries (in Italy, for instance). The proposal of introducing some elements of non-Euclidean geometry in high schools could be very interesting; but we run the risk of introducing it only by giving some theorems. On the contrary, it is important to give students an idea of the epistemological meaning of the non-Euclidean revolution: i.e., the overcoming of the old conception of geometry.

A historical analysis allows teachers to understand why a certain concept is difficult for the student

The concept of function, one of the main concepts in high school mathematics courses, can be considered as a unifying idea occurring in various chapters of scientific knowledge. The different names that it assumes—operation, correspondence, relation, transformation—reflect the historical circumstances in which it appeared in the fields of mathematics, of physics, of logic.

While two centuries ago functions were thought of as formulas that described the relation between two variables involving algebraic expressions (in Euler's meaning), the modern definition of function is not so limited. It is not necessary that functions are represented by a graph or by a formula, and the corresponding sets do not only consist of numerical elements.

In modern mathematics courses a function from X to Y (where X and Y are sets of real or complex numbers) is defined to be a rule that assigns to each element x of the set X a unique element $y = f(x)$ of the set Y. Sometimes the function f is defined in terms of the set of all pairs $(x, f(x))$, a subset of the Cartesian product set $X \times Y$.

In mathematics education it is often forgotten that the concept of function was the result of a long train in mathematical thought developing slowly. On the contrary, at the school level this concept is in general introduced very early as a basis on which other concepts are introduced. But is this basis really well understood? What happened during its long history?

The idea of function can be seen in ancient mathematics. For instance ratios can be seen as functional relations. But it must be clear that Greek geometry was concerned more with form than with variation, so that the function concept in itself was not developed. There was in Greek geometry no idea of a curve as corresponding to a function.

The notion of function arose as the necessary mathematical tool for the quantitative study of natural phenomena,[5] begun by Galileo (1564–1642) and Kepler (1571–1630). Its development was based in the expressive possibilities provided by the modern algebraic notation created by Viète (1540–1603) and, especially, by the analytic geometry introduced by Descartes (1596–1650) and Fermat (1601–1665). Descartes stated that an equation in two variables, geometrically represented by a curve, indicates a dependence between variable quantities. Relationships between these quantities were often described by means of equations. However, these geometrical variables were viewed primarily as being associated with the curve itself, rather than with each other.

We can see here, as, for example, much earlier in Nicole Oresme's (1323–1382) *Treatise on the Configurations of Qualities and Motions,* a notion of a functional relationship between variables, but the several variables associated with a curve were not generally viewed as depending upon some single "independent" variable.

In the last part of the seventeenth century Leibniz (1646–1716) introduced the word "function" into mathematics precisely as a term designating the various geometrical quantities associated with a curve; they were the "functions" of the curve. Then, as increased emphasis was placed on the formulas and equations relating the functions of a curve, attention came naturally to be focused on their roles as the symbols appearing in these equations, that is, as variables depending only on the values of other variables and constants in equations (and thus no longer depending explicity on the original curve). This gradual shift in emphasis led ultimately to the definition of a function given by Euler (1707–1793) at the beginning of the *Introductio in Analysin Infinitorum* (1748): a function of variable quantity is an analytical expression composed in any way from the variable quantity and from numbers or constant quantities. In present day terminology, we can say that Euler's definition included just the analytic functions, a restricted subset of the already small class of continuous functions. As far as mainstream mathematics is concerned, the identification of functions with analytical expressions would remain unchanged for all the eighteenth century.

Euler's *Introductio* was the first work in which the function concept played a central and explicit role. It was the identification of functions, rather than curves, as the principal objects of study that permitted the arithmetization of geometry and the consequent separation of infinitesimal analysis from proper geometry.

Euler later (1755) gave a broader definition that is virtually equivalent to modern definitions of functions: if some quantities so depend on other quantities that if the latter are changed the former undergo change, then the former quantities are called functions of the latter. This denomination is of the broadest nature and comprises every method by means of which one quantity can be determined by others. If, therefore, x denotes a variable quantity, then all quantities which depend upon x in any way or are determined by it are called functions of it [$y = f(x)$].

This definition was revised and extended in the nineteenth century into a definition which deeply changed the nature and meaning of the term "function." (Ponte, 1992). The definition in terms of arbitrary sets was reached only in the second half of that century. In retrospect, it is pertinent to remark that whereas the idea of variability had been banned from Greek mathematics because it led to Zeno's paradoxes, it was precisely this concept which, revived in the later Middle Ages and represented geometrically, led in the seventeenth century to the calculus. Nevertheless, as the culmination of almost two centuries of discussion on the basis of the new analysis, the very aspect which had led to its rise was in a sense again excluded from mathematics with the so-called 'static' theory of the variable which Weierstrass developed.

At school level the set theory approach—that is a static approach—in which a function is a 'triad' (domain, codomain, correspondence between their elements) can blur the *dynamism* inherent in the function idea as dependence between two variables quantities. On the other hand, it could happen that where the aspect of 'dependence' is dominant, the function domain gets to assume a minor role. Moreover, when students use only Euler's definition, misconceptions which lead to stereotypes arise. The students develop 'prototype examples' of the function concept, such as: a function is like $y = x^2$, or a polynomial, or $1/x$, or a sine function (Baker, Tall, 1991).

I think that we could take from the rich history of the concept of a function the ideas for different approaches in introducing this concept at school level: the different levels of representing a function become necessary for its comprehension; the different notations are complementary, but they do not involve the same degree of difficulty.

References

Baker, M., Tall, D.: 1991. 'Students' Mental Prototypes for Functions and Graphs', in F. Furinghetti (ed.) *Proc. XVth PME* , Assisi (Italy), I, pp. 104–111.

Boyer, C. B.: 1959, *The History of the Calculus and its Conceptual Development,* Dover Publications, New York.

Deledicq, A. (Ed.),: 1991, Math & malices (le magazine de mathèmatiques pour tous), n. 1, ACL-Editions, Paris.

Edwards, C. H.: 1979. *The Historical Development of the Calculus,* Springer-Verlag, New York.

Euler, L.: 1748 (1922), *Introductio in Analysin Infinitorum,* vol. 1, *Opera Omnia,* Ser. 1, vol. 8, Leipzig.

Euler, L.: 1755 (1913), *Instituiones Calculi Differentialis,* Opera Omnia, Ser. 1, vol, 10, Leipzig.

Giusti, E.: 1980, *Bonaventura Cavalieri and the Theory of Indivisibles,* Edizioni Cremonese, Bologna.

Glasersfeld, E. von: 1991, 'A Constructivist's view of Learning and Teaching', International Conference on Psychology of Learning and Mathematics Learning, (preprint in English), *L'Educazione Matematica,* April 1992, 27–39.

Grugnetti L.: 1988, 'The Role of the History of Mathematics in an Interdisciplinary Approach to Mathematics Teaching', in L. Bazzini & H. G. Steiner (eds.) *Proc. of the first Italian-German bilateral symposium on didactics of mathematics,* 9–23.

Grugnetti, L. and Jaquet, F.: 1996, Senior Secondary School Practices (13–18 years), *International Handbook of Mathematics Education* (A. Bishop et al. eds.), The Kluwer International Handbooks of Mathematics Education, 615–645.

Lombardo Radice, L. & Mancini Proia, L.,: 1979 *Il metodo matematico,* Principato, Milano.

Pepe, L.: 1990, 'Storia e Didattica della matematica', *L'Educazione Matematica,* Suppl. al n.2, 23–24.

Ponte, J. P.: 1992, 'The History of the Concept of Function and Some Educational Implications', *The Mathematics Educator,* (3) 2, 3–8.

Weil, A.: 1978, History of Mathemathics: why and how, in *Proc. of the International Congress of Mathematicians,* Helsinki I, 227–236.

Endnotes

[1] See, at least, *Les obstacles épistémologiques et les problémes en mathématiques* by G. Brousseau, in the Proceedings of 28th CIEAEM meeting, 1976, Louvain-la Neuve and A. Serpinska's works on the epistemological obstacles concerning the concept of the limit in: Recherches en Didactiques des Mathématiques, 6, 5, 1985, pp. 5–67; Proceedings of the 37th CIEAEM's Meeting, Leiden, 1986, pp. 73–95.

[2] Among the different proposals there are the well known French works of some IREM that, in particular, are concerned with a critical analysis of "originals" directly in classrooms (see. for example, *Pour une perspective historique dans l'enseignement des mathématiques, Mathématiques: approche par des textes historique (tome 2),* Paris VII, Janvier 1990).

[3] English translations of brief but illustrative sections of these two lengthy works may be found in Struik's *A Source Book in Mathematics, 1200–1800,* Cambridge, MA: Harvard University Press, 1969, pp. 209–219.

[4] See E. Giusti, *Bonaventura Cavalieri and the theory of Indivisibles,* Edizioni Cremonese, 1980.

[5] See J. P. Ponte (1992).

Euclid versus Liu Hui: A Pedagogical Reflection[1]

Wann-Sheng Horng
National Taiwan Normal University

I. Introduction

It has been argued that ancient mathematics of the West and the East might have contributed complementary aspects, namely, the structural and the operational, to mathematics. These aspects are particularly revealing in the works of Euclid and Liu Hui (third century CE).[2] In this article, the author will compare the epistemology and methodology reflected in the methods of these two mathematicians in finding the greatest common divisor of two natural numbers, in their circle measurements, and in their proofs of the formula for the volume of pyramid. Pedagogical comments will be made to show the contrast between Euclid's structural (or theoretical) and Liu Hui's algorithmic (or constructive) approaches to the mathematical problems in question.

II. Arithmetica vs. "Suan Shu" (Logistica)[3]

In the elementary school mathematics curriculum, "arithmetic" usually refers to the computational procedures, or algorithms, for real numbers. This aspect of mathematics was called *logistica* (a word related to our "logistics") by the ancient Greeks (Cf. Bunt et al., 1988, p. 75). For them, arithmetic (or *arithmetica,* a word related to our discipline "number theory"), on the other hand, was a study of the abstract mathematical properties of numbers, chiefly the natural numbers. Greek enthusiasm about arithmetic, among other mathematical disciplines, was witnessed by Plato:

> [T]his knowledge of the kind [i.e., mathematics] for which we are seeking, has a double use, military and philosophical; for the man of war must learn the art of number or he will not know how to array his troops, and the philosopher also, because he has to rise out of the sea of change and lay hold of true being, and therefore he must be an arithmetician (Republic, Book VII, p. 525).

In fact, arithmetic, along with geometry, astronomy and harmonics formed the quadrivium in the Classical World. No wonder Euclid would devote Books VII, VIII and IX of his *Elements* to arithmetic, since this text was probably designed for teaching mathematics in ancient Greece.

However, side by side with the tradition of classical Greek geometry, which is known from the works of Euclid, Archimedes, Apollonius, and Pappus, a more popular tradition existed, a tradition of arithmetical and geometrical problems with numerical solutions, similar to the problems we find in Egyptian, Babylonian and Chinese collections (Van der Waerden, 1983, p. 154). This is very well illustrated by Heron's *Metrika.* Whereas Euclid operates with the line segments, polygons, circles, and solids themselves, without ever using words like "length", "area", or "volume", Heron is mainly concerned with the numerical values of areas and volumes. This indeed points to the fact that Heron was dealing with mathematics in terms of its operational aspect.

Heron's treatment of mathematics was no exception in the history of mathematics. In addition to his Babylonian and Egyptian predecessors, Heron also had Chinese counterparts. Ancient Chinese mathematics is well known for its algorithmic aspects which are devoted primarily to solving practical problems. This may explain in part why the *Jiu Zhang Suan Shu* (Nine Chapters on Mathematics, first century CE) introduces the rules for addition, subtraction, multiplication, and division of fractions from the very beginning. Indeed Chapter 1 of the *Jiu Zhang Suan Shu* was written to calculate the area of plane figures whose linear dimensions are often given as fractions.

It should be noted that the rules of addition, subtraction, multiplication and division given in the *Jiu Zhang Suan Shu* all agree with those we learn at school. More interesting is the method for simplifying fractions:

> If both [the denominator and numerator] can be halved, then halve them. When both cannot be halved, set down the numerals of the denominator and numerator, and subtract the smaller from the larger. Continue to diminish mutually through subtractions ("Geng Xiang Jian Sun") to seek a pair of equal numbers ("Deng Shu"). Use [this number called] the "Deng Shu" to reduce [the fraction] (Qian, 1963, pp. 94–95. Here the translation follows Lam and Ang, 1992, p. 56).

The text gives two examples, namely Problems 5 and 6 of Chapter 1. The latter is to simplify 49/91. We are instructed first to lay out

$$\begin{matrix} 49 & 49 & 7 & 7 & 7 & 7 & 7 & 7 \\ 91 & 42 & 42 & 35 & 28 & 21 & 14 & 7 \end{matrix}$$

then to divide the numerator and denominator by the "equal number" 7 in order to get the answer 7/13. According to Van der Waerden, since it is quite easy to get the answer in this case, "the mention of the algorithm is not a logical or didactical necessity; it is just an addition by a systematically-minded teacher, who wanted to teach a never-failing method" (Van der Waerden, 1983, p. 38).

Even so, an explanation of how the algorithm works made by Liu Hui in his commentary to the *Jiu Zhang Suan Shu* is equally impressive:

> To reduce by using the "Deng Shu" means to divide [the denominator and numerator] by it. As the mutually subtracted numbers are all multiples ("Chong Die") of the "Deng Shu", this is the reason why the "Deng Shu" is used to reduce [the fraction] (Qian, 1963, p. 95. Here too the translation follows Lam and Ang, 1992, p. 56).

Whether or not this argument is convincing, one must admit that Liu Hui did tell us the reason why the algorithm ("Geng Xiang Jian Sun") worked successfully in simplifying the fraction. Meanwhile, due to his argument that "the mutually subtracted numbers are all multiples of the 'Deng Shu'", one cannot help thinking of his structural concern in the context of algorithmic mathematics.

In contrast, Euclid was primarily interested in finding a common measure (divisor) of two [natural] numbers. For example, Proposition 1 of Book VII reads:

> Two unequal numbers being set out, and the less being continually subtracted in turn from the greater, if the number which is left never measures the one before it until an unit is left, the original numbers will be prime to one another (Heath, 1956, vol. 2, pp. 296–297).

Immediately following Proposition 2 there is the method of finding the greatest common measure of two numbers not prime to one another:

> Given two numbers not prime to one another, to find their greatest common measure (Heath, 1956, vol. 2, pp. 298–300).

These two propositions comprise the so-called Euclidean algorithm. It should be noted that the Euclidean algorithm is just the same as the Chinese "Geng Xiang Jian Sun" despite the fact that the latter awaited Liu Hui's explanation. Indeed the Greeks as well as the Chinese reduce the pair of numbers (m, n) by alternative subtractions until they become equal, and then divide m and n by the resulting common divisor.

That concepts like prime number and numbers prime to one another never occurred in traditional Chinese mathematics may be because the ancient Chinese never paid any attention to mathematical problems which could be associated with number theory or arithmetic in the Greek sense. Nor did they discover the relation between the greatest common measure and the least common multiple. In fact, Chapter 4 of the *Jiu Zhang Suan Shu* gives the concept of the least common multiple in order to deal with the addition of fractions. However, the concept of the "Deng Shu" is never mentioned in the computational procedures. What is more surprising is that Liu Hui, the most important commentator of the *Jiu Zhang Suan Shu* in ancient China, also says nothing in this context. In other words, it seems that the structural relation between the greatest common measure and the least common multiple never entered Liu Hui's mind even though he paid much attention to the concept of the "Deng Shu" (the greatest common measure).

On the other hand, Books VII, VIII and IX of Euclid's *Elements* are devoted to number theory. A theoretical framework therefore is provided for dealing with the properties of number *per se,* for example, definitions of prime number and composite number etc. (Heath, 956, vol. 2, pp. 277–278). What most concerns Euclid is the structural aspect of (natural) numbers including the relation between prime and composite numbers as well as that of the greatest common measure and the least common multiple (Heath, 1956, vol. 2, pp. 336–341). No wonder Euclid stresses the significant role prime numbers play in number theory. Under such circumstances, it is not surprising to see that he even tries to count how many prime numbers there are (Heath, 1956, vol. 2, p. 412). Moreover, Euclid's proofs of Propositions VII-1 and VII-2 deserve our special attention. This is because even in the case involving an algorithm, say, the Euclidean algorithm, he still appeals to the argument of *reductio ad absurdum* (reduction to absurdity) (Cf. Heath, 1956, vol. 2, pp. 296–300). He is primarily concerned with the structural aspect of mathematics even when he is doing something very algorithmical.

III. Circle Measurement

In Problems 31 and 32 of Chapter 1 in the *Jiu Zhang Suan Shu,* we are asked to calculate the areas of circular fields. For example, Problem 31 reads:

> A circular field has circumference 30 "Bu" (1 "Bu" is about 126 cm in 1st-century China) and diameter 10 "Bu". What is the area of the field? (Qian, 1963, p. 103)

In solving this problem, the *Jiu Zhang Suan Shu* provides four formulas for the area S of the circle T. The first of them is: $S = (C/2)(D/2)$, where C, D are the circumference and diameter of the circle respectively (Qian, 1963, pp. 103–108).[4]

In order to show why this formula can help compute the area of a circle, Liu Hui suggests that from the beginning we can regard $C/2$ and $D/2$ respectively as two sides of a rectangle, say R. If we can transform the circle T into the rectangle R, preserving the area, then the proof is accomplished (see Figure 1). This is Liu Hui's strategy (Qian, 1963, p. 103). In fact, it is very natural for Liu Hui to adopt this strategy, for he has already proved the area formulas for the triangle and trapezoid by transforming each of them into rectangles (Cf. Qian, 1963, pp. 101–102). What he uses is the so-called In-Out Principle ("Yi Ying Bu Xu"; see Figure 2). Since the circle is a curved plane figure, therefore, in addition to the use of the In-Out Principle, Liu Hui apparently realizes that he needs something more. This may explain why he also counts on the method of limit to accomplish the proof of the formula: $S = (C/2)(D/2)$.

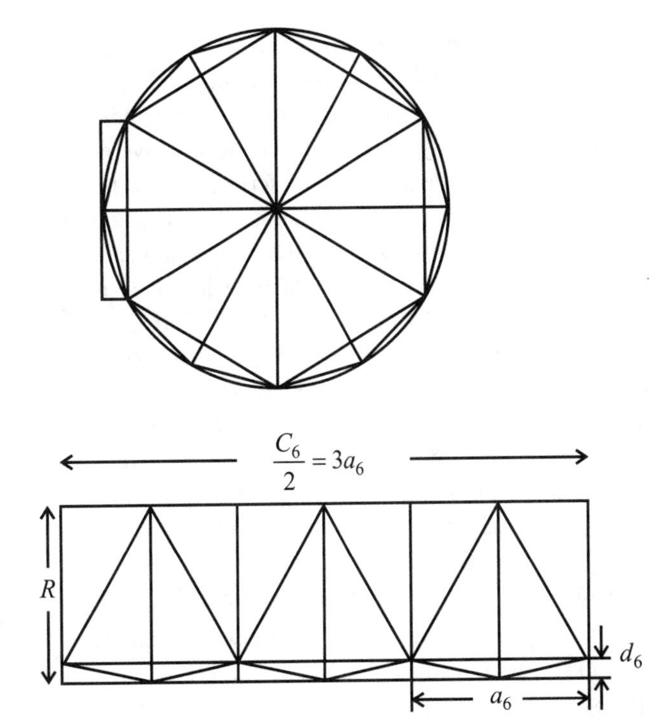

FIGURE 1

In his proof, Liu Hui inscribed a regular hexagon within a circle. By successively doubling the number of sides—a method known as the "Ge Yuan" (circle division) method, he believed it was possible to reach a polygon which coincided with the circle and therefore "exhausted" its area (Qian, 1963, pp. 103–104; Lam and Ang, 1986). There is no doubt that Liu Hui would allow the cutting to be infinite, for "if one makes the cut finer, the loss [of the

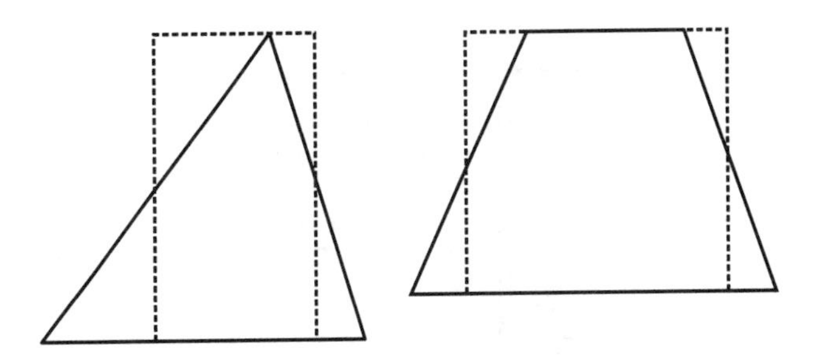

FIGURE 2

area of the inscribed regular polygon with respect to the area of the circle] will be smaller" (Qian, 1963, pp. 103–104). Here Liu Hui's conceptions of infinity and indivisible point are also clearly reflected in his quantitative estimation of the loss of the area of the inscribed regular polygon with respect to the area of the circle. Yet then why did Liu Hui need to emphasize: "cut the circle again and again till it cannot be cut any more; then the inscribed regular polygon will coincide without any loss"? Perhaps to Liu Hui exhausting the cutting to reach an eventual geometric entity was meaningful only if he had a formula to prove. This certainly was not the case with the area of a segment of a circle ("Hu Tian") in Chapter 1 of the *Jiu Zhang Suan Shu* in which no exact formula of the "Hu Tian" was given (Cf. Qian, 1963, pp. 108–110; Horng, 1995).

Now let us turn to Euclid. The "area formula" Euclid presents in his *Elements* is Proposition 2 of Book XII:

> Circles are to one another as the squares on the diameters (Heath, 1956, vol. 3, pp. 371–373).

The proposition implies that he only knows that the ratio of a circle to the square of its diameter is a constant. However, he never mentions what the ratio is. Perhaps what he had in mind is actually $\pi : 4$. Yet a ratio, say $a : b$, in his *Elements* is not treated as a number a/b *per se* (Kline, 1972, p. 73). Therefore Euclid should not be credited with giving a formula for the area of a circle.

In order to prove this theorem, Euclid applies the principle of exhaustion, a legacy of Eudoxus. He puts this principle as Proposition 1 of Book X in his *Elements* (Heath, 1956, vol. 3, p. 14). The method does not really suggest that the inscribed polygons of the circle are going to "exhaust" the circle. Instead it allows us to "approximate" the circle with the inscribed polygons to whatever extent we like. His proof of the theorem begins by assuming that the result is not true—an argument based on a *reductio ad absurdum* (Katz, 1992, pp. 86–87).

It should be noted that "[t]he method of exhaustion, although equivalent in many respects to the types of argument now employed in proving the existence of a limit in the differential and the integral calculus, does not represent the point of view involved in the passage to the limit" (Boyer, 1949, p. 35). In fact, in his proof of the formula for the circle area Euclid never allows the inscribing of polygons in the circle literally to be carried out to an infinite number of steps. Meanwhile, although the inscribed polygon could be made to approach the circle as nearly as possible, it could never become the circle, for this would imply an end in the process of subdividing the circumference.

It seems quite enough at this point for us to compare Liu Hui and Euclid's treatments of the circle measurement. In contrast with Euclid's proof, Liu Hui's argument is more intuitive and illuminating in that every stage of his proof shows clearly his goal. Therefore, one cannot but agree that underlying his methodology Liu Hui had "the belief in a balanced employment of rigorous argument and heuristic reasoning with the aim of achieving better understanding" (Siu, 1993). In fact, for certain audiences (Horng, 1994), Liu Hui's proof is even more attractive than that of Archimedes' formula:

> The area of any circle is equal to the area of a right triangle in which one of the legs is equal to the radius and the other to the circumference (Fauvel and Gray eds., 1987, pp. 148–150).

Archimedes' formula is equivalent to the Chinese one, namely both have the same form of $(1/2)C \cdot r$, where C and r are the circumference and radius respectively. Following Euclid, Archimedes also uses the method of exhaustion to give his proof (Cf. Fauvel and Gray eds., 1987, pp. 148–150). Unlike Euclid however, Archimedes knows the formula for the area of a circle.

Even so, just like Euclid's case, each step of Archimedes' proof hints at nothing about how the formula $(1/2)C \cdot r$ was obtained in a constructive manner.[5] Judged by this standard, Liu Hui's treatment of circle measurement is more heuristic than Archimedes' despite the fact that Archimedes in his book *The Method* shows his enthusiasm for his method of discovery.

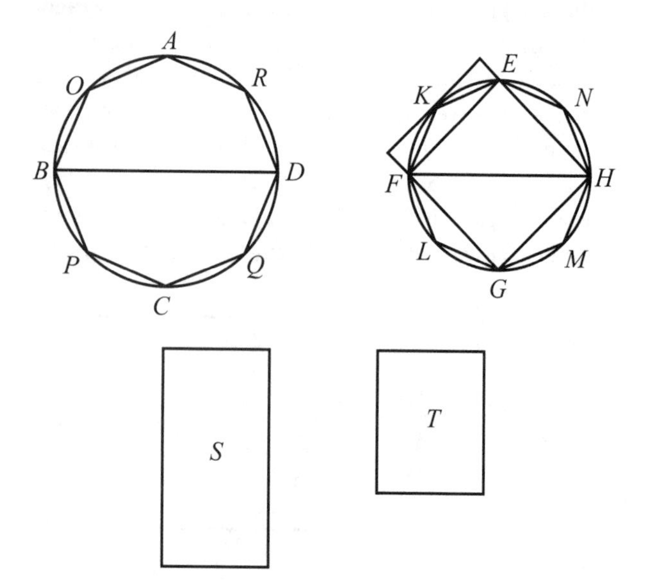

FIGURE 3

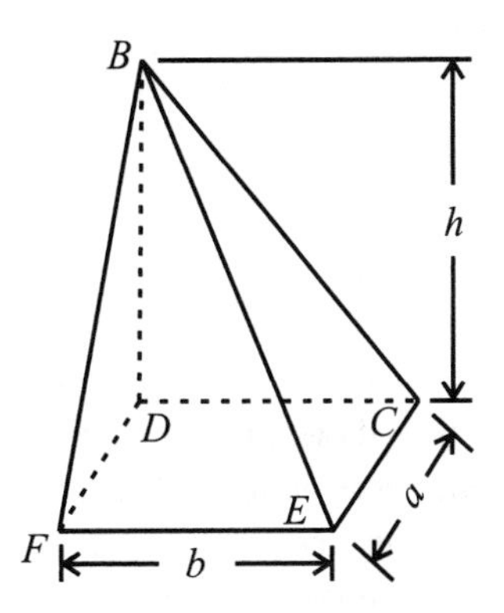

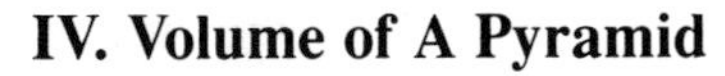
FIGURE 4

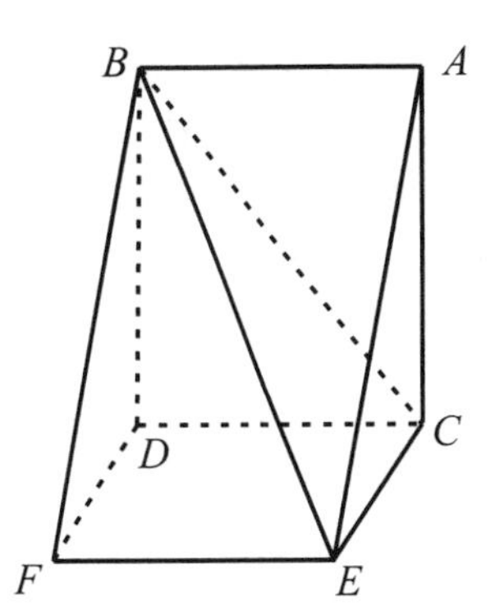

FIGURE 5

IV. Volume of A Pyramid

As is well known, any proof of the formula for the volume of a pyramid must use infinitesimal considerations in one form or another. It is not surprising to see, just as in the case of circle measurement, that Liu Hui also makes use of a limit process in deriving the volume of the pyramid, "Yang Ma", which has a rectangular base and one lateral edge perpendicular to the base. In fact, Liu Hui's argument was given in his commentary on Problem 15 of Chapter 5 in the *Jiu Zhang Suan Shu*:

> Given a "Yang Ma" which has a length of 5 "Bu" and a width of 7 "Bu" [for its rectangular base] as well as a height [i.e., lateral edge perpendicular to the base] of 8 "Bu", find its volume (Qian, 1963, p. 166).

Let us begin with a brief summary of Liu Hui's derivation, "translated" into modern language and symbolism. Let a "Yang Ma" be given with dimensions a, b and h as shown in Figure 4. A tetrahedron "Bie Nao", i.e., a pyramid with right-triangular base and with one lateral edge perpendicular to the base, could be chosen to fit together with the "Yang Ma" forming a wedge "Qian Du" (see Figure 5). Here the "Yang Ma" is $BDFEC$ and the "Bie Nao" is $BACE$. Set c as the volume of the "Qian Du" $ABDCEF$; y as the volume of the "Yang Ma" $BDFEC$; and p as the volume of the "Bie Nao" $BACE$. Since Liu Hui has already proved that $c = (1/2)abh$, in order to prove that $y = (1/3)abh$ it is sufficient to show that $y = 2p$. Now make the division as shown in Figures 6 and 7. It is clear that the sum of the volumes of the two "Qian Du" pieces of the "Bie Nao" is exactly one-half of the sum of the volumes of the one box and two "Qian Du" pieces of the "Yang Ma". Thus it remains to show that the two smaller "Bie Nao" pieces of the "Bie Nao" together represent half the volume of the two smaller "Yang Ma" pieces of the "Yang Ma".

These smaller "Bie Nao" and "Yang Ma" can again be divided up in the same way as shown in Figures 6 and 7. This division again yields some parts whose volumes have the desired ratio, and which altogether form $3/4$ of the volumes remaining in the first plus the remainders,

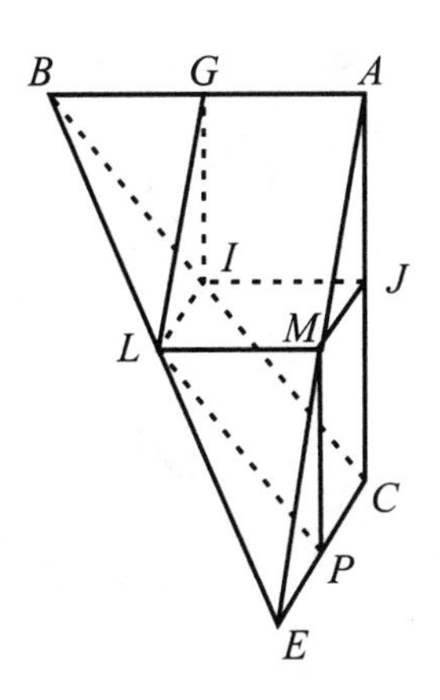

FIGURE 6

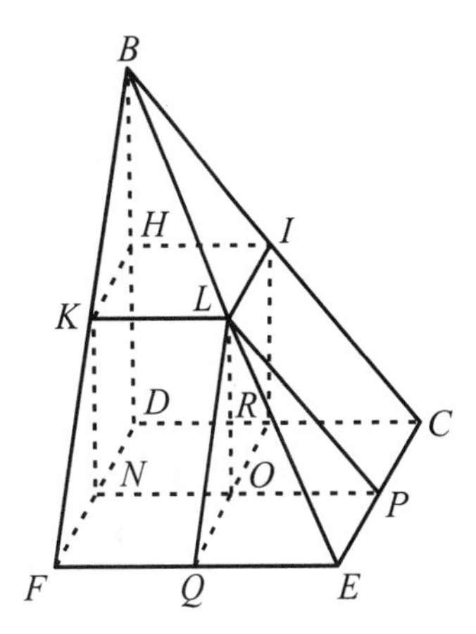

FIGURE 7

namely, four smaller "Bie Nao" and four smaller "Yang Ma". The still undetermined remainders are only $1/4$ of $1/4$ of the original "Qian Du". These can again be divided up in the same way. Eventually, "to exhaust the calculation," as Liu Hui puts it, "halve the remaining breadth, length, and height; an additional three-quarters can thus be determined" (Qian, 1963, p. 168).

But how did Liu Hui actually carry the process to the limit? He says:

> The smaller they are halved, the finer ["Xi"] are the remaining [dimensions]. The extreme of fineness is called "subtle" ["Wei"]. That which is subtle is without form ["Xing"]. When it is explained in this way, why concern oneself with the remainder? (Qian, 1963, p. 168; here the translation follows Wagner, 1979)

This passage makes it clear that Liu Hui has an idea of carrying the process to the limit (Van der Waerden, 1983, p. 202). For Liu Hui it is possible to exhaust the process of halving to reach a state of "Wei", a concept of indivisible point which he might have drawn from the pre-Qin period thinkers like the Mohists. In fact, coupled with his (quantitative) estimation of the remainders, Liu Hui must have used the argument of "exhausting the halving" to convince his readers that his demonstration really works, as does his proof of the formula for the area of the circle.

In addition, at each stage of his halvings, Liu Hui is able to hint at the so-called "Liu Hui`s Principle"—"Yang Ma": "Bie Nao" $= 2:1$ (Cf. Wu, 1982) which leads eventually to his proof of the volume formula of the "Yang Ma". In other words, Liu Hui's method is constructive in a sense that each step of the argument repeatedly points to the goal to be obtained.

As in his circle measurement, Euclid also uses the method of exhaustion to prove his volume formula of a pyramid. In addition, the argument of *reductio ad absurdum* is adopted apparently because, just as in the case of his circle measurement, his formula for the volume of a pyramid is expressed as a proportion, namely Proposition XII-5:

> Pyramids which are of the same height and have triangular bases are to one another as the bases (Heath, 1956, vol. 3, pp. 386–388).

Again he still needs Proposition X-1 to approximate the pyramid with its inscribed prisms. Prior to that it is also necessary for him to provide Proposition XII-3:

> Any pyramid which has a triangular base is divided into two pyramids equal and similar to one another, similar to the whole and having triangular bases, and into two equal prisms; and the two prisms are greater than the half of the whole pyramid (Heath, 1956, vol. 3, p. 378).

and Proposition XII-4:

> If there be two pyramids of the same height which have triangular bases, and each of them be divided into two pyramids equal to one another and similar to the whole, and into two equal prisms, then, as the base of the one pyramid is to the base of the other pyramid, so will all the prisms in the one pyramid be to all the prisms, being equal in multitude, in the other pyramid (Heath, 1956, vol. 3, p. 382).

In fact, in his proof of Proposition XII-5 Euclid takes away from the pyramid two prisms which together are more than half the pyramid (see Figure 8). From the remainder he again takes away more than half, and so on, until the remainder is less than any assigned volume (Cf. Heath, 1956, vol. 3, pp. 386–388). Van der Waerden has noted its parallel in methodology to Liu Hui's proof of the "Yang Ma": Liu Hui takes away three quarters from the prism ("Qian Du"), and from the remainder he again takes away three quarters, and so on, until the remainder is completely negligible (Van der Waerden, 1983, p. 203). On the other hand, Crossley and Lun comment that "Euclid did not need to use non-constructive methods in his derivation of the volume of the pyramid" (Crossley and Lun, 1994). This is because, as they suggest, if for given pyramids P, P' Euclid considered $P-$(prisms in P) and $P'-$(prisms in P') and repeated the subdivisions, then he could have constructive convergence (to zero) on both cases.[6]

With the resemblance of Liu Hui and Euclid's methods in hand, one is tempted to agree with Van der Waerden that Liu Hui was influenced by Greek sources (Van der Waerden 1983, p. 203). Whatever the case, Liu Hui's

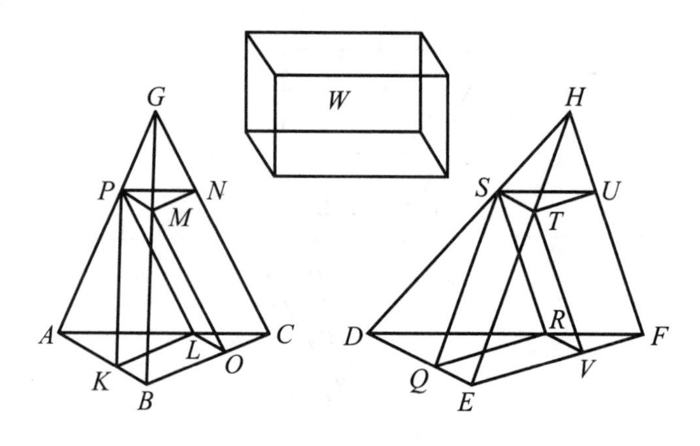

FIGURE 8

method indeed provides a direct avenue which help us to understand how the volume of the "Yang Ma" is two times that of the "Bie Nao". In contrast, due apparently to the argumentation of *reductio ad absurdum,* the indirect feature of Euclid's approach is evident in his proof of the formula for the volume of a pyramid, just as in his proof of the formula for the area of a circle. Meanwhile, one has to admit that Euclid does not say much about what the volume of the pyramid is. He simply puts it as: "Any prism which has a triangular base is divided into three pyramids equal to one another which have triangular bases." (Proposition 7 of Book XII, cf. Heath 1956, vol. 3, p. 394). But nowhere in his *Elements* does Euclid mention any formula for the volume of the prism. As to the reason, we would like to speculate in the next section.

V. Operational vs. Structural

It seems quite clear, from the examples discussed above, that in contrast with the structural features of the *Elements,* the *Jiu Zhang Suan Shu* is a mathematical classic which is basically operational. Therefore, if the structural and the operational aspects of mathematics can be regarded to be complementary, then so are the learning of the *Elements* and the *Jiu Zhang Suan Shu.* In fact, even "[i]n Euclid, the role of dialectic is to justify a construction—i.e., an algorithm" (Davis and Hersh, 1981, p. 182).

For Euclid indeed, the definitions, postulates and common notions of the *Elements* are those suggested by common sense by which geometrical construction could be made. Besides, his geometry never loses contact with spatial intuition. For example, Definition 4 of Book I is about a straight line which is reminiscent of the experience of the ancient Egyptian rope-stretcher:

> A straight line is a line which lies evenly with the points on itself (Heath, 1956, vol. 1, p. 153).

It has also been suggested that in the *Elements* Proposition X-2 as well as Proposition X-3 were originally used as a method of approximating the length of the diagonal of a square rather than in order to prove its incommensurability with the side (Heath, 1956, vol. 3, pp. 17–22; Lloyd, 1984, pp. 106–108). These two together with Proposition VII-1 and VII-2, i.e., the Euclidean algorithm, are called "anthyphairesis", a term showing that a clear practical bearing can be located among these propositions (Lloyd, 1984, pp. 106–108).

Now let us briefly trace Euclid and his *Elements* back to a period in which the Greek philosophers Parmenides, Zeno, Plato and Aristotle dominated mathematical studies, at least at the academic level. Then we can better explain why Euclid was only interested in the structural aspect of mathematics. On the other hand, since Liu Hui also showed a very clear concern about philosophical issues in his commentary to the *Jiu Zhang Suan Shu,* perhaps it is not irrelevant for us to explore the interaction between philosophy and his mathematical study as well.

Euclid's use of the method of exhaustion and the argument of *reductio ad absurdum* suggests that the *Elements* is based on "refined intuition" rather than "naive intuition" (Boyer, 1949, p. 47). In so doing, the infinitesimal is excluded from the demonstrations of geometry as shown in the proofs of Propositions XII-2 and XII-5. The reason may lie in Euclid's response to the Eleatic philosophers Parmenides and Zeno. In fact, Zeno's famous Achilles paradox is directed against the thesis that space and time are infinitely divisible since, according to his explanation of the infinitesimal, it is impossible to conceive intuitively the sum of infinitely many infinitesimals (Boyer, 1949, p. 24). Apparently "it was the Eleatics who provided the first clear statement of the key thesis that serves as the epistemological basis for any abstract inquiry such as mathematics, namely the insistence on the use of reason (as opposed to the senses) as the criterion" (Lloyd, 1984, p. 110). On the other hand, granted that "[a] number is a multitude composed of units" (Definition 2 of Book VII in the *Elements*) (Heath, 1956, vol. 2, p. 177), the ratio of the diagonal and the side in a square (i.e., the square root of 2) is simply not a number at all. This may have compelled Euclid to incorporate Eudoxus's theory of proportion into Book V of the *Elements*—an indication that Euclid was forced to deal with number and magnitude separately since number in the Greek sense is not enough to deal with incommensurable magnitudes (i.e., irrational numbers in the modern sense). Consequently, Euclid was obliged to express the area of a circle and the volume of a pyramid in proportional forms.

Indeed, incorporation of Eudoxus's principle of exhaustion as well as the argument of *reductio ad absurdum* into the *Elements* is only one piece of evidence showing that Euclid had accomplished the grand synthesis of Greek philosophy and mathematics. By taking seriously into account the Eleatic challenge, Euclid, among other Greek mathematicians, adopted the hypothetico-deductive format in order to secure the certainty of mathematical knowledge (Lloyd, 1984, pp. 70–72, 110, 118). This may well explain why the *Elements* reflects primarily the structural aspects of mathematics.

The role of Liu Hui in the history of Chinese mathematics is parallel to that of Euclid in the history of Greek mathematics. In fact, just as Euclid concludes prior Greek mathematics with a hypothetico-deductive structure of his own, Liu Hui establishes a conceptual framework to analyze the underlying logical interconnections of algorithms in the *Jiu Zhang Suan Shu* (Cf. Guo, 1995, pp. 301–322). Unfortunately, as a commentator, Liu Hui was obliged to confine his study of mathematics to explaining methods, formulas and algorithms of the *Jiu Zhang Suan Shu*. In so doing, Liu Hui, just like other ancient Chinese scholars, was not expected to break the format of the text or to reorganize its content into a new text (Cf. Wagner, 1979). Under such circumstances, he did not have a chance to build a theoretical structure of his own. It comes as no surprise that his mathematics retains the algorithmic features on which the practical concern of Chinese mathematics could lean—since for the ancient Chinese, mathematics is one of the six arts in solving practical problems.

Even so, in dealing with mathematics, Liu Hui, just like Euclid, does not count on "naive intuition" but "refined intuition". In the case of the "Geng Xiang Jian Sun", although his structural concern had to be fitted into the algorithmic jacket designed by the *Jiu Zhang Suan Shu*, Liu Hui's explanation not only helps justify the method but also enhances understanding of the structural aspect. In the cases of circle measurement and the volume of a pyramid, Liu Hui's explanations of their formulas are constructive in the sense that they provide a direct avenue for us to attain the final goal. It should be noted that Liu Hui's argument always tells us how to proceed in the next step and where to go for the goal. Moreover, his proofs of the formulas for the area of a circle and the volume of a pyramid not only offer justifications but explain the ways to re-discover the formulas. In other words, by contrast with Euclid's sole concern about the structural aspect of mathematics, Liu Hui emphasizes both the operational and the structural.

Now that we have established that Liu Hui was trying to add something structural to the *Jiu Zhang Suan Shu*, a basically operational text, can we explain his epistemology and methodology in the socio-cultural context of third-century China? As argued above, we can trace the influence on Euclid's *Elements* of the philosophies of the Eleatics, Plato and Aristotle. Yet how about Liu Hui's mathematics?

Indeed, Liu Hui's concern about the argumentation of mathematics had something to do with the heritage of several schools of thought, especially the sophists and the Mohists in the pre-Qin period (Cf. Wagner, 1979; Guo, 1995, pp. 323–348; Horng, 1995). Especially striking is the resemblance of Liu Hui's epistemology to that of the Mohists. In his treatments of circle measurement and the volume of a pyramid, Liu Hui apparently adopted the Mohists' definition of an indivisible point and thereby applied his own principle of exhaustion. Besides, Liu Hui also justified his propositions essentially in terms of arguments related to the concept of an indivisible point. In so doing, Liu Hui might have followed the Mohists in their studies of argumentation (Cf. Guo, 1995, pp. 267–300). However, if Chinese mathematics was basically devoted to the empire's practical affairs such as calendar-reform, agriculture, and so on, then how could Liu Hui justify his own mathematical study which doubtless reflected something beyond a mere practical flavor? It should be noted that unlike in Plato's philosophy, mathematics simply has no place to show its epistemological power in Confucian philosophy. This means that mathematical training has nothing to do with Chinese sagehood (Cf. Horng, 1991). Under such circumstances, mathematics was always regarded as an inferior art which scholars only learned for leisure use. If that is the case, why has Liu Hui bothered with something more?

Although almost nothing about his life is known to us, Liu Hui was active in the late third century, for he wrote his commentary to the *Jiu Zhang Suan Shu* in 263 CE. That is a period when Confucian philosophy lost its status as a state orthodoxy which, as mentioned above, never respected mathematics highly as an academic discipline. Thus Liu Hui should have been comfortable in studying mathematics for its own sake under a more flexible intellectual background (Cf. Horng, 1992). This in turn might have led him to pay attention to issues of methodology with the source materials available thanks to the re-appearance of pre-Qin texts like the Mohist at the time (Cf. Guo, 1995, pp. 267–301, 323–348). Therefore, just as in Euclid's case, Liu Hui's mathematics can be better understood in terms of socio-intellectual factors.

VI. Conclusion: A pedagogical reflection

First of all, in mathematical learning one should be aware that "the terms 'operational' and 'structural' refer to inseparable, though dramatically different, facets of the same thing" (Sfard, 1991). In other words, these two aspects are to be regarded as "the different sides of the same coin". In this connection, therefore, their duality rather than dichotomy is emphasized (Sfard, 1991).

If that is indeed the case, we would like to recommend Liu Hui's approach to mathematics for school teachers. This is principally because one can learn from Liu Hui "both what and how to do" or "both rules and reasons" (Cf. Skemp, 1995, pp. 203–213, 221–225). In the case with finding the greatest common divisor of two natural numbers, the "Geng Xiang Jian Sun" method suggests that a series of "reciprocal subtractions" would be performed with counting rods ("Suan Chou") on the counting board in the manner shown in Section II above (Cf. Lam and Ang, 1992, pp. 54–56). According to Van der Waerden, the algorithm manifests itself even without Liu Hui's explanation (Cf. citation in Section II above). However, to our students who learn the same procedure in terms of manipulation of Hindu-Arabic numerals, Liu Hui's explanation reminds them to check the procedure if an error occurs in the final result. Of course, the Euclidean algorithm now familiar to school teachers and students is easy to manipulate too. Yet the "Geng Xiang Jian Sun" method and the accompanying explanation by Liu Hui are still more illuminating in "both rules and reasons". Nevertheless, the Euclidean algorithm should not be ignored because it is at the more advanced stage of concept formation. For example, after students have become familiar with how to find the greatest common divisor and in turn how to simplify fractions, they should be encouraged to explore what the concept of divisor *per se* is all about. For them then the structural aspect of the natural numbers may become a genuine issue. In other words, maybe now the students are ready for some features of elementary number theory.

In the cases of circle measurement and the volume of a pyramid, we also think that it is appropriate to introduce to students Liu Hui's approach at the first stage of their learning. This is because Liu Hui`s method shows a very heuristic flavor that is lacking in Euclid's solely structural concern. Moreover, Liu Hui`s approach also solicits the readers to share his methods with models with which they are familiar in their daily lives.[7] Above all, his instruction may serve to encourage teachers not to sever mathematics teaching at elementary levels from the ordinary life experience of students. Nevertheless, Euclid's treatments of circle measurement and the volume of a pyramid deserve a place in geometry teaching not only for their standard of rigor but also for their emphasis on the structural relations between geometrical entities. These treatments would help brighter students come to realize that exploring geometrical entities for their own sake is worthwhile. This would in turn help them to appreciate the beauty of mathematics—an intellectual satisfaction that all qualified teachers like to share with their classes.

As a concluding remark, let me write a few words about how the history of mathematics should be related to the pedagogy of mathematics. In concluding their study concerning the comparison of the logic employed by Euclid and Liu Hui in proving the formula for the volume of a pyramid, Crossley and Lun comment that:

> Nowadays we use the abstract approach of Euclid and the algorithmic and practical approach of Liu Hui together. In this way our mathematics can provide sound proofs and accurate calculations at the same time. These are essential for writing good computer programs. They also reflect how beneficial it is to use the ideas of our ancestors from the West and the Middle Kingdom together (Crossley and Lun, 1994).

However, the argument of this article has tried to go one step further. I hope that it has successfully shown how a comparison of mathematical methods developed in different civilizations could be beneficial to mathematical teaching. In fact, the contrast of Euclid and Liu Hui should have suggested that meaningful teaching can be accomplished by approaching mathematics from different epistemological and methodological perspectives. Such a pedagogy is even more fruitful than the usual one in classrooms in its multicultural bearing. In this respect, historians of mathematics around the world all have something to contribute to the teaching enterprise. We look forward to seeing a closer cooperation of historians and teachers in the setting of mathematics education.

Endnotes

[1] The early version of this article has been presented, in an abridged form but with the same title, to the conference, HEM Braga 96, July 24–30.

[2] According to Cullen, "Liu Hui does not use any word which corresponds to Greek *apodeixis* or English `proof'" (Cullen, 1995). Nevertheless, if by a proof we mean not only to justify but also to enhance understanding, then comparison in methodology should be legitimate at least in the connections of HPM. In this article, therefore, the author will use the term "proof" in a broad sense—it refers not merely to Euclid's rigorous deductive demonstration but to Liu Hui's explanation as well.

[3] The term "Suan Shu", appearing often in ancient China, did not refer to mathematics in the modern sense. Literally it indicates a deep-rooted emphasis on calculation and algorithm. For example, the *Jiu Zhang Suan Shu* bearing the term "Suan Shu" in its title is the paradigmatic example in ancient China (Cf. Siu, 1996).

[4] The other three formulas for the circle are (in modern symbolic terminology): a). $(C \cdot D)/4$; b). $(3 \cdot D^2)/4$; c). $C^2/12$. Liu Hui is quite certain that the first is equivalent to $(C/2)(D/2)$, while the last two are exact only under the conditions that $\pi = 3$ (Cf. Qian, 1963, pp. 106–108).

[5] In explaining how Archimedes might have come up with his formula for the circle, Johann Kepler wrote that "I divide the circumference into as many parts as there are points on it, which is an infinite number. We consider each of them to be the base of an isosceles triangle with altitude r" (Quoted in Bero, 1995). In other words, Kepler offers an explanation of this formula in a constructive manner.

[6] Crossly and Lun also compare Euclid and Liu Hui's methodology in this connection: "The methodology of Euclid was more

abstract and general than that of Liu Hui but this general approach will sometimes give non-constructive proofs which do not allow direct calculations" (Crossley and Lun, 1994).

[7] For example, in proving the formulas of rectilinear solids in Chapter 5 of the *Jiu Zhang Suan Shu,* Liu Hui cut up the solid in question into at most four standard blocks, a "Li Fang" (cube), a "Qian Du", a "Yang Ma" and a "Bie Nao", each with length, breadth, and height 1 "Chi" (about 21 cm at Liu Hui's time). Then he manipulates the formulas for the parts to arrive at the formula for the whole (Cf. Li and Du, 1987, pp. 71–73).

References

Baron, Margaret E.: 1989, *The Origins of the Infinitesimal Calculus,* Dover Publications, Inc., New York.

Bero, Peter: 1995, "Volume Calculation in the Manner of the 17th Century", in IREM Montpellier (ed.), *History and Epistemology in Mathematics Education,* IREM de Montpellier.

Boyer, Carl B.: 1949, *The History of the Calculus and Its Conceptual Development,* Dover Publications, Inc., New York.

Bunt, Lucas N. H., Phillip S. Jones and Jack D. Bedient: 1988, *The Historical Roots of Elementary Mathematics,* Dover Publication, Inc., New York.

Crossley, J.N. and A.W.C. Lun: 1994, "The Logic of Liu Hui and Euclid as Exemplified in Their Proofs of the Volume of a Pyramid", *Philosophy and the History of Science: A Taiwanese Journal* 3(1), 11–27.

Cullen, Christopher: 1995, "How Can We Do the Comparative History of Mathematical Proof? Proof in Liu Hui and the Zhou Bi", *Philosophy and the History of Science: A Taiwanese Journal* 4(1), 59–94.

Davis, Philip J. and Reuben Hersh: 1981, *The Mathematical Experience,* The Harvest Press, New York.

Fauvel, John and Jeremy Gray eds.: 1987, *The History of Mathematics: A reader,* The Open University, London.

Guo Shuchun: 1995, *Gu Dai Shu Xue Tai Dou Liu Hui* (*Liu Hui—One of the Greatest Masters of Mathematics in Antiquity World*), Ming Wen Book Co., Taipei.

Heath, Thomas L.: 1956, *The Thirteen Books of the Elements* (three volumes), Dover Publications, Inc., New York.

Horng, Wann-Sheng: 1991, "Kongzi Yu Shu Xue (Confucius and Mathematics)", in Wann-Sheng Horng, *Kongzi Yu Shu Xue* (Collected Essays on the History of Mathematics), Ming Wen Book Co., Taipei, pp. 1–12.

——: 1992, "Chong Fang *Jiu Zhang Suan Shu* Ji Qi Liu Hui Zhu (Revisit the *Jiu Zhang Suan Shu* and Its Commentary by Liu Hui)", *Mathmedia* 16(2) (Taiwan), 29–36.

——: 1994, "San Ge Gong Shi Ji Yuan Mian (Three Formulas for the Area of a Circle", *Science Monthly* (Taiwan) 25(7), 539–544.

——: 1995, "How Did Liu Hui Perceive the Concept of Infinity: A Revisit", *Historia Scientiarum* Vol. 4-3, 207–222.

Katz, Victor J.: 1993, *A History of Mathematics: An Introduction,* HarperCollins College Publishers, New York.

Kline, Morris: 1972, *Mathematical Thought from Ancient to Modern Times,* Oxford University Press, New York.

Lam Lay Yong and Ang Tian-Se: 1986, "Circle Measurements in Ancient China", *Historia Mathematica* 12, 325–340.

——: 1992, *Fleeting Footsteps,* World Scientific, Singapore.

Li Yan and Du Shiran: 1987, *Chinese Mathematics: A Concise History,* Clarendon Press, Oxford, London.

Lloyd, G. E. R.: 1984, *Magic, Reason and Experience,* Cambridge University Press, London.

Plato: *Republic* (translated by B. Jowett), Vintage Books, New York.

Qian Baocong: 1963, *Suan Jiang Shi Shu Dian Jiao* (*Annotation on the Chinese Ten Classics of Mathematics*), Zhong Hua Book Co., Beijing.

Sfard, Anna: 1991, "On the Dual Nature of Mathematical Conceptions: Reflections on Processes and Objects as Different Sides of the Same Coin", *Educational Studies in Mathematics* 22, 1–36.

Siu, Man Keung: 1993, "Proof and Pedagogy in Ancient China: Examples from Liu Hui's Commentary on *Jiu Zhang Suan Shu*", *Educational Studies in Mathematics* 24, 345–357.

——: 1995, "Mathematics Education in Ancient China: What Lesson Do We Learn from It?" *Historia Scientiarum* Vol. 4-3, 223–232.

——: 1996, "Ancient Chinese Mathematics", in *Proceedings of HEM Braga* 1996 vol. 1, pp. 54–65.

Skemp, Richard R.: 1995, *The Psychology of Learning Mathematics,* Jiu Zhang Press, Taipei. A Chinese translation based on the revised 2nd edition (1987) published by Lawrence Erlbaum Associates Inc., Publishers.

Van der Waerden, B. L.: 1983, *Geometry and Algebra in Ancient Civilizations,* Springer-Verlag, Berlin/Heidelberg/New York/Tokyo.

Wagner, Donald B.: 1979, "An Early Chinese Derivation of the Volume of a Pyramid", *Historia Mathematica* 6, 164–188.

Wu Wenjun: 1981, "Chu Ru Xiang Bu Yuan Li (In-Out Principle)", in Wu Wenjun (ed.), *Jiu Zhang Suan Shu Yu Liu Hui* (*Jiu Zhang Suan Shu* and Liu Hui), Beijing Normal University Press, Beijing, pp. 58–75.

Glossary

Bie Nao	鼈臑
Bu	步
Chi	尺
Chong Die	重疊
Deng Shu	等數
Ge Yuan	割圓
Geng Xiang Jian Sun	更相減損
Hu Tian	弧田
Jiu Zhang Suan Shu	九章算術
Li Fang	立方
Liu Hui	劉徽
Qian Du	塹堵
Suan Sha	算術
Wei	微
Xi	細
Xing	形
Yang Ma	陽馬
Yi Ying Bu Xu	以盈補虛

The Long Tradition of History in Mathematics Teaching: An old Italian Case

Fulvia Furinghetti
University of Genova

Introduction

The use of history in mathematics teaching is a matter of discussion very much alive both in the world of historians and of mathematics educators. It may be encouraging for researchers and teachers involved in this kind of study to know that there is a long tradition behind the experiences and the discussions carried out at present; evidence of this fact is offered by the old mathematical journal for students of secondary school *Il Pitagora* (*Pythagoras*) which we present in this paper. The interest of this journal in connection with the debate on the use of history in mathematics teaching mainly relies on two facts:

- it is one of the few (the only?) journals of its time which had the declared aim of introducing the history of mathematics in classroom practice
- the great majority of the authors who published articles in it were teachers and thus their suggestions really came from a practical knowledge of classroom life.

The Journal *Il Pitagora*

The editor and the life of the journal (1895–1918)

The first issue of the journal *Il Pitagora* appeared in January 1895 in Avellino, a little town in Southern Italy near Naples. The founder and editor was Gaetano Fazzari (1856–1939), a prominent secondary mathematics teacher of the period. According to the customs of those times the journal belonged to the editor and thus followed the events of his life: for example, starting from 1899 it was published in Palermo (Sicily), since Fazzari was appointed as a teacher in a high school of this town.

The last issue of the journal appeared in December 1918. One of the reasons it stopped appearing was the critical situation in Italian society in that period with strikes and riots which made the regular publication of this and other mathematics journals very difficult. The main reason for the journal ceasing to appear was, though, the First World War, as is expressed in the following passage by the editor (1916–17, v.1, s.2, a.22, p. 144)[1]: "This is the last issue of the volume of *Il Pitagora* published in the present year [1917] during the glorious war of our Italy, while the young boys, who last year were students in high school, are heroically fighting for the freedom and the redemption of peoples."

The founder and editor Gaetano Fazzari was born in Calabria, a region of Southern Italy which in the past was part of the important Greek colony known as *Magna Grecia* (Great Greece). Important philosophers and mathematicians of the classical period (Pythagoras, for one) lived

IL PITAGORA

GIORNALE DI MATEMATICA

PER

GLI ALUNNI DELLE SCUOLE SECONDARIE

Premiato con medaglia d'argento all'Esposizione Nazionale di Torino del 1898

Pubblicato per cura

DI

GAETANO FAZZARI

Prof. di matematica nel R. Liceo Umberto I di Palermo

Anno VII. — 1900-01

PALERMO

CASA ED. «ERA NOVA»

1901

there. Fazzari taught in the Classical Lyceum, a type of Italian high school with a strong humanistic orientation, where Latin and ancient Greek are the most important subjects. All his life he kept a strong penchant for the classical world; his work in history mainly concerns the Greek authors and the translation of Archimedes' writings. He authored research papers in mathematics and the history of mathematics, as well as articles on the teaching of mathematics and on the use of history in mathematics teaching. Fazzari had regular contacts with the international scientific world, received publications in his fields of interest from many countries, prepared reports for the first ICMI meetings, and was a member of an important mathematical association of those times which still exists (*Circolo Matematico di Palermo*). These facts affect the style of the journal *Il Pitagora* which, in spite of its secluded origin, does not have a provincial flavor.[2]

The editorial line

The subtitle of *Il Pitagora* is "Mathematics journal for secondary school students." This makes it one of the first Italian mathematical journals for students of this school level. (Note that, according to the Italian school system, secondary means until the age of 19.) The aim of the editor, stated in the circular letter of presentation written in November 1894, is "to instill the love for the mathematical disciplines into the minds of the young people attending secondary school." To attain this aim he warmly asks his colleagues' collaboration; with a few exceptions, the authors of the articles published in the journal were, indeed, school teachers.

The articles concern various mathematical subjects (algebra, arithmetic, geometry, logic, probability), with a prevalence of algebra, arithmetic and geometry, the subjects most treated in school. The level of the presentation is elementary, but in some problems (e.g., that of the measurement of a circle that we shall present in the following) advanced topics such as integration and plane algebraic curves of degree greater than two are introduced. Except for a few cases, the articles are quite short and at the students' level. As a general impression we can say that the initial intention of addressing secondary students is respected with a good balance of different styles and topics. This is the value of *Il Pitagora,* since other journals born with the same purpose degenerated into a rather boring list of exercises. However it seems that the most efficient way to employ the journal was to use it as a source for developing a given topic under the teacher's guidance. (We will give an example of this in the following.) For this reason we consider the teachers as the principal readers and as mediators.

What makes the journal singular in the international panorama of the period is the role ascribed to the history of mathematics in pursuing the editor's aim ("to instill the love . . . "). In the circular letter of presentation to potential subscribers of the journal, the editor presents it as a "tournament open to the young minds of the students attending secondary schools. [The journal will contain:] questions and exercises, with the best solutions of the young readers; puzzles and (scientific) paradoxes, *historical, biographical and bibliographical notions,* questions and answers by students, *original excerpts from ancient authors* [emphasis is mine]." His hope is to help students to overcome the difficulties in doing mathematics, of which, as a teacher, he is perfectly aware.

In the first issue of the journal the intention of the editor is made evident (1895, a.I, p. 1): "Being convinced that it is a great advantage for young people to know the history of mathematics, *Il Pitagora* will attempt to present the studies on which the efforts of the scholars of this discipline are concentrating all over the world in recent times.[3] And who cannot see how much this is useful not only from the pedagogical, but also and first of all from the scientific and historical point of view? As a matter of fact nobody can doubt that any erudition, according to the cognitive process, has to be a history before being a science. Thus erudition, as a center which links the criticism to hermeneutics, is the first means for understanding the Greek and Latin Classics. Without any doubt erudition is very important for human knowledge, since it makes the centuries present and associates us to past generations. Moreover no young people having attended our high school should be ignorant of the place occupied in the history of civilizations by Thales, Pythagoras, Euclid, Archimedes, ... and particularly by the Italian mathematicians who from the XIII to the XVI century were the followers of Greek science, not only extending it, but spreading it all over Europe."

The presentation of the editor gives an idea of the underlying educational philosophy of the journal, which may be summarized in the following points:

- students do not naturally love mathematics and have an inborn difficulty in learning it
- puzzles and history are the 'devices' with which the teacher can arouse interest for mathematics
- to know the facts and how things have evolved is the first step for understanding things.

In this view of history we can recognize the Italian orientation favorable to historicism which strongly influenced the school and the culture. The position of the journal may be criticized; nevertheless the editor's merit is to have clearly stated it to his interlocutors. In the light of later research in mathematics education, the remedies to the students' difficulties proposed by the editor may appear quite naïve and simplistic, but we have to acknowledge that he has conceived and developed a project consistent with his ideas about mathematical culture. The way he proposed for solving the mathematics teaching/learning problems is one of the possible answers to the awkwardness of mathematics teachers facing the difficulties of their profession. In the same period other teachers proposed very different solutions to this problem. For example, in the U. K. the Perry movement proposed to emphasize applications in the teaching of mathematics. The limit we can recognize to Fazzari's project is that he trusted too much in the possibility of transferring to students the pleasure he felt in doing mathematics, based on the enjoyment of the 'aesthetics' of this discipline.

The philosophy underlying *Il Pitagora* is strongly influenced by the role ascribed to history in mathematics teaching by the Italian historian Gino Loria (1862–1954), whose ideas, clearly stated in (Loria, 1890), are widely reported as a *manifesto* in the letter of presentation of the journal. One of the main points of Loria's view is that the history of mathematics is an efficient means for promoting links among the various subjects taught in school; in this way it contributes to widening and deepening the culture of the student. These ideas of Loria were shared by other researchers; see for example (Heppel, 1893). This last paper shows a very pragmatic approach to the issue, by stating very clearly the basic restrictions and limitations under which history may be advantageously employed in teaching mathematics, as shown in the following passage (Heppel, 1893, p.19):

> I. The History of Mathematics should not form a separate subject of education, but be strictly auxiliary and subordinate to Mathematics teaching.
> II. Only those portions should be dealt with which are of real assistance to the learner.
> III. It is not to be made a subject of examination.

In the journal *Il Pitagora* we do not find such a position clearly expressed, but, as we will see in our 'virtual' didactic unit, the choice of the historical topics to be presented in the classroom appears really aimed at promoting the students' learning and understanding.

A 'virtual' didactic unit from the journal *Il Pitagora*

To determine the effectiveness of the proposals coming from the journal, we unfortunately have no records on what teachers of the beginning of the century were doing in their school practice, nor if and how they applied the suggestions and the stimuli coming from the journal. What we can do is to think of a 'virtual' teacher who is a regular reader of the journal and imagine what he/she could have done on finding inspiration from the articles published there. Under this condition we have constructed a 'virtual' didactic unit using seven articles of the journal *Il Pitagora*.[4]

The topic we shall develop in this 'virtual' didactic unit is *the problem of the measurement of a circle,* which is treated in many issues of the journal from various points of view. The topic has a particular didactic character since it can be treated with different degrees of depth, using elementary to advanced concepts. In fact, Felix Klein (1849–1925), who was very concerned with the problem of linking the different levels of mathematical instruction, chose it as one of the topics to be developed at the Göttingen meeting of the German Association for the Advancement of the Teaching of Mathematics and the Natural Sciences (Klein, 1962). Moreover we think that the historical analysis of this problem is particularly suitable to give students an element very important for their mathematical learning, that is *the sense of what solving a problem means.* As a matter of fact we have often observed that students have three beliefs when facing a problem. First, a problem always has a solution; second, the solution is unique; and third, the way of solving the problem is unique. In addition, they are not aware that to be solvable or not is linked to the context in which the problem is set and the means available. Since solving problems has to be the core of mathematical activity (in the classroom as well as in research) we think that

1. Survey on the historical development of the problem "measurement of a circle"	The symbol π. Number of digits of π. The transcendence of π.
2. How to find values of π	Through elementary constructions; Through advanced constructions; Through mechanical machines; Experimentally
3. An approach to sources	The propositions I, II, and III of Archimedes' Measurement of a circle

Table 1
The plan of the 'virtual' didactic unit

Researcher	Year	Number of Digits of π
Metius	1556	6
Viète	1579	9
Viète	1610	35
Romanus	1613	15
Ludolph von Ceulen	1615	32
Grienberger	1630	39
Matsumura	1633	7
Sharp	1699	71
Lagny	1719	112
Vega	1794	136
Thibaut	1822	156
Dahse	1844	200
Clausen	1847	248
Richter	1853	330
Rutheford	1853	440
Shanks	1853	530
Shanks	1874	707

Table 2
The approximated values of π (data from the journal *Il Pitagora*)

giving clear ideas on the nature of this activity is an aim to be pursued in teaching.

We have planned the didactic unit having this aim in mind. We address it to high school students. The material provided by the articles of the journal *Il Pitagora* is organized according to the schema of Table 1. All our suggestions are based on materials and ideas contained in articles of the journal, to which we add a few comments. Here we can simply hint how the sections 1, 2, 3 can be developed; the reader can find the complete treatment in the works quoted in the references and in other books on the history of mathematics.

1. Survey on the historical development of the problem 'measurement of a circle'

There are reports of experiences in the classroom which show that it may be efficacious to introduce students to a given problem through an initial very descriptive phase. In our case this phase is necessary to give an idea of how important this problem was in the development of mathematical speculation.

At the beginning we consider the history of the symbol π. It appeared for the first time in the *Synopsis palmariorum* (1706, London) of William Jones (1674–1749); thanks to Leonhard Euler (1707–1783) it was definitely accepted. Afterwards the various approximations of π and the related number of digits are presented. The survey begins with the Holy Bible, the Egyptians and the Babylonians and arrives up to the nineteenth century. Table 2 summarizes the main attempts of researchers looking for algorithms approximating π from the sixteenth century onwards. It may be amusing for students to quote the funny lines of verse from the famous book *Récréations mathématiques* (E. Lucas, Paris, 1891, v.II, p.155) in which each word has the number of letters equal to the corresponding decimal digit of π until the thirtieth. These two different pieces of information (the table and the lines) show the two main aspects of the problem, its impact in the mathematical community and in the common people *via* school experience.

The foregoing presentation of decimal digits not only has the function of surprising and amusing the students, but it is also a means for introducing them to the transcendence of π, instilling no doubt that the number in question is 'special'. Thus, when the narrative survey arrives at the eighteenth century the interest of the students shifts from the problem of the pure computation of digits to the problem of guessing what the nature of π may be. The school level we are considering does not allow us to present the proofs involved, but students are able to understand the statements of the theorems and become acquainted with the transcendence of π.

2. How to approximate the value of π

Through elementary constructions. After this quite descriptive survey, which consists of giving pieces of infor-

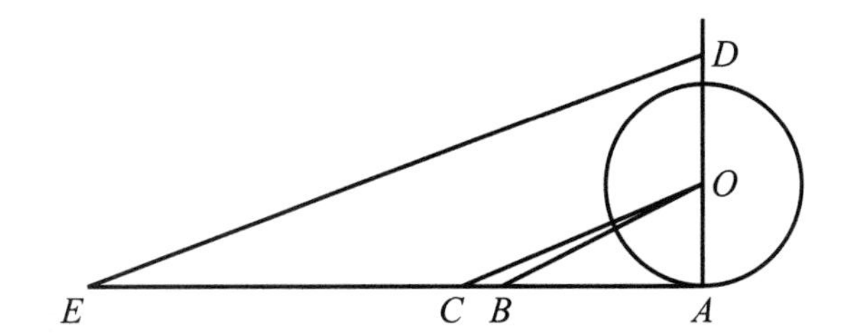

FIGURE 1
The construction of π by Specht

mation on the role played by the problem in the development of mathematical culture, the students need to work actively on it. We present two elementary constructions (I and II) which have the aim of approximating the value of π, working in a field (Euclidean geometry) which is quite familiar to students.

I. In a circle of radius 1 the sum of the sides of the equilateral triangle and of the square inscribed is $\sqrt{3}+\sqrt{2} = 3.1462\ldots$; thus we can take a segment with length equal to this sum as the half of the circumference with an error less than 0.005.

II. In (Specht, 1828) there is the following construction (see Figure 1). On the tangent at A to a given circle with radius r take the segments $AB = (2+1/5)r$ and $BC = (2/5)r$. On the diameter through A take $AD = OB$, and draw DE parallel to OC (E is its intersection with the tangent at A). The length of AE gives an approximation of the length of the circumference with an error less than two millionths of the radius.

Proof. $AD = OB = r\sqrt{1+(11/5)^2} = r\sqrt{146}/5$, since $AE : AD = AC : AO = 13/5$, $AE = (13/5)AD = 2(3.1415919\ldots)r$, so that, if $r = 1$, the difference $AE/2 - \pi$ is about 0.0000007. This means that for a circle with the radius of 7000 kilometers (note that the mean radius of the earth is 6366 kilometers) this construction gives the circumference with an error of approximately 10 meters.

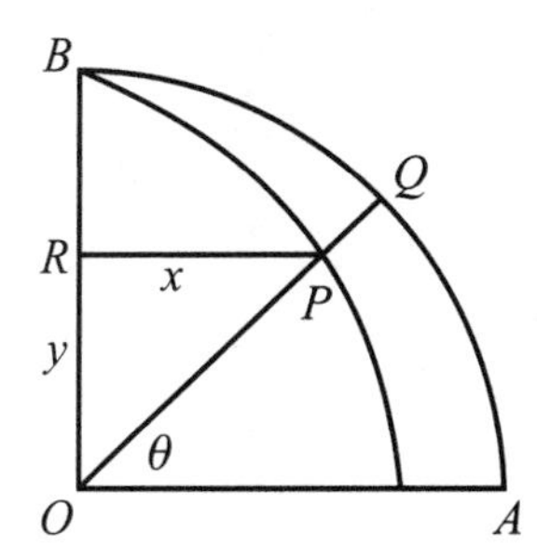

FIGURE 2
The generation of the quadratrix of Dinostratus

Through advanced constructions. The method presented is based on the curve, called the quadratrix of Dinostratus (fourth century BCE), described as follows (see Figure 2).

Let a point Q starting at A describe the circular quadrant AB with uniform velocity, and let a point R starting at O describe the radius OB with uniform velocity. Let Q and R start simultaneously and simultaneously reach the point B. Let the point P be the intersection of OQ with a line perpendicular to OB drawn from R. The locus of P is the quadratrix. Letting $\angle QOA = \theta$, and $OR = y$, we find that y/θ is equal to $2r/\pi$ (r is the radius of the circle). The curve has equation $x = y\cot\theta = y\cot(\pi y/2r)$, and intersects the x axis at the point with abscissa $x = 2r/\pi$ obtained as y approaches 0. Thus if we can construct the curve, we can construct π. For this purpose, it is sufficient to represent the branch of the curve described by the point P, as in Figure 2. In Figure 3 (taken from the journal), all the branches of the curve $x = y\cot(\pi y/2r)$ are drawn; this figure also contains the circumference with center O which is related to the dynamic construction, but is not part of the curve. With a convenient choice of the axes X and Y the curve of Figure 3 can be obtained as a graph of a real function of one real variable. At present in the classroom, we can use the computer to show the dynamic construction of the branch that gives π (see Figure 2) as well as to draw the complete curve of Figure 3.

Through mechanical machines. An article of the journal considers also processes based on non-elementary means such as algebraic curves of degree greater than two or integral curves. We present them, even if they are beyond the usual mathematical programs of high school, since they give interesting insights on the use of mechanical machines in fashion at those times.

We consider the integraph designed and realized by the Polish Bruno Abdank-Abakanowicz (1852–1900) based on the kinematic principle that if a wheel rotating round an

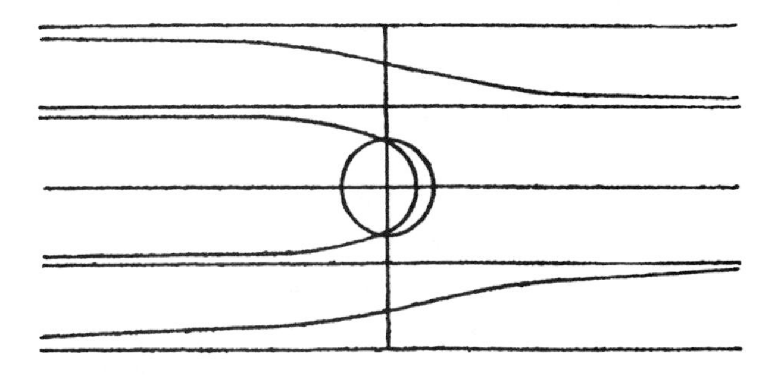

FIGURE 3
The complete quadratrix of Dinostratus

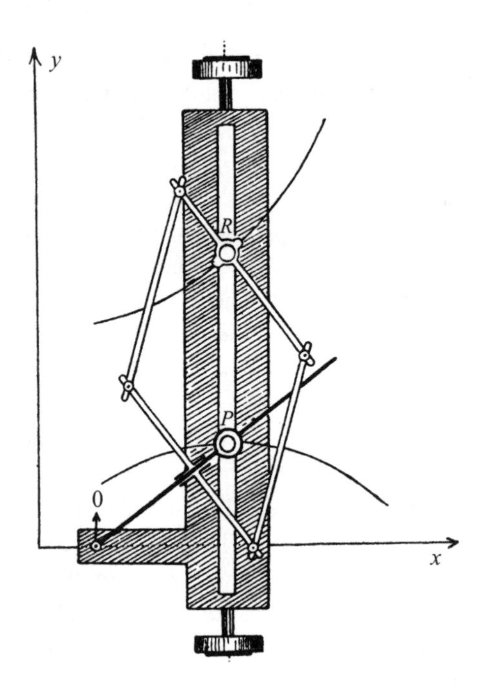

FIGURE 4
The Abdank-Abakanowicz integraph

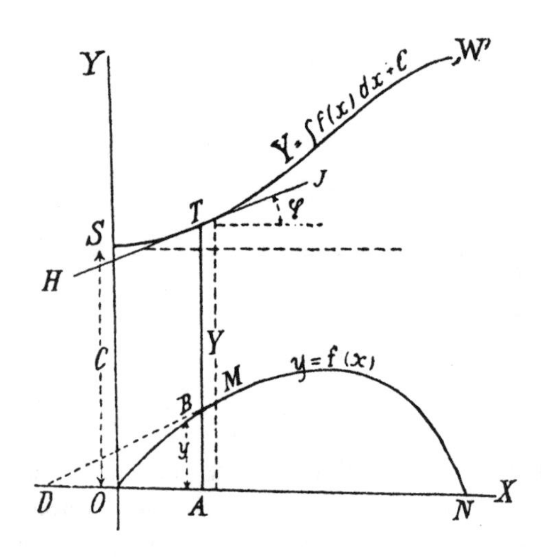

FIGURE 5
The use of the integraph for integration

axis parallel to a given plane rolls on the plane itself, the tangent to the curve described by the point of contact lies on the plane of the wheel. This instrument enables us to trace the *integral curve*

$$Y = F(x) = \int f(x)\,dx$$

when we have given the *differential curve*

$$y = f(x).$$

For this purpose, we move the linkwork of the integraph so that the *guiding point* (P) follows the differential curve; the *tracing point* (R) will then trace the *integral curve* (see Figure 4[5]). We simply indicate the mathematical principle on which the integraph is based (see Figure 5 taken from the journal). Let B be any point of the differential curve $y = f(x)$, Construct the right-angled triangle with the vertices $B(x, y)$, $A(x, 0)$ and $D(x-1, 0)$. The tangent of the angle BDA is equal to y. Thus the hypotenuse BD is parallel to the line tangent to the integral curve at the point $T(X, Y)$ corresponding to the point $B(x, y)$. For a fuller description of this instrument we refer to (Abdank-Abakanowicz, 1889). Some old books of calculus and analysis published in the first decades of the twentieth century present this and other integraphs.[6]

To construct π, we consider the differential curve $x^2+y^2 = r^2$, and its integral curve $Y = \int \sqrt{r^2 - x^2}\,dx = (r^2/2)\arcsin(x/r) + (x/2)\sqrt{r^2 - x^2}$, which consists of a family of equal branches. The points where it intersects the OY axis have ordinates $0, \pm r^2\pi/2, \ldots$; the points where it intersects the lines $x = \pm r$ have ordinates $0, \pm r^2\pi/4, r^2 3\pi/4, \ldots$. If we take $r = 1$ the ordinates of the intersection points give π and its multiples. In Figure 6 (taken from the journal *Il Pitagora*) there is the construction that the author says is taken from Klein's book *Leçons sur certaines questions de géométrie élémentaire* (Nony, Paris, 1896). An English version of this construction is in (Klein, 1962).

Experimentally. We take an 'experimental' method for finding the area of the circle from the treatise *Admirandis Archimedis Syracusani monumenta* by Franciscus Maurolycus (1494–1575), published posthumously (Palermo, 1685). In this book there are also versions of Archimedes'

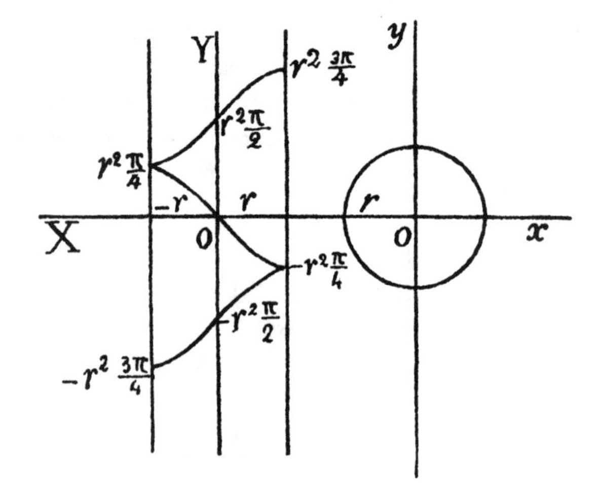

FIGURE 6
The construction of Klein for finding π

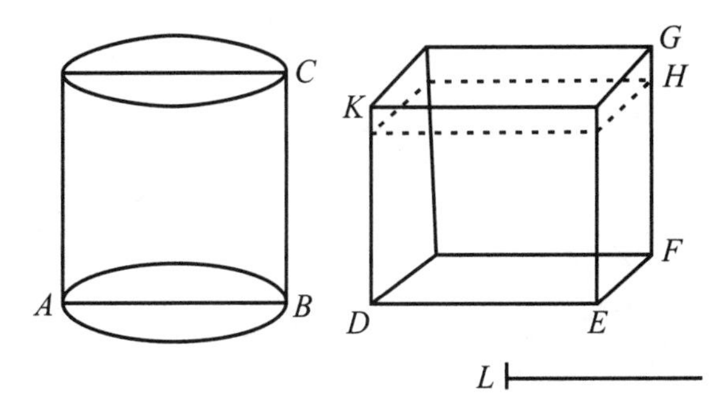

FIGURE 7
The area of the circle according to the method of Maurolycus

work on the measurement of a circle, Hippocrates' method for squaring the circle ('Hippocratis tetragonismus'), and a method by Maurolycus himself ('Maurolyci tetragonismus').

On page 39 we find the problem 'Modus alius quadrandi circulum' in which the author gives the following method. We construct an empty cylinder and an empty cube (see Figure 7). The cylinder has the circle as base and height equal to the diameter d of the circle; the cube has side equal to the diameter d. We fill the cylinder with water and then pour it into the cube. The product of the depth x of the water in the cube and the side d of the cube equals the area of the base of the cylinder.

Proof. The volume of the water in the cube is xd^2, the volume of the same water in the cylinder is given by the area of the circle multiplied by d; then xd^2 = area of circle multiplied by d and the area of the circle is xd.

3. An approach to sources

This part asks the students to come to the core of the problem by studying the work of a chief character in the history of π, Archimedes (287–212 B.C.). Due to the importance of this author the presentation of the topic has to be preceded by some notes on his life and works. Afterwards students have to discuss the following propositions taken from the *Measurement of a circle* (see Heath, 1921).

I. The area of a circle is equal to that of a right-angled triangle in which the perpendicular is equal to the radius, and the base to the circumference, of the circle.

II. The area of a circle is to the square of its diameter as 11 to 14.

III. The ratio of the circumference of any circle to its diameter is less than $3\ 1/7$ but greater than $3\ 10/71$.

The Archimedean proofs of these propositions may be found in (Archimedes, 1558; Heath, 1897 and 1912; Ver Eecke, 1921) as well as in many treatises on the history of mathematics; here we give only the general idea of the processes involved to point out the strong didactic character. In proposition I the area of the circle is approximated by inscribing and circumscribing successive regular polygons with a number of sides continually doubled (starting from squares). The proof is interesting from the pedagogical point of view, since it introduces a general method applicable to other situations, viz. the method of exhaustion, which rests on a principle stated in the following enunciation of Euclid X, 1, see (Heath, 1956):

> Two unequal magnitudes being set out, if from the greater there be subtracted a magnitude greater than its half, and from that which is left a magnitude greater than its half, and if this process be repeated continually, there will be left some magnitude which will be less than the lesser magnitude set out.

About proposition II, in proving it we observe that it depends on the result of proposition III. As pointed out in (Knorr, 1993, p.153), "the writing as we know it has been altered to its detriment through editorial revisions and scribal confusions." The correct order of the two propositions can be discussed in class as a further occasion for understanding the topic.

The proof of proposition III is the most interesting; it offers a good example of an algorithm for performing approximate calculations. The method requires the approximate calculation of the perimeter of two regular polygons of 96 sides, one of which is circumscribed about, and the other inscribed in, the circle. In the process of accomplishing this, Archimedes gives the following approximation without explanations:

$$265/153 < \sqrt{3} < 1352/780.$$

The discussion of this value offers students a good occasion to reflect on this type of computation. Students can work on the problem with programmable pocket calculators or a computer and improve Archimedes' algorithm. The fact that history may be a good approach to problems of approximation has already been discussed in (Furinghetti, 1992).

All these proofs have to be considered, according to the distinction discussed in (Hanna, 1990), not only proofs which prove, but also proofs which have a strong explanatory character. For these reasons they are a recurring theme in school. Proposition III has been developed for the classroom in recent times. For example, this experiment was carried out in a French Lyceum, as described in (Bühler, 1990). In this paper this and other ideas presented in the

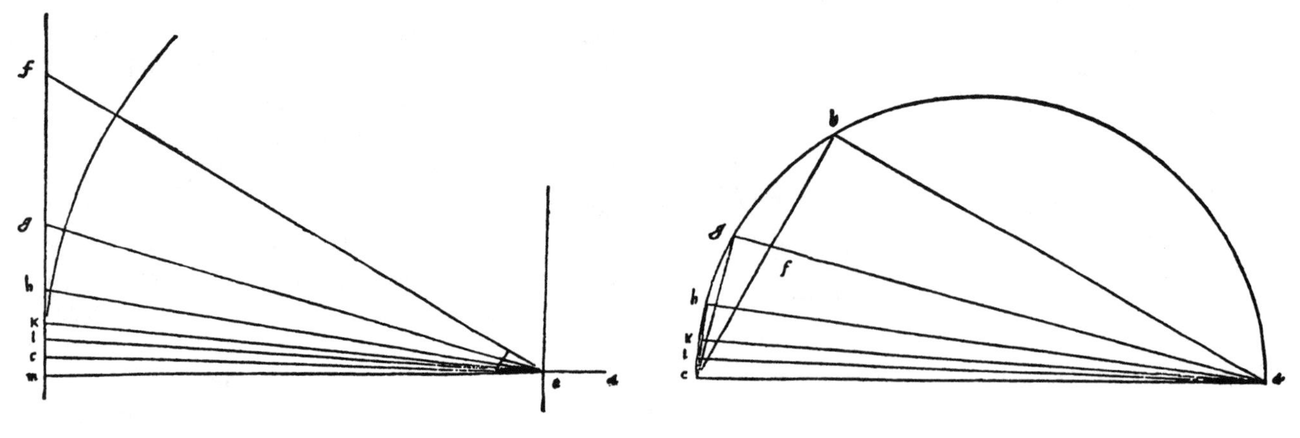

FIGURE 8
The figures of the proposition III from *Archimedis opera non nulla* ... (1558)

journal *Il Pitagora* are developed, including that of involving the colleague who teaches ancient Greek in the translation of Archimedes' excerpts, since the students are studying this language. Other papers present experiences in the classroom based on Archimedes' measurement of a circle, integrated with works of successive authors. In (Führer, 1991) it is recognized that the history of π gives the feeling that "good ideas seem to have a touch of eternity" (p.27).

Conclusions

We think that our 'virtual' didactic unit satisfies the requirement we have stated at the beginning of enriching the students' ideas on what solving a mathematical problem means. The articles of the journal provide us with materials that foster a gradual presentation from the phase of a general survey of the problem to the phase in which one really goes to the core of the problem and perceives the main concepts (*big mathematical ideas*) involved in it. The study of the three Archimedean propositions allows us to introduce students to a method transferable from this case to other cases. The concept of solving a problem is extended to use means different from the usual ones and to consider different aspects (theoretical, existential, computational,. . .). These facts foster a flexibility in learning which is advocated by many educators. It is curious to observe that our didactic unit based on history leads in a natural way to the use of computing machines for implementing algorithms or drawing curves. It is interesting also to observe that with our didactic unit we can give to students an idea of what mathematics development has been, reproducing some aspects of it in the classroom.

To analyze the actual contribution of the journal to the discussion on the use of history in mathematics teaching we have to consider how this use may be carried out in school practice. First, we must observe that, in spite of the declared intentions of the editor of addressing students, the historical materials proposed in the journal can reach the students' minds only *via* a careful mediation of the teacher. The discussion of how this mediation can be performed leads us to the core of the discussion concerning whether and how to use history in mathematics teaching. In (Furinghetti, 1997) we have summarized our opinion on this point by advocating the *integration* of history in mathematics teaching. For us this integration means to pursue the mathematical objectives of the classroom activity through history or, as (Katz, 1997) put it, to use the history of mathematics to make learning better for the students.

Students can feel themselves participants in the cultural project of the development of mathematics only if they actively live their personal mathematical experience. The main stages of this process of *integration* are sketched in Figure 9.

On the ground of the preceding observations on our 'virtual' didactic unit we can say that the journal reached its scope of providing students, or, to be precise, their teachers with resources suitable for *integrating* history in the teaching of mathematics. Thus the first two stages of the process are fulfilled. We lack important elements for judging the effectiveness of the proposals that we have read in the journal, viz. the kind of mediation made by the teachers of those days when working in the classroom and the analysis of the students' reactions. This means that the last three stages have to be fulfilled. Would a modern teacher experiment with our 'virtual' didactic unit and write the missing part of the *Il Pitagora*'s project?

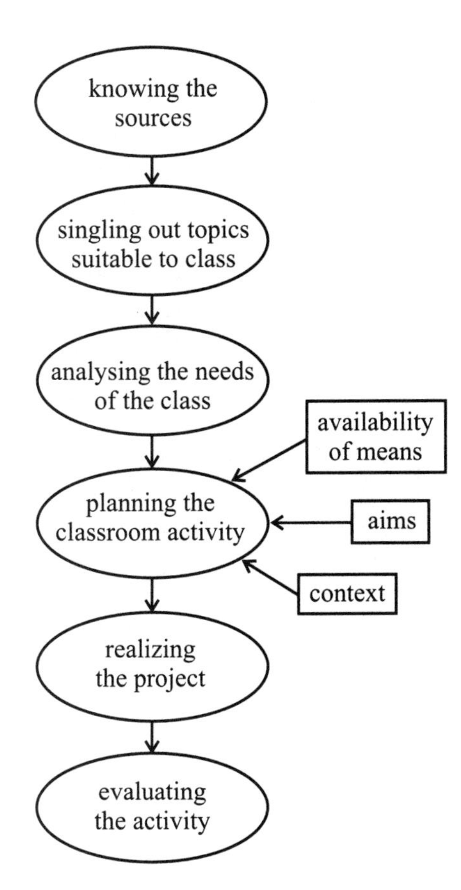

FIGURE 9
The process of introducing history in the teaching of mathematics

Acknowledgment. I would like to thank Victor Katz (Washington, DC) for valuable criticism and for polishing the English of the present paper.

References

Abdank-Abakanowicz, B.: 1889, *Die Integraphen,* Teubner, Leipzig (There is also a French edition).

Archimedis opera non nulla a Federico Commandino Urbinate nuper in latino conversa, et commentariis illustrata: 1558, Venezia.

Bühler, M.: 1990, "Reading Archimedes' *Measurement of a circle*," in J. Fauvel (editor), *Mathematics through History,* Q.E.D. Books, York, 43–58.

Führer, L.: 1991, "Historical stories in the mathematics classroom," *For the Learning of Mathematics,* vol. 11, no. 2, 24–31.

Furinghetti, F.: 1992, "The ancients and the approximated calculation: some examples and suggestions for the classroom," *The Mathematical Gazette,* vol. 76, no. 475, 139–142.

Furinghetti, F.: 1997, "History of mathematics, mathematics education, school practice: case studies linking different domains," *For the Learning of Mathematics,* vol. 17, no. 1, 55–61.

Hanna, G.: 1990, "Some pedagogical aspects of proof," *Interchange,* v.21, n.1, 6–13.

Heath, T. H.: 1897 and 1912: *The Works of Archimedes* and *The Method of Archimedes,* Cambridge University Press, Cambridge, reprinted in one volume (Dover, New York, 1956).

Heath, T. H.: 1921: *A History of Greek Mathematics,* v.II, Oxford University Press, London.

Heath, T. L.: 1956: *The Thirteen Books of Euclid's Elements,* v.III, Dover Publications Inc., New York (reedition of the second edition published by Cambridge University Press in 1926).

Heppel, G.: 1893, "The use of history in teaching mathematics," *Nineteenth general report of the Association for the improvement of geometrical teaching,* W. J. Robinson, Bedford, 19–33.

Katz, V.: 1997, "Some ideas on the use of history in the teaching of mathematics," *For the Learning of Mathematics,* vol. 17, no. 1, 62–63.

Kidwell, P.: 1996, "American mathematics viewed objectively: the case of geometric models," in R. Calinger (editor), *Vita Mathematica,* MAA Notes no. 40, 197–207.

Klein, F.: 1962, "Famous problems of elementary geometry," in F. Klein, W. F. Sheppard, P. A. MacMahon & L. J. Mordell *Famous Problems and Other Monographs,* Chelsea, New York, i–xiii and 1–92 (translation by W. W. Beman & D. E. Smith in 1897).

Knorr, W. R.: 1993, *The ancient tradition of geometric problems,* Dover, New York.

Krazer, A. (editor): 1905, *Verhandlungen des dritten internationalen Mathematiker-Kongresses,* B. G. Teubner, Leipzig.

Loria, G.: 1890, "Rivista bibliografica: Treutlein, P. 'Das geschichtliche Element in mathematischen Unterrichte der höheren Lehranstalten', Braunsweig, 1890," *Periodico di matematica,* a.V, 59–61.

Maurolycus, F.: 1685, *Admirandis Archimedis Syracusani monumenta,* Palermo.

Specht, C.: 1828, "Annäherungs—Construction des Kreis—Umfangs und Flächen—Inhalts," *Journal für die reine und angewandte Mathematik,* vol. III, 83.

Ver Eecke, P.: 1921, *Archimède: les oeuvres complètes,* (French translation from the 2nd edition of Heiberg), Desclée & de Brouwer, Paris-Bruges.

Endnotes

[1] The original is in Italian. Here and elsewhere the translations of the passages originally in Italian are by the author of the present paper.

[2] A presentation of the historical content of *Il Pitagora* is in the dissertation (supervisor F. Furinghetti) Demontis, F.: 1995, *La storia come strumento didattico nella rivista Il Pitagora,* Dipartimento di Matematica dell'Università di Genova.

[3] In that period studies in history of mathematics were flourishing all over the world with important editions of classic authors and the birth of the first journals explicitly dedicated to history of mathematics.

[4] With the expression 'didactic unit' we mean a series of lessons on a given topic. The articles used are published in (1895, a.I, 26), in 1899 (a.V, 1° sem., 14–15), in 1904–05 (a.XI, 144), in

1908–09 (a.XV, 101–102), in 1898 (a.IV, 2° sem., 95–96), in 1902–03 (a.IX, 31–32; 47–51), in 1912–13 (a.XIX 5–23).

[5] This figure is taken from Tacchella, G.: 1957, "Calcolo meccanico," in Berzolari, L., Vivanti, G. & Gigli, D. (editors), *Enciclopedia delle matematiche elementari,* vol. I, p. I, 411–442. This article is contained in a famous book for teacher training, first published in 1930.

[6] We have to remember that in those days the use of physical, concrete helps for doing mathematics was becoming fashionable, in connection both with the growing interest in pedagogical problems in teaching mathematics and the raising of positivist philosophy. In this connection we recall that in the International Conference of Mathematicians held in Heidelberg in 1904 there was an exhibition of about 300 mathematical models, tools and computing machines (including that of Leibniz), as described in the third part of the proceedings (Krazer, 1905, 717–755). For a discussion on the connection between pedagogical reforms and concrete objects see (Kidwell, 1996).

Problem Solving from the History of Mathematics

Frank J. Swetz
The Pennsylvania State University

"Where can I find some good problems to use in my classroom?" is a question I am often asked by mathematics teachers. My answer is simple: "The history of mathematics."

Since earliest times, written records of mathematical instruction have almost always included problems for the reader to solve. Mathematics instruction was certainly considered an activity for self-involvement. The luxury of a written discourse and speculation on the theory of mathematics appeared fairly late in the historical period with the rise of Greek science. Records from older civilizations such as Babylonia, Egypt, and China reveal that mathematics instruction was usually incorporated into a list of problems whose solution scheme was then given. Quite simply, the earliest known mathematics instruction concerned problem solving—the doing of mathematics. Obviously, such problems, as the primary source of instruction, were carefully chosen by their authors both to be useful and to demonstrate the state of their mathematical art.[1] The utility of these problems was based on the immediate needs of the societies in question and thus reflect aspects of daily life seldom recognized in formal history books. Such collections of problems are not limited to ancient societies but have appeared regularly throughout the history of mathematics.

In this literature of mathematics, thousands of problems have been amassed and await as a ready reservoir for classroom exercises and assignments. The use of actual historical problems not only helps to demonstrate problem-solving strategies and sharpen mathematical skills, but also:

- imparts a sense of the continuity of mathematical concerns over the ages as the same problem or type of problem can often be found and appreciated in diverse societies at different periods of time;
- illustrates the evolution of solution processes—the way we solve a problem may well be worth comparing with the original solution process, and
- supplies historical and cultural insights of the peoples and times involved.

In the following discussion, I will survey 28 problems that readily lend themselves to classroom use.

A Sample Problem Solving Situation

During an in-service workshop on problem solving conducted for twenty experienced secondary school teachers,[2] the participants were given the following homework exercise:

> A circle of radius 4 units is inscribed in a triangle.
> A point of tangency of the circle with a side of

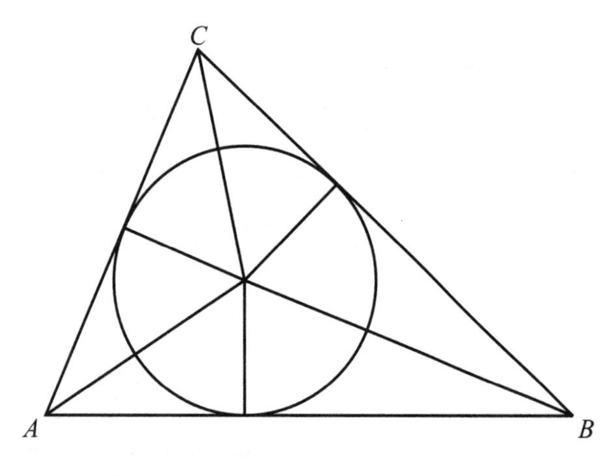

FIGURE 1

the triangle divides the side into lengths of 6 and 8 units. What are the lengths of the remaining two sides of the triangle?

Further, they were constrained from using trigonometric techniques. (An analysis of a diagram for the situation reveals three pairs of congruent right triangles upon which a repeated use of trigonometric functions would supply the correct answer.) See Figure 1.

The problem appears simple but without alternate problem solving strategies, the teachers were frustrated in their solution attempts. On the following day, only one teacher had solved the problem. Working with areas, she determined those of the known components of the triangle, two pairs of congruent right triangles 4×8 and 6×4, then let the length of the unknown segments of the required two sides of the triangle be x and used Heron's formula:

Given a triangle with sides of length a, b, c, the area of the triangle will be given by:

$$A = \sqrt{s(s-a)(s-b)(s-c)}$$

where s is the semi-perimeter of the triangle. Thus she obtained, $\sqrt{(x+14)(6)(8)x} = 32 + 24 + 4x$ and solving the equation, she found $x = 7$ and the required sides are 15 and 13 units long. Upon demonstrating this solution process to the other workshop participants, several questions arose: initially, "What is Heron's formula?" as the majority of the teachers had never heard of it; second, "Who is Heron?" (Alexandrian mathematician, circa 75 CE). Finally, someone asked where this particular problem originated. All the teachers were amazed to find that it came from 15th century Italy. A follow-up assignment was the derivation of Heron's formula.

In this instance much useful mathematical introspection and learning took place. A problem, usually considered an exercise in trigonometry, was resolved through a consideration of area—a more concrete approach to a solution that was popular for hundreds of years. The issue of another formula or method for determining the area of a triangle arose—most teachers and students only know the prescription that the area of a triangle is one-half the product of its base and altitude. In an applied problem, the measure of an altitude may remain elusive and, in itself, may be the subject of further investigation. Land surveyors determine areas of triangular plots by working with the sides of the triangles in question and seldom consider altitudes. A problem-solving approach recognized from the history of mathematics thus supplies relevance for contemporary needs.

A Survey of Historical Problems, Some General Impressions

Mathematical historians generally now concede that several ancient peoples probably preceded the Greeks in obtaining and using the correct mathematical relationships between the sides of a right triangle, the relationship commonly known as the Pythagorean Theorem.[3] A concern with right triangle theory and applications is evident in many old problems.

> A reed stands against a wall. If it moves down 9 feet (at the top), the (lower) end slides away 27 feet. How long is the reed? How high is the wall? (Babylonia, 1600–1800 BCE)[4]

> An erect (vertical) pole of 30 feet has its base moved out 18 feet. Determine the new height and the distance the top of the pole is lowered. (Egypt, 300 BCE)

> A bamboo shoot 10 feet tall has a break near the top. The configuration of the main shoot and its broken portion forms a triangle. The top touches the ground 3 feet from the stem. What is the length of the stem left standing erect? (China, 300 BCE)

> A spear 20 feet long rests against a tower. If its end is moved out 12 feet, how far up the tower does the spear reach? (Italy, 1300 CE)

While the information given is similar within this sequence of problems, the required results are different. If, for each problem, x represents the desired unknown, the solution situations are quite different as shown in Figure 2.

The "bent bamboo" problem provides an example of mathematical borrowing and transmission across cultures. It is known to have first appeared in China as the thir-

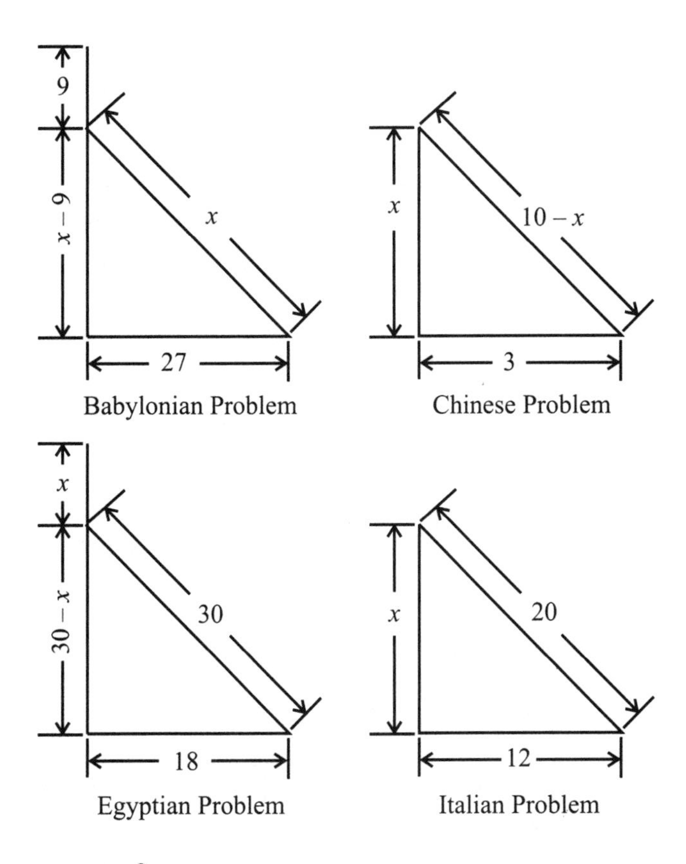

FIGURE 2

teenth problem in the ninth chapter of the *Jiuzhang suanshu* [*Nine Chapters on the Mathematical Art*] (300 BCE–200 CE).[5] Later it reappears in the Sanskrit mathematical classic *Ganita—Sara* [*Compendium of Calculation*] compiled by Mahavira (c. 850 CE). Finally, it found its way to Europe in Philippi Calandri's "Arithmetic" of 1491.[6] See Figure 3. Such illustrations can be made into overhead transparencies and shown to a class. Students can be placed in the position of mathematical archeologists and asked to decipher the problems from the diagrams and the limited, recognizable, information given. Archaic units of measure, for example, *braccia,* can be converted to their modern equivalents. Their findings can then be compared to the original problems.

While these are simple applications of the "Pythagorean" theorem, more complex and imaginative situations were also considered:

> A tree of height 20 feet has a circumference of 3 feet. There is a vine which winds seven times around the tree and reaches the top. What is the length of the vine? (China, 300 CE)
>
> (Given a vertical pole of height 12 feet.) The ingenious man who can compute the length of the pole's shadows, the difference of which is known to be nineteen feet, and the difference of the hy-

Calandri's Problems

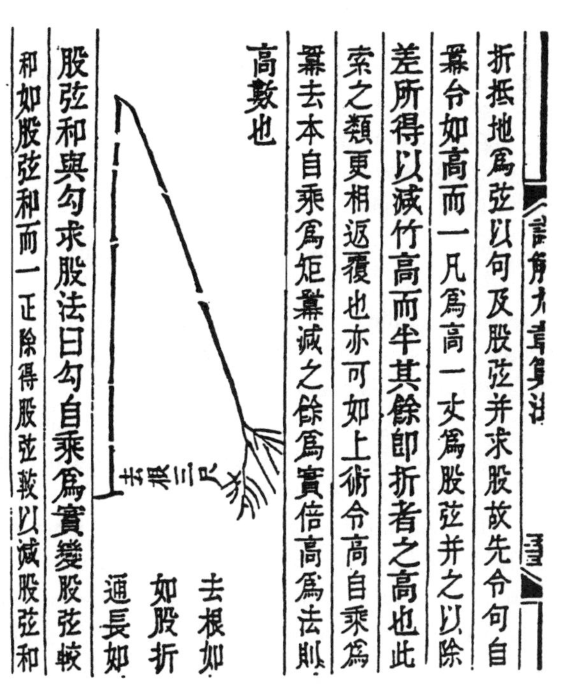

Bent Bamboo

FIGURE 3

potenuses formed, 13 feet, I take to be thoroughly acquainted with the whole of algebra as well as arithmetic. (India, 1000 CE)

This second problem provides a true computational challenge!

Problem similarities are not only limited to mathematical contexts. Societal concerns also allow for comparisons. Early societies' concern with food, particularly grain, is evident from the context of many problems. Ancient peoples used grain as a currency and paid wages and taxes with grain.

> Suppose a scribe says to thee, four overseers have drawn 100 great baskets of grain, their gangs consisting, respectively, of 12, 8, 6 and 4 men. How much does each overseer receive? (Egypt, 1600 BCE)

> From a certain field, I harvest 4 wagons of grain per unit of area. From a second field, I harvest 3 wagons of grain per unit area. The yield of the first field was 50 wagons more than that of the second. The total area of both fields is known to be 30 units. How large is each field? (Babylonia, 1500 BCE)

> The yield of 3 sheaves of superior grain, 2 sheaves of medium grain, and 1 sheaf of inferior grain is 39 baskets. The yield (from another field) of 2 sheaves of superior grain, 3 sheaves of medium grain, and 1 sheaf of inferior grain is 34 baskets. (From a third field) the yield of 1 sheaf of superior grain, 2 sheaves of medium grain, and 3 sheaves of inferior grain is 26 baskets. What is the yield of superior, medium, and inferior grains? (China, 300 BCE)

Early society's concern with grain, its harvesting, storage and distribution reflects their agricultural status. Problems 41–60 of the Rhind papyrus (c. 1650 BCE) concern the geometry of grain storage.[7] One chapter of the Chinese *Jiuzhang suanshu* is entitled "Millet and Rices." It concerns the use of proportions in the distribution of these grains. The use of proportion was a powerful mathematical tool in early problem solving.

Throughout the ages, the division and pricing of foodstuffs has provided the basis for many problems. Fibonacci, writing for a 13th century Italian audience, demonstrates that, mathematically speaking, "things are not what they may seem."

> There were two men, of whom the first had 3 small loaves of bread and the other 2; they walked to a spring, where they sat down and ate. A soldier joined them and shared their meal, each of the three men eating the same amount. When all the bread was eaten the soldier departed, leaving 5 bezants to pay for his meal. The first man accepted 3 of the bezants, since he had three loaves; the other took the remaining 2 bezants for his 2 loaves. Was the division fair? (Italy, 1202).

Insights into the use and rewards for labor can be found in many cultures:

> It is known that the digging of a canal becomes more difficult the deeper one goes. In order to compensate for this fact, differential work allotments were computed: a laborer working at the top level was expected to remove 1/3 sar of earth in one day, while a laborer at the middle level removed 1/6 sar and at the bottom level, 1/9 sar. If a fixed amount of earth is to be removed from a canal in one day, how much digging time should be spent at each level? (Babylonia, 1500 BCE)

> Now there is a city wall of upper width 2 *zhang*, lower width of 5 *zhang*, 4 *chi*, height 3 *zhang*, 8 *chi*, and length 5,550 *chi* to be constructed. The work capacity of a person in autumn is 300 *chi*. Find the manpower needed. (China, c. 200 CE)[8]

> Warner receives $2.50 a day for his labor and pays $.50 a day for his board; at the finish of 40 days, he has saved $50. How many days did he work and how many days was he idle? (United States, 1873).

Compensation for work performed reveal social injustices, societal inequities and the fact that our forefathers worked with indeterminate situations:

> If 100 bushels of corn be distributed among 100 people in such a manner that each man receives 3 bushels, each woman 2, and each child 1/2 bushel, how many men, women and children were there? (England, 800 CE)

> If 20 men, 40 women and 50 children receive $350 for seven weeks work, and 2 men receive as much as 3 women or 5 children, what sum does a woman receive for a week's work? (England, 1880)

In a 1000 years, has the status of a woman's labor really improved?

Alcuin of York compiling problems in about the year 800, notes the need for military conscription:

> A king recruiting his army, conscripts 1 man in the first town, 2 in the second, 4 in the third, 8

> in the fourth, and so on until he has taken men from 30 towns. How many men does he collect in all?

Apparently in Victorian England the practice was still taking place, although at a less dramatic level.

> The number of disposable seamen at Portsmouth is 800; at Plymouth 756; and at Sheerness 404. A ship is commissioned whose complement is 490 seamen. How many must be drafted from each place so as to take an equal proportion? Hamblin's *Treatise of Arithmetic* (1880).

Military needs have been a persistent theme in historical problem solving.[9]

Geometric situations also provided a challenging setting for problems:

> Given a triangular piece of land having two sides 10 yards in length and its base 12 yards, what is the largest square that can be constructed within this piece of land so that one side lies along the base of the triangle? (Al-Khwarizmi's *Algebra*, c. 820 CE)

> A circular field of land can contain an equilateral triangle of side 36 feet. What is the size of the field? (Egypt, 300 BCE)

> Given a right triangle with legs 8 and 15 feet, respectively. What is the largest circle that can be inscribed in this triangle? (China, 300 BCE)

In examining collections of old problems, it is interesting to note that there have always existed purely intellectual, riddle-type, problems such as the "three sisters" problem from ancient China:

> Now there are three sisters. The eldest returns every 5 days, the second returns once every four days and the youngest returns once every three days. Find the number of days before the three sisters meet together. (c. 200 CE)

or the famous "goat, wolf and cabbage" problem from Western mathematics:

> A wolf, a goat and a cabbage must be transported across a river in a boat holding only one besides the ferryman. How must he carry them across so that the goat shall not eat the cabbage, nor the wolf the goat? (800 CE)[10]

Mathematical authors have always provided both challenges and fun in their problem selections.

Problems as a Mathematical Testament

While problems can supply much information about the societies and times in question, they also illustrate the mathematical needs and ingenuity of our ancestors. From examining a problem's contents, students are often amazed to discover that before the Christian era, people were solving systems of linear equations and applying iterative algorithms to compute square and cube roots of numbers to a good degree of accuracy.[11] At various periods of history certain problems dominate the mathematical environment. While ancient Chinese and Egyptian mathematics focused on utilitarian needs, Greek mathematicians were busy with such non-utilitarian concerns as:

1. The duplication of the cube
2. The trisection of an angle
3. The quadrature of the circle

These problems were to be solved with the use of a straightedge and compass alone. It was over two thousand years before this feat was proved impossible, but yet, in the interim search for solutions, many useful mathematical discoveries were made, including a theory of conic sections and the development of cubic, quadratic, and transcendental curves. Out of this particular legacy emerged a series of geometric problems that can challenge and fascinate modern students. In the search to achieve a quadrature of the circle, a theory of lunes developed and problems like those that follow resulted.[12]

> Given a semicircle with diameter AB, arc ADB (a quadrant) is inscribed in the semicircle. The region bounded between the semicircle and the arc is called a lune. Show that the area of the lune $ACBD$ is equal to the area of the inscribed triangle ACB where $\overline{AC} \cong \overline{CB}$. (See Figure 4.)

Problems become more complex and further removed from reality, as shown by a consideration of the arbelos and its properties, a curve studied by Archimedes.[13]

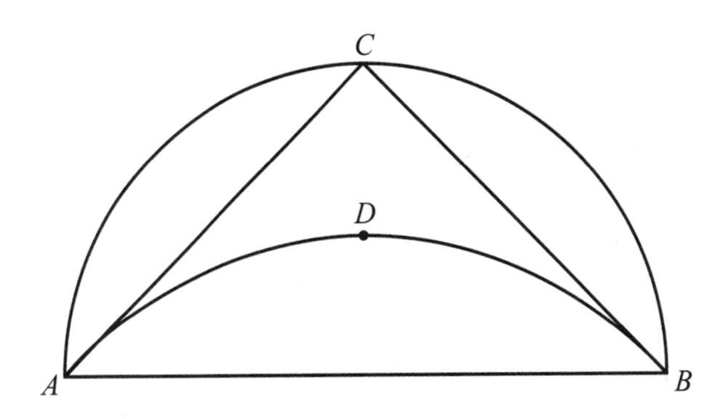

FIGURE 4

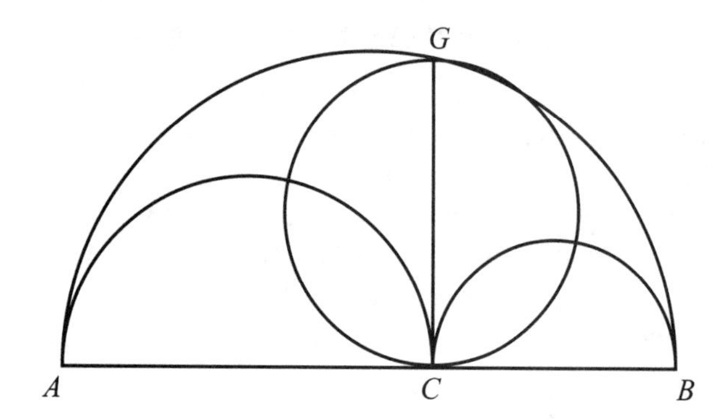

FIGURE 5

Let A, C, and B be three points on a straight line. Construct semicircles on the same side of the line with AB, AC, and CB as diameters. The region bounded by these three semicircles is called an arbelos. At C construct a perpendicular line to AB intersecting the largest semicircle at point G. Show that the area of the circle constructed with CG as a diameter equals the area of the arbelos. (See Figure 5.)

Given an arbelos packed with circles $C_1, C_2, C_3, \ldots$, as indicated, show that the perpendicular distance from the center of the nth circle to the line ACB is n times the diameter of the nth circle. (See Figure 6.)

During the Edo period of Japanese history (1603–1867), Japan remained isolated from the Western world. Little Western knowledge entered the island empire. However, it was a period of strong mathematical activity with a broad spectrum of common people posing and solving rather intricate problems involving geometric figures. Most of these problems concerned the packing of circles into certain specified plane figures.[14] Two elementary problems of this genre are:

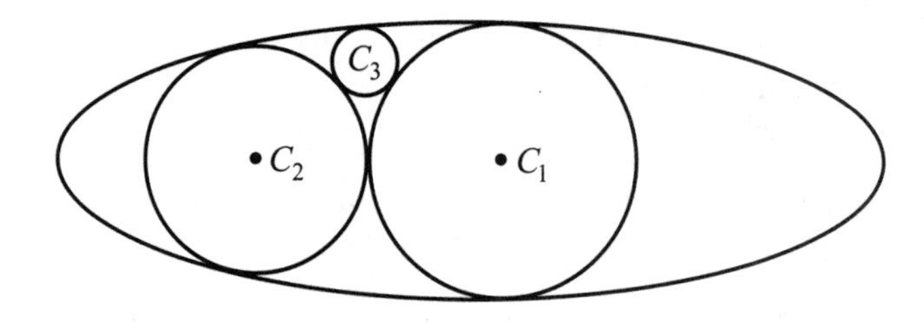

FIGURE 7

Given three circles C_1, C_2, C_3 with radii of r_1, r_2, and r_3 respectively. The circles are inscribed in an ellipse with major axis of length a and minor axis of length b. Circles C_1 and C_2 touch each other externally, and each also touches the ellipse at two distinct points. The circle C_3 touches the ellipse internally and also touches each of the other two circles externally. Find r_3 in terms of a, b, and r_1. (See Figure 7.)

Given triangle ABC with sides of known lengths a, b, c respectively. Three circles of radii r_1, r_2, r_3 are in mutual contact and are inscribed in the triangle. What are the radii of the circles in terms of the given sides? (See Figure 8.)

Such problems, when solved, were inscribed on wooden tablets and presented at the local Shinto temple. Today they are known as "Japanese Temple Problems."

Both the classical problems of ancient Greece and the temple problems of a reclusive seventeenth century Japan, possess a certain aesthetic that while different is also strangely similar. Geometric relations and perceptions of space have always fascinated people across time and cultures. It was well into the nineteenth century before the Greek classical problems were shown to be unsolvable using straightedge and compass, but their legacy has left us with much interesting mathematics. The temple problems

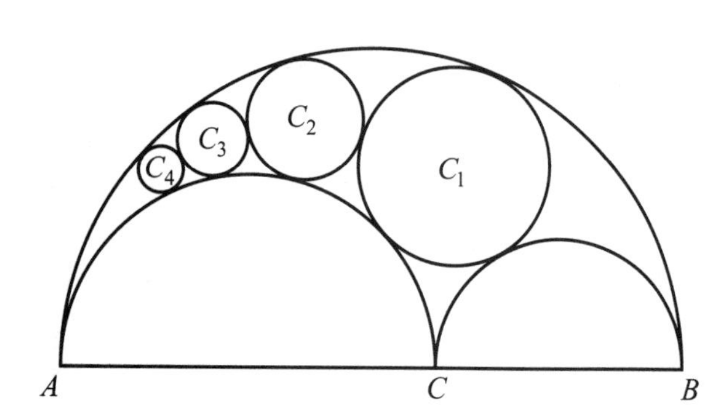

FIGURE 6

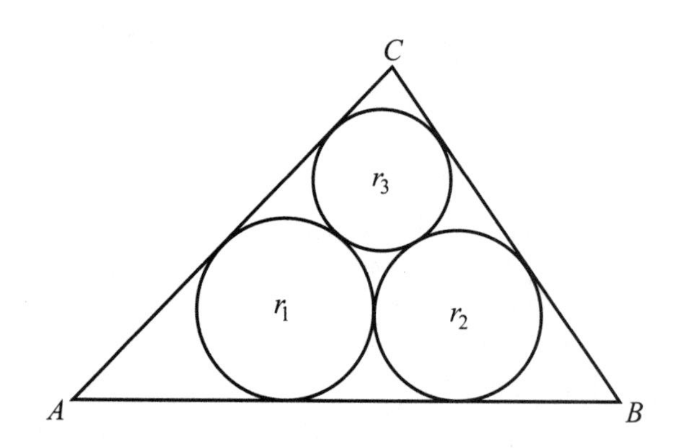

FIGURE 8

of Japan still remain little known outside their country of origin—they wait to be explored!

Conclusion

The history of mathematics contains a wealth of material that can be used to inform and instruct in today's classrooms. Among these materials are historical problems and problem solving situations. While for some teachers, historical problem solving can be a focus of a lesson, it is probably a better pedagogical practice to disperse such problems throughout the instructional process. Teachers who like to assign a "problem of the week" will find that historical problems nicely suit the task. Ample supplies of historical problems can be found in old mathematics books and in many survey books on the history of mathematics. These problems let us touch the past but they also enhance the present. Their contents reveal the mathematical traditions that we all share. Questions originating hundreds or even thousands of years ago can be understood, appreciated, and answered in today's classrooms. What a dramatic realization that is!

Notes and References

[1] The emphases of textual problems have varied for different periods of time and have reflected the instructional intent of their authors. Texts written in the bureaucratic contexts of ancient Egypt and China considered contemporary societal problems and employed contemporary data. Their problems discussed and reflected the needs of the times and thus provide valid historical insights. Similarly the abaci manuscripts of the late middle ages reflect the economic and commercial climate of southern Europe. See, for example, Warren Van Egmond, *The Commercial Revolution and the Beginnings of Western Mathematics in Renaissance Florence 1300–1500* (Ph.D. thesis, Department of History and Philosophy of Science, Indiana University, 1976). Other textbook authors, primarily interested in the theoretical aspects of mathematics, have devised problems whose realistic application is open to question.

[2] The workshop was given in 1995 by the author to a select group of teachers from Southeast Asia.

[3] See discussion in Victor Katz, *A History of Mathematics* (New York: Harper Collins, 1993) pp. 26–31; R.C. Buck, "Sherlock Holmes in Babylon," *American Mathematical Monthly* 87 (1980) pp. 335–345.

[4] Assume that the reed initially stands even with the top of the wall.

[5] For a more complete discussion and analysis of this problem see Frank Swetz and T.I. Kao, *Was Pythagoras Chinese? An Examination of Right Triangle Theory in Ancient China* (University Park, PA: Pennsylvania State University Press, 1997) pp. 44–45.

[6] The migration of this problem is discussed in Vera Sanford, *The History and Significance of Certain Standard Problems in Algebra* (New York: Teachers College Press, 1927) p. 77. The problems illustrated from Calandri's *Arithmetic* are:

> A tower is 40 braccia high and at its base runs a river which is 30 braccia wide. I want to know how long a rope which runs from the top of the tower to the other side of the river will be?
>
> There is a tree on the bank of a river which is 60 braccia high and the river is 30 braccia wide. The tree breaks at a point such that the top of the tree touches the opposite bank of the river. I want to know how many braccia broke off and how high the stump was?

[7] As given in the translation: A.B. Chace, *The Rhind Mathematical Papyrus* (Washington, DC: The National Council of Teachers of Mathematics, 1978 [reprint of 1927, 1929 editions]).

[8] The units of measure employed are the *chi*, the Chinese "foot" and the *zhang*, where 1 *zhang* = 10 *chi*. See Lam Lay Yong and Ang Tian Se, *Fleeting Footsteps: Tracing the Conception of Arithmetic and Algebra in Ancient China* (Singapore: World Scientific, 1992) p. 84. The unit *chi* is a measure of both length and volume depending on context. Thus the work capacity of 300 *chi* should be understood to mean 300 cubic *chi*. The accompanying answer determines the volume to be constructed as 7,803,300 (cubic) *chi* for which 26,011 men will be required.

[9] Books were devoted to military problem solving, for example, Leonard and Thomas Digges, *Arithmetical Militare Treatise*, named *Stratioticos* (London, 1572.) The concept of "drumhead trigonometry" owes its origins to military needs. See D.E. Smith, *History of Mathematics* vol. II (New York Dover Publications, 1958) p. 355.

[10] The problem apparently originated in Alcuin of York's *Propositiones ad acuendos juvenes* (800 CE) See translation and discussion in David Singmaster and John Hadley, "Problems to Sharpen the Young," *The Mathematical Gazette* 76 (1992) pp. 102–126. A discussion of the same problem in different contexts and cultures is given in Marcia Ascher, *Ethnomathematics* (Pacific Grove, CA: Brooks/Cole Publishing Company, 1991) pp. 109–112.

[11] An analysis of Yale Babylonian Collection tablet 7289 reveals that its ancient scribes calculated $\sqrt{2} = 1.414213562$. See discussion in H.L. Resnikoff and R.O. Wells, *Mathematics in Civilization* (New York; Holt, Rinehart and Winston, 1973) pp. 76–78.

[12] Leonardo da Vinci experimented with the mathematics of lunes. See Herbert Wills, *Leonardo's Dessert* (Reston, VA: National Council of Teachers of Mathematics, 1985).

[13] For a fuller discussion of the arbelos see L. Raphael, "The Shoemaker's Knife," *Mathematics Teacher* (April, 1973) pp. 319–323.

[14] These problems are considered in Yoshio Mikami, *The Development of Mathematics in China and Japan* (New York: Chelsea Publishing Company, 1974 [reprint of 1913 edition]). An extensive discussion of problems and their solutions is given in H. Fukagawa and D. Pedoe, *Japanese Temple Geometry Problems* (Winnipeg, Canada: The Charles Babbage Research Centre, 1989).

Part III

Teaching a Particular Subject Using History

Second Degree Equations in the Classroom: A Babylonian Approach*

Luis Radford
Université Laurentienne, Canada

and

Georges Guérette
Conseil de l'éducation de Sudbury, Canada

In this paper, we present a teaching sequence whose purpose is to lead the students to reinvent the formula that solves the general quadratic equation. Our teaching sequence is centered on the resolution of geometrical problems related to rectangles using an elegant and visual method developed by Babylonian scribes during the first half of the second millennium BCE. Our goal is achieved through a progressive itinerary which starts with the use of manipulatives and evolves through an investigative problem-solving process that combines both numerical and geometrical experiences. Instead of launching the students into the modern algebraic symbolism from the start—something that often discourages many of them— algebraic symbols are only introduced at the end, after the students have truly understood the geometric methods. The teaching sequence has been successfully undertaken in some high school classrooms.

1. The Babylonian Geometric Method

Before explaining the teaching sequence it is worthwhile to mention briefly some of the features of the Babylonian geometric method. The method to which we are referring was identified by J. Høyrup who called it *Naive Geometry*.[1] In order to show the method, let us discuss one of the simplest Babylonian problems, namely, problem 1 of a tablet preserved at the British Museum and known as BM 13901.

The statement of the problem, which seeks to find the length of the side of a square, is the following:

> The surface and the square-line I have accumulated: 3/4.

As in most of the cases, the scribe states the problem using a very concise formulation. He is referring to the surface of a square, and the square-line means the side of the square. Thus, the problem is to find the side of the square, knowing that the sum of the area of the square and the side is equal to 3/4. The method of solution is not fully explained in the text. Indeed, the text shows only a list of instructions concerning a sequence of calculations that allows one to get the answer.

The instructions, as they appear in the tablet, are the following:

> 1 the projection you put down. The half of 1 you break, 1/2 and 1/2 you make span [a rectangle, here a square], 1/4 to 3/4 you append: 1, make 1 equilateral. 1/2 which you made span you tear out inside 1: 1/2 the square-line. (Høyrup, 1986, p. 450)

Of course, the Babylonians should have had a method on which the numerical calculations were based. For some time it was believed that the Babylonians somehow knew our formula to solve second-degree equations. However, this interpretation has been abandoned because of the multiple intrinsic difficulties that it implies, one of them being the well known lack of algebraic symbols in Babylonian mathematical texts and the related impossibility for the Babylonians to handle complex symbolic calculations without an explicit symbolic language (details in Radford 1996 and Radford in press).

Based on a philological and textual analysis of the Babylonian texts, J. Høyrup suggested that the solution of problems (such as the preceding one) was underlain by a geometrical configuration upon which the oral explanation was based. In the case of the previous problem, the scribe thinks of an actual square (Fig. a). However, the side is not seen as a simple side (Fig. b) but as a side provided with a canonical projection that forms, along with the side, a rectangle (Fig. c). The duality of the concept of side is based on a metrological equality: the length of the side and the area of the rectangle that it forms along with its canonical projection have the same numerical value (see Høyrup 1990a). Keeping this in mind and coming back to problem 1, BM 13901, it appears that the quantity $3/4$ refers then to the total area of Fig. 1. Next, the scribe cuts the width 1 into two parts and transfers the right side to the bottom of the original square (see Fig. 2).

Now the scribe completes a big square by adding a small square whose side is $1/2$ (Fig. 3). The total area is the $3/4$ (that is, the area of the first figure) plus $1/4$ (that is, the area of the added small square). It gives 1. The side of the big square can now be calculated: that gives 1; now the scribe subtracts $1/2$ from 1, he gets $1/2$: this is the side of the original square.

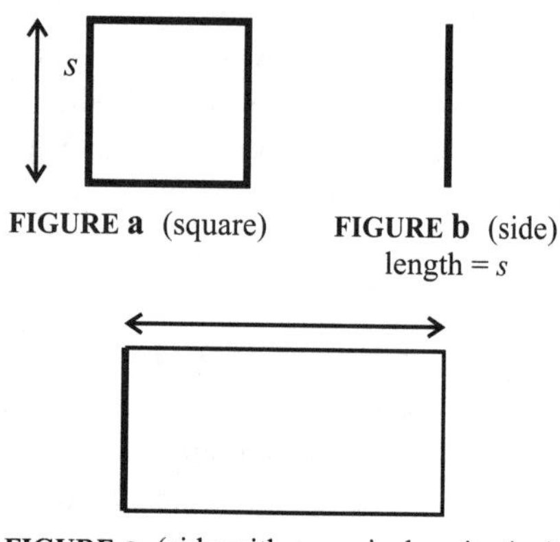

FIGURE a (square) **FIGURE b** (side) length = *s*

FIGURE c (side with canonical projection) area = *s*

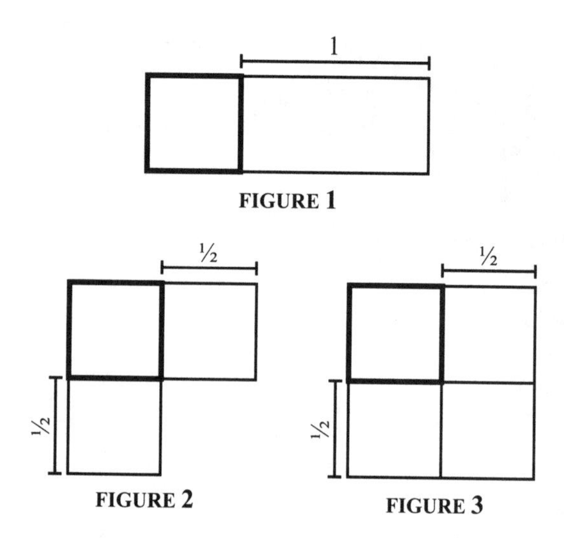

FIGURE 1

FIGURE 2 **FIGURE 3**

This is the same type of transformation that seems to be the basis of the resolution of many problems found in a medieval book, the *Liber Mensurationum* of Abû Bekr (probably ninth century), whose Arabic manuscript was lost and which we know of through a twelfth century translation by Gerard of Cremona (ed. Busard, 1968). In fact, many of these problems are formulated in terms similar to those of the *Naive Geometry*.

Let us consider an excerpt from one of the problems of the *Liber Mensurationum* (problem 41; Busard 1968, p. 95):

> And if someone tells you: add the shorter side and the area [of a rectangle] and the result was 54, and the shorter side plus 2 is equal to the longer side, what is each side?

As in the case of Babylonian texts, the steps of the resolution given in the *Liber Mensurationum* indicate the operations between the numbers that one has to follow.[2] In all likelihood, the calculations are underlain by a sequence of figures like figures 4 to 10 hereinafter. The initial rectangle is shown in Fig. 4. The shorter side, x, (placed at the right of the figure) is provided with a projection equal to 1 (Fig.5), so that the length of the side is equal to the area of the projected rectangle, as in the case of the Babylonian problem discussed above. Given that $y = x + 2$, the base 'y' (bottom of Fig. 5) can be divided into two segments 'x'

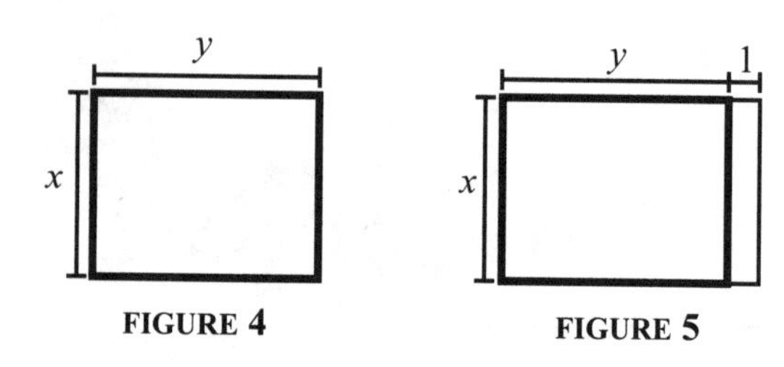

FIGURE 4 **FIGURE 5**

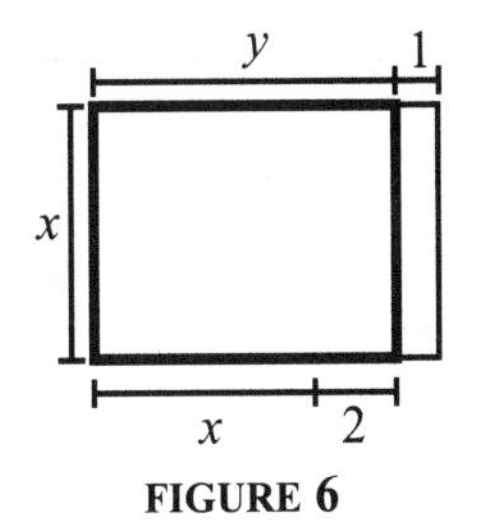

FIGURE 6

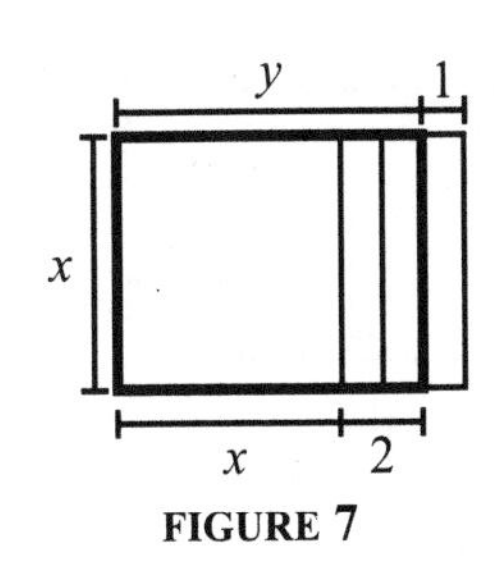

FIGURE 7

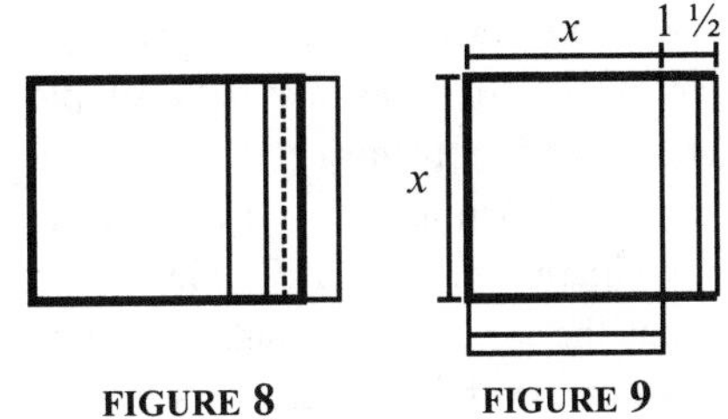
FIGURE 8

FIGURE 9

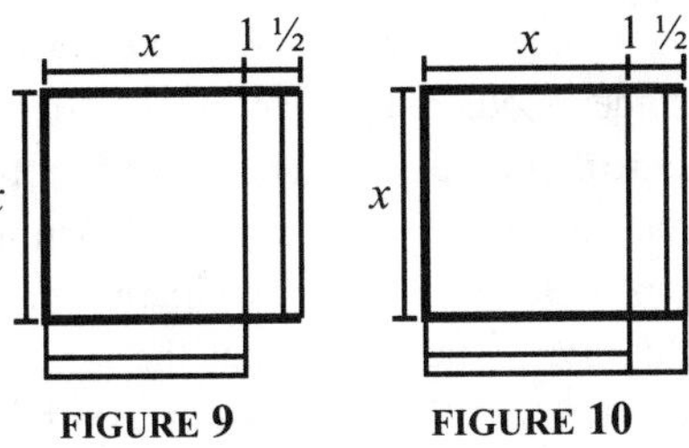

FIGURE 10

and '2' (see Fig. 6). Now two small rectangles are placed inside the original rectangle, as shown in Fig. 7.

The next step is to divide into two the set of the three equal rectangles (Fig. 8); one of these parts (that is, a rectangle and a half) is placed at the bottom of the remaining figure. As a result of this transformation, we now have Fig. 9, which is *almost* a square.

The key idea in the resolution of these types of problems (and which appears in an explicit manner in Al-Khwarizmi's *Al-Jabr* (ed. Hughes 1986)) is to complete the current figure (Fig. 9) in order to get a square. The completion of the square (Fig. 10) is achieved then by adding in the right corner a small square whose area is equal to $\left(1\frac{1}{2}\right)^2 = 2\frac{1}{4}$. The final square then has an area equal to $54 + 2\frac{1}{4} = 56\frac{1}{4}$, so that its side is $\sqrt{54\frac{1}{4}} = 7\frac{1}{2}$. The shorter side, x, of the original rectangle is then equal to $7\frac{1}{2} - 1\frac{1}{2} = 6$, so the longer side is 8.

We are not going to discuss here the historical arguments that support the reconstruction of the procedures of resolution for problems such as the preceding one in terms of the *Naive Geometry* (see Høyrup 1986 or Høyrup 1990b). We shall limit ourselves to simply indicating that the explicit appearance of these procedures in Al-Khwarizmi's work leaves no doubt that these procedures were well-known in the ninth century in certain Arabic milieus.

2. The Teaching Sequence

In this section we shall present the teaching sequence that we have developed in order to introduce the students to second degree equations and that culminates with the reinvention of the formula to solve these equations. The sequence is divided into 5 parts (whose duration may vary according to the students' background).

For each part of the sequence:

(i) we give the indications of the different steps to follow in the classroom;

(ii) we include an item called *particular comments,* which, through concrete examples, intends to shed some light on the issues of the teaching sequence according to our classroom experience.

Part 1. The introduction to the *Naive Geometry*

In part 1, the students are presented with the following problem:

> What should the dimensions of a rectangle be whose semi-perimeter is 20 and whose area is 96 square units?

Working in cooperative groups, the students are asked to try to solve the problem using any method. After they complete the task, the teacher, returning to the geometrical context of the problem and using large cardboard figures on the blackboard, shows them the technique of *Naive Geometry.*

This can be done through the following explanation: If you take a square whose side is 10, then its area is 100 (Fig.11). One must therefore cut out 4 square units of the square whose side is 10 (Fig. 11) to obtain a figure whose area is 96. This can be achieved (and that is the key idea of the resolution) by cutting out of the big square a smaller square whose side is 2 (see Fig. 12). In order to obtain a rectangle one cuts the rectangle shown by the dotted line in Fig. 13 and places it vertically on the right (Fig. 14). The sought-after sides then measure 12 units and 8 units.

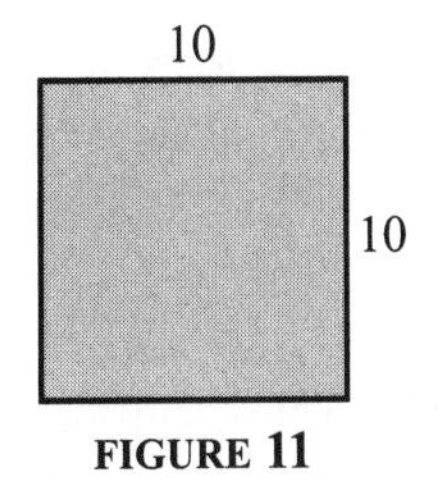

FIGURE 11

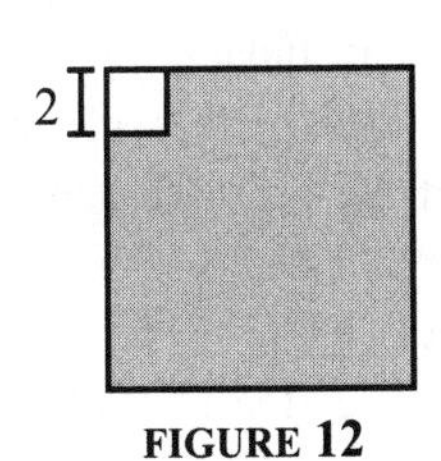

FIGURE 12

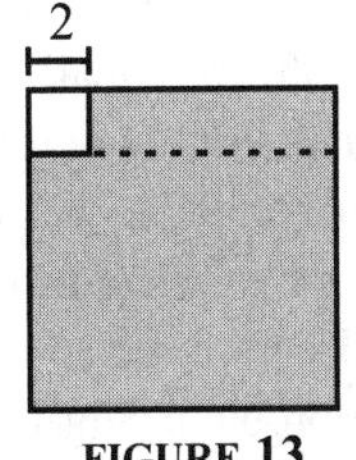

FIGURE 13

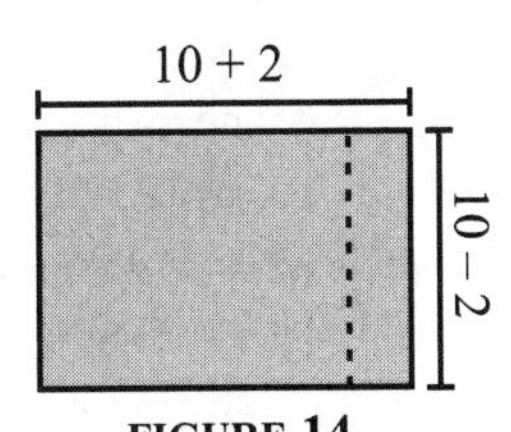

FIGURE 14

Once the technique is presented, the teacher gives the students other similar problems. In order to avoid a simple repetition, the parameters of the problem (i.e., the area and the semi-perimeter) may be chosen as follows: area of the rectangle $= 30$ and semiperimeter $= 12$. Problems like this are particular in that the area of the small square to be removed (Fig. 12) is not a perfect square. This led the students to reflect about the *Naive Geometry* technique on a deeper level.

In order to help students achieve a better understanding, the teacher asks them to bring a written description to class the next day outlining the steps to follow to solve this type of problem. They may be told that the written description or "message" should be clear enough to be understood by any student of another class of the same grade.

Particular comments.

(1) The idea of asking the students to solve the problem 1 using any method is simply to get them exploring the problem. As expected, usually, they use a trial-and-error method. Other students choose a rather numerical-geometric method; they choose a square of side equal to 10 (a solution motivated by the fact that the number 10 is half of the semi-perimeter 20). A less usual strategy is to take the square root of 96. In the last two strategies, when they try to justify their answer (sometimes at the request of the teacher), they realize that it is incorrect. The teacher may then ask for ideas about direct methods of solution (something that excludes trial-and-error methods).

(2) The geometrical resolution of this problem, a problem that can actually be found in a numerical formulation in Diophantus' *Arithmetica* (c. 250 CE) (Book 1, problem 27), is far from evident to the students. As we have quite often noticed, when we first show the *Naive Geometry* approach in the classroom, the visual seductive geometrical particularity of the resolution awakens a genuine interest among the students.

(3) In one of our sessions, when confronted with the problem of area $= 30$ and semi-perimeter $= 12$, one group of students started assuming, according to the technique, that the sides were each equal to 6 (which meets the requirement of semi-perimeter $= 12$). Given that the area of this square is equal to 36, they realized that they needed to take away 6 square units. In order to avoid irrationals, they cut out a rectangle whose sides were equal to 2 and 3. Then they realized that in doing so it is not possible to end with a rectangle, as required by the statement of the problem. Fig. 15b shows the non-rectangular geometric object to which one is led when one takes away a 2×3 rectangle (Fig. 15a) instead of a square whose sides are equal to $\sqrt{6}$). Then, they became aware that a square of area equal to 6 has to be removed and that they had to take away a side of length equal to $\sqrt{6}$.

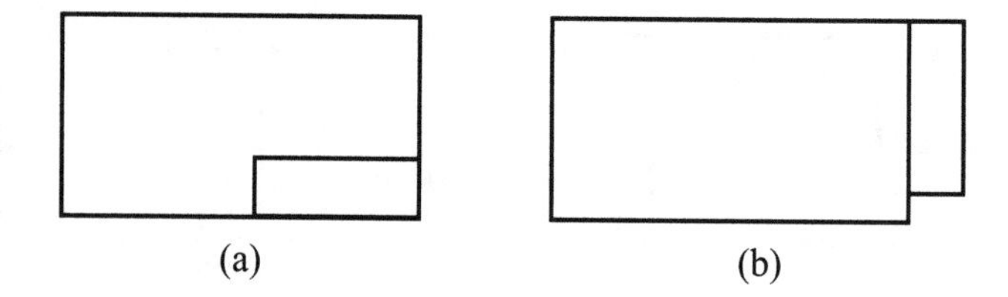

FIGURE 15

Part 2.

This part begins with a discussion of the messages containing all the steps required in the resolution of the problems seen in part 1. Working in cooperative groups, the students have to discuss and come to an agreement about the points which could cause a conflict or could give way to an improvement. When all the group members are in agreement, the teacher can choose one student of each cooperative group to present the work to the other groups. This allows certain students to better understand. Following this, the students are asked to *pose* problems themselves with the following restriction: the sides of the sought-after rectangle have to be expressed in whole numbers; then, as a second exercise, the sides of the sought-after rectangle do not have to be expressed in whole numbers. The students may even be asked to find fractional answers. A few of these problems would be used in the test at the end of the chapter.

Particular comments. We want to stress the fact that many students have some problems in writing the message in general terms, that is, without referring to particular numbers for the semi-perimeter and the area. Although they were not explicitly asked to write the message using letters (i.e., using an algebraic language), it was very hard for many of them to express, at a general level, the actions that they were able to accomplish when working with manipulatives (concrete rectangles on paper that they produced themselves using scissors) or drawings.

Despite the intrinsic difficulty of this task, it is important that students try to use the natural language to express their ideas. This will simplify the transition to the abstract symbolic algebraic language.

Part 3.

In part 3, the students are presented with a problem that requires a different use of the Naive Geometry technique.

The problem, inspired by that of Abû Bekr seen at the end of section 1, is the following:

> Problem 2: The length of a rectangle is 10 units. Its width is unknown. We place a square on one of the sides of the rectangle, as shown in the figure. Together, the two shapes have an area of 39 square units. What is the width of the rectangle?

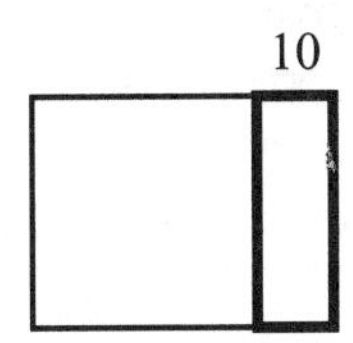

The teacher asks the students to solve the problem using similar ideas as the ones used to solve problem 1. If students do not succeed in solving the problem by the *Naive Geometry* technique, the teacher may show the new problem-solving method as follows: Using large cardboard figures placed on the blackboard, the teacher cuts the initial rectangle vertically in two (Fig. 17), then takes one of the pieces and glues it to the base of the square (Fig. 18). Now the students notice that the new geometrical form is almost a square. The teacher then points out that the new form could be completed in order to make it a square. In order to do so, a small square, whose side is 5 (Fig. 19), has to be added. The small square has an area equal to 25. Thus the area of the new square (Fig. 19) is equal to $39 + 25 = 64$. Its side is then equal to 8. From Fig. 18 it follows that $x + 5 = 8$, which leads us to $x = 3$. Next, other similar problems are given to the students to solve in groups.

As in part 1, the students are asked to work on a written description or message of the steps to follow in order to solve this type of problem.

Particular Comments. The students soon realize that the problem-solving procedure used in problem 1 does not apply directly to problem 2. Here, the central idea (and it is important that the teacher emphasizes it) is the completion of a square, something that will be very important when the students work with algebraic symbols later.

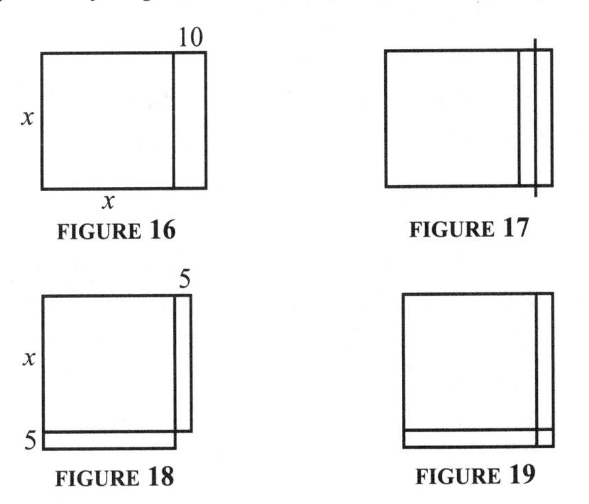

FIGURE 16 FIGURE 17

FIGURE 18 FIGURE 19

Part 4.
As in part 2, the students' written descriptions or messages are discussed. After this, the teacher asks them to pose some problems requiring a specific condition on the sides of the rectangle:

(i) the sides of the rectangle have to be expressed in whole numbers;
(ii) the sides of the rectangle have to be expressed in fractional numbers;
(iii) the sides of the rectangle have to be expressed in irrational numbers.

Part 5. Reinventing the formula.
In this part, the students will keep working on a problem of the same type as in parts 3 and 4. The difference is that *concrete numbers* are given neither for the base of the rectangle nor for the area that the two shapes cover together. The goal is to help students reinvent the formula that solves quadratic equations.

In order to do so, the teacher explains to the students that s/he is interested in finding a formula which will provide one with the answers to the problems seen in parts 3 and 4.[3] The teacher may suggest that they base their work on the written message produced in step 4 and to use letters instead of words. To facilitate the comparison of the students' formulas in a next step, the teacher may suggest using the letter "b" for the base of the rectangle and "c" for the area of both shapes (see Fig. 20). The equations are discussed in co-operative groups. The final equation is

$$x = \sqrt{c + \left(\frac{b}{2}\right)^2} - \frac{b}{2}.$$

The teacher may then proceed to translate the geometric problem into algebraic language: if the unknown side is 'x', then the area of the square is x^2 and that of the rectangle is bx; thus the sum of both areas is equal to c, that is: $x^2 + bx = c$. Now, in order to link equations to the formula, the teacher gives some concrete equations (like $x^2+8x = 9$,

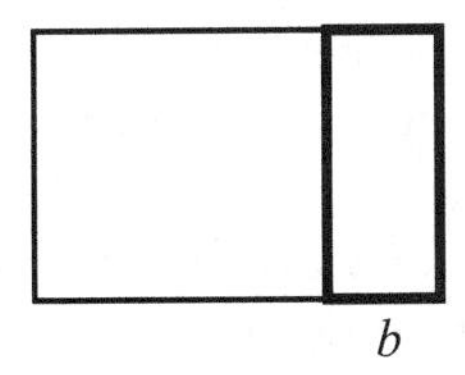

FIGURE 20

$x^2 + 15x = 75$) and asks the students to solve them using the formula.

The next step is to give the students the equation $ax^2 + bx = c$ and ask them to find the formula to solve this equation. The students might note that if this equation is divided by a (we suppose that $a \neq 0$), then we are led to the previous kind of equation. It suffices then to replace 'b' by 'b/a' and 'c' by 'c/a', in the previous formula, which gives the new formula

$$x = \sqrt{\frac{c}{a} + \left(\frac{b}{2a}\right)^2} - \frac{b}{2a}.$$

The last step is to consider the general equation $ax^2 + bx + c = 0$ and to find the formula that solves it. The formal link with the previous equation $ax^2 + bx = c$ is clear: we can rewrite this equation as $ax^2 + bx - c = 0$. Thus, in order to get the equation $ax^2 + bx + c = 0$ we need to replace 'c' by '$-c$' and to do the same in the formula.

When we replace 'c' by '$-c$' in the formula we obtain the general formula:

$$x = \sqrt{-c + \left(\frac{b}{2a}\right)^2} - \frac{b}{2a}.$$

Of course, this formula is equivalent to the well-known formula:

$$x = \frac{-b + \sqrt{b^2 - 4ac}}{2a},$$

where in order to obtain all the numerical solutions one needs to consider the negative square root of $b^2 - 4ac$. This leads us to the formula

$$x = \frac{-b \pm \sqrt{b^2 - 4ac}}{2a}.$$

Particular comments.

(1) Usually, the students are able to provide the formula that solves the equation $x^2 + bx = c$ and to use it to solve concrete equations (as those mentioned above). That they can produce such a formula and realize the amount of work that the formula saves is appreciated very much by the students. This gives them a 'practical' sense of the formula.

(2) However, many students need some time in order to abandon the geometrical context and to limit themselves to the numerical use of the formula. Further, there are many students who prefer to keep thinking in terms of the *Naive Geometry* technique. The geometrical versus numerical preference may be caused by the specific students' own kind of rationality (something that is referred to in the educational field by the unfelicitous expression "styles of learning", an expression that hides more things than it explains!). Some students have the impression that they no longer understand if they merely use a formula. *Understanding,* for many of them, does not seem to mean simply 'being able to do something'.

(3) Most of the students are able to find the formula which solves the equation $ax^2 + bx = c$. However, some students may experience some difficulties. The main problem is that here, as in the subsequent steps, the geometrical context is being progressively replaced by a symbolic one.

(4) To end these comments, we want to stress the fact that our approach cannot avoid or conceal the problems that are specific to the mastering and understanding of algebraic symbols (see section 4 below). Our approach aims to provide a useful context to help the students develop a meaning for symbols. It is worthwhile to mention that the use of manipulatives and geometric techniques in order to derive the formula were appreciated by our high-school students. A girl, for instance, said: "I better understand with the drawings, I find this a lot more interesting and fun than the other mathematics."

3. About the duration of the teaching sequence

The teaching sequence that we discussed here may vary in time depending on the background of the students and their familiarity with classroom research activities. In one of the first times that we undertook it, we allowed a period of 80 minutes to each step. However, it is possible to reduce the time and the steps of the sequence. A variant of the sequence, that we undertook in an advanced mathematics high school course, consisted of steps 3 to 5. This can be done in two periods of 80 minutes each.

4. A concluding (theoretical) remark about the use of symbols in mathematics

The passage from numbers to letters does not consist of a simple transcription, as we have noted. In fact, the symbol must, in part 5, summarize the numerical and geometrical experiences developed in parts 3 and 4. The encapsulation of these experiences includes a stage of generalization and of reorganization of the actions which opens up on a much more ample description of mathematical objects.

Of course, the new semiotic category (that is, the category in which the algebraic symbolism is embedded) offers new challenges to the students (see the students' reinvention of the formula shown in Fig. 21) as it did for past mathematicians (see Radford, in press). For instance, operations with symbols need to be provided with new meanings. In this sense abstraction does not seem to proceed to a *detachment* of meanings or to more "general" ideas. Indeed, contrary to a general interpretation, abstraction does not mean to take away some features of a given object but to *add* new ones and to be able to focus our attention on the features required by the context. This was suggested by the transference from geometric to syntactic algebraic symbolism and vice-versa that students showed when solving second-degree equations at the end of the sequence (for example, $2x^2 + 12x - 64 = 0$), after having reinvented the formula.

These considerations lead us to the following intriguing idea: abstraction is a contextually based operation of the mind.

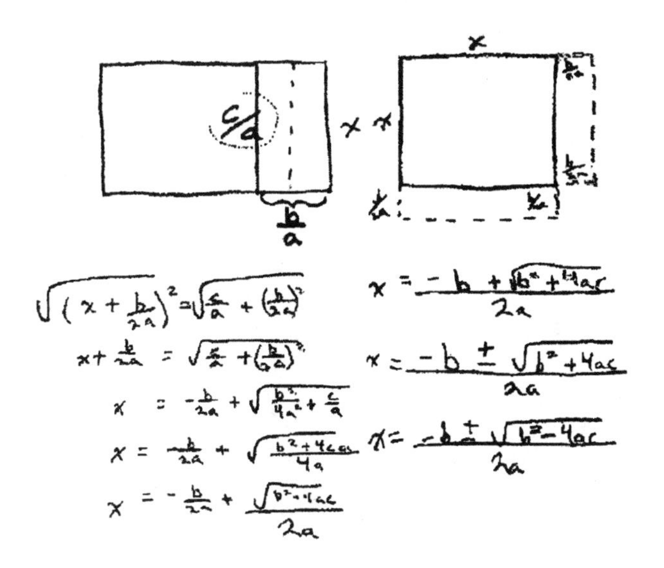

FIGURE 21

References

Busard, H. (1968) "L'Algébre au Moyen Âge: Le "Liber Mensurationum" d'Abû Bekr," *Journal des savants,* Avril-juin, 65–124.

Høyrup, J. (1986) "Al-Khwarizmi, Ibn-Turk, and the Liber Mensurationum: on the Origins of Islamic Algebra," *Erdem* 2 (Ankara), 445–484.

Høyrup, J. (1990a) "Dynamis, the Babylonians, and Theaetetus 147c7–148d7," *Historia Mathematica,* vol. 17, 201–222.

Høyrup, J. (1990b) "Algebra and Naive Geometry. An Investigation of Some Basic Aspects of Old Babylonian Mathematical Thought," *Altorientalische Forschungen,* vol. 17, 27–69, 262–354.

Hughes, B. (1986) "Gerard of Cremona's Translation of Al-Khwarizmi's *Al-Jabr*: A Critical Edition," *Mediaeval Studies,* No. 48, 211–263.

Radford, L. (1996) "The Roles of Geometry and Arithmetic in the Development of Algebra: Historical Remarks from a Didactic Perspective," in: *Approaches to Algebra,* eds. N. Bednarz, C. Kieran and L. Lee, Kluwer Academic Publishers, pp. 39–53.

Radford, L. (in press) The Historical Origins of Algebraic Thinking, in: *Perspectives in School Algebra,* eds. R. Sutherland, T. Rojano, A. Bell and R. Lins, Dordrecht/Boston/London: Kluwer.

Endnotes

* This article is part of a research project granted by FCAR No. 95ER0716, Québec, and Laurentian University Research Funds, Ontario.

[1] His main work on Naive Geometry is (Høyrup 1990b).

[2] The solution is given in the Liber Mensurationum as follows:

> The way to find this is that you add two [to one], such that you have 3. Now you take half which is one and a half and multiply that by itself so you get two and a quarter. So, add 54 to this and you get 56 and a quarter; take the root and subtract 1 and a half; you are left with 6 and that is the smaller side; add to it 2 and you will have the longer side, that is, 8. However, there is a method to find this according to the people of the al-gabr...

[3] The students are already familiar with the concept of formula: not only have they seen formulas in mathematics, e.g., the formulas for the areas of regular geometrical figures, but in the sciences as well.

Anomalies and the Development of Mathematical Understanding

Janet Heine Barnett
University of Southern Colorado

A number of historians of mathematics have identified anomalies as a major force shaping the development of mathematics. Wilder [1980], for example, classifies paradoxes and inconsistencies as components of 'hereditary stress', "the most important force exerted from the mathematical community upon the individual mathematical mind". Several of the famous *Ten 'laws' concerning patterns of change in the history of mathematics* identified by Crowe [1975] also point to the role of anomalies in both delaying and advancing mathematical developments.[1] In his critique of Crowe's laws, Mehrtens [1976] remarks not only on the role played by anomalies in the history of mathematics, but also proposes the concept of an 'anomaly' as a valuable tool for the historian in assessing historical developments.

In view of the importance assigned to anomalies by historians, it is reasonable to expect this concept to be of some importance to individual student learning as well. Underlying this expectation is the Piagetian belief that there are parallels between the way individuals construct mathematical knowledge (psychologically) and the way humanity constructed such knowledge (historically). This proposal views historical analysis as a critical factor in making instructional and curricular decisions, even when the actual implementation of these decisions might not involve the use of history in the mathematics classroom. Thus, the history of mathematics is seen as a lens through which to view 'mathematical understanding' in order to develop and assess pedagogical principles for the mathematics classroom.

This paper offers such an analysis of the role played by anomalies in developing mathematical understanding. We begin with overviews of three historical episodes in which the resolution of anomalies played a central role. The primary purpose of these overviews is to identify pedagogical principles that could be applied to the teaching of any mathematical topic by highlighting certain aspects of the episodes in question. The specific examples considered (incommensurable magnitudes, non-Euclidean geometries, and Cantor's infinite sets) were chosen to draw particular attention to the role of intuition in generating and resolving anomalies. In the closing sections, we relate these aspects of the historical developments to principles underlying current learning theory, and offer specific examples of how these ideas might be implemented in classroom instruction.

Incommensurables

We do not know exactly how or when the existence of magnitudes which cannot be measured by a common unit was discovered, due to the lack of primary source material concerning pre-Euclidean Greek mathematics. The discovery was certainly known by 399 BCE when Thaetetus (417–369 BCE) informs us in Plato's dialogue[2] of that name:

> Theodorus here was proving to us something about square roots, namely, that the sides of squares representing three square feet and five square feet are not commensurable in length with the line representing one foot, and he went on in this way, taking all the separate cases up to the root of seventeen square feet.[3]

It is interesting that Plato takes $\sqrt{3}$ as the starting point of Theodorus' work, a detail which suggests that the incommensurability of $\sqrt{2}$ was known prior to Theodorus' research. This interpretation is consistent with the hypothesis that incommensurables were discovered in the context of finding a common measure for the side and diagonal of a unit square. A general outline for a proof equivalent to our standard proof of the irrationality of $\sqrt{2}$ is given by Aristotle in several places, including the following:

> the diagonal of the square is incommensurate with the side, because odd numbers are equal to evens if it is supposed to be commensurate.[4]

Another hypothesis concerning the initial discovery suggests that the first known pair of incommensurable magnitudes was the side and diagonal of the regular pentagon,[5] a figure associated with the pentagram which served as a Pythagorean recognition symbol. Von Fritz [1945] argues that the pentagon discovery is apparent almost at first sight"[6] from the fact that the diameters of the pentagon form a new regular pentagon in the center. The formal proof results from applying the repeated subtraction process of the Euclidean algorithm to the magnitudes. In this instance, this repeated subtraction procedure (known as *antenaresis*) results in a pattern of remainders that establishes the impossibility of a 'least' magnitude.[7]

One point on which most scholars agree is that the existence of incommensurable magnitudes represented an unexpected and unwelcome discovery for the early Pythagorean philosophy that 'all is number'. The severity of the shock is suggested by the well-known story of Hippasus being drowned at sea for revealing the discovery.[8] Aristotle describes the Pythagorean belief in number as the substance of the physical universe as follows:

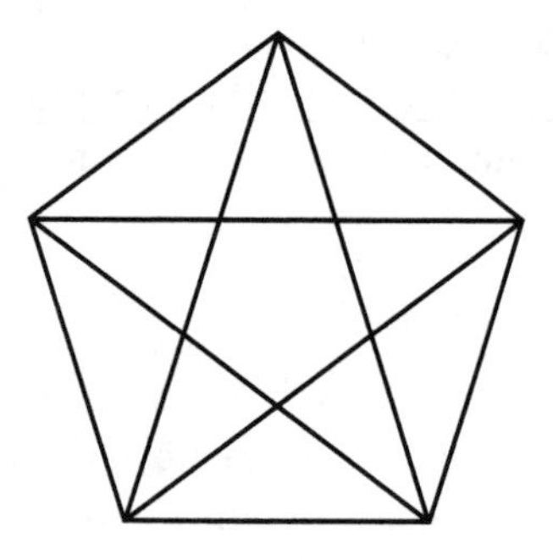

> in numbers they seemed to see many resemblances to the things that exist and come into being...; since, again, they saw that the modifications and the ratios of the musical scales were expressible in numbers; since, then, all these things seemed in their whole nature to be modeled on numbers,... they supposed the elements of numbers to be the elements of all things.[9]

This intuition that everything can be counted, and hence assigned a number, is easily applied to lengths by designating a unit of measure which can be replicated to cover the given length. For two or more magnitudes, the intuition that one can find some (possibly) small unit that can be used as a common measure is reasonable based on our experiences with physical measurements. In fact, the need to prove the existence of such a common measure would not have occurred to the Babylonians and Egyptians, in part because they were more empirically oriented than the Greeks, but also because of the strength of the intuition in question.

Given the requirement of a proof in Greek mathematics, the discovery of incommensurables was inevitable. Although it is questionable whether this discovery formed a mathematical crisis,[10] commensurability did play a key role in Greek geometrical proofs; the discovery could not be ignored. In this instance, the required 'resolution' was achieved by Eudoxus (408– 355 BCE) who developed a theory of proportions for continuous *magnitudes* as a companion to the existing theory of proportional *numbers.* As presented by Euclid in Book V of the *Elements,* the Eudoxean Theory of Proportion[11] provided the necessary logical foundation for the existing theory of similar rectilinear figures, as well as a basis for the limiting process of the Method of Exhaustion used to establish area and volume formulae.

While the details of the Eudoxean resolution of the incommensurability anomaly are of clear mathematical interest,[12] other aspects of the episode may be of greater interest to the mathematics educator. Foremost among these is the high level of mathematical activity which appears to have taken place in that period, including the work of Hippocrates of Chios (c. 420 BCE) on lunules, the study of irrationals begun by Theodorus (c. 400 BCE) in the late fifth century BCE and continued by Thaetetus in the fourth, and on-going efforts to solve the well-known geometrical problems of circle quadrature, angle trisection and cube duplication[13] throughout the fifth and fourth centuries. The mathematicians associated with Plato's Academy in the fourth century BCE were especially productive, both before and after Eudoxus' arrival at the Academy in the middle of that century. Whatever the reasons for this intense level of activity, it is clear that the existence of the anomaly did not halt mathematical production, or even restrict its scope to problems that avoided incommensurables.

Another interesting aspect of the post-resolution period is the failure of the Eudoxean theory of magnitudes to replace the Pythagorean theory of proportional numbers, which appears in Book VII of Euclid. The resolution of the incommensurable 'crisis' led not to a more general conception of number, but rather to a strictly enforced distinction between continuous magnitudes and discrete number. This distinction was in part due to the authority of Aristotle, who was the first to define it. It is also indicative of the reluctance of the mathematical community to cast off a theory that has served it well. In this case, the techniques for working with numbers were considerably less cumbersome than those for working with magnitudes, leading to even greater reluctance to discard the old theory in favor of a new, more general theory.[14]

Finally, but perhaps most surprisingly, it is worthwhile to note the rapidity with which the old intuition of 'commensurability' came to be replaced by the new mathematical intuition developed by Eudoxus. Looking to Aristotle once more, we read that:

> it seems wonderful to all who have not yet seen the reason, that there is a thing which cannot be measured even by the smallest unit. ... [yet] there is nothing which would surprise a geometer so much as if the diagonal turned out to be commensurable.[15]

The anomaly had thus been completely reversed; it would now be more surprising, at least to the initiated, if all magnitudes were commensurable. This is precisely the goal of mathematics educators: to move students from their current 'naive' understanding to new and deeper intuitions. Understanding this phenomenon of building new intuitions will be the focus of our remaining historical examples.

Non-Euclidean Geometry

The anomaly, which precipitated the development of non-Euclidean geometries, was not a foundational 'crisis' of the kind often associated with the discovery of incommensurables. The truth of the parallel postulate was never questioned by Euclid's contemporaries, nor by the mathematicians who investigated it over the next twenty centuries. Rather, efforts to derive the parallel postulate as a proposition, or at least to re-state it in a more self-evident form, were based on a more subtle dissatisfaction with its role in the formal axiomatic system of the *Elements*. First, there is the fact that Euclid's parallel postulate is strikingly complicated, especially in contrast to the simplicity and self-evident characteristic of the first four postulates:

Euclid's Postulates:[16]

1. To draw a straight line from any point to any point.
2. To produce a finite straight line continuously in a straight line.
3. To describe a circle with any center and distance.
4. That all right angles are equal to one another.
5. That, if a straight line intersecting two straight lines make the interior angles on the same side less than two right angles, the two straight lines, if produced indefinitely, meet on that side on which the angles are less than two right angles.

The parallel postulate itself is not employed in a proof until Proposition I-29. It is also the converse of Proposition I-17. These facts suggested to many that the parallel postulate might itself be a theorem, and not a postulate.

Attention to this anomaly began as early as the first century BCE;[17] investigations of the postulate were carried out by Posidonius (first century BCE), Ptolemy (2nd century CE), ibn Sina (980–1039), Levi ben Gerson (1280–1344), Omar Khayyam (1048–1131), Nasir Eddin al-Tusi (1201–1274), and J. Wallis (1616–1703), to name but a few.[18] A significant step towards the development of a non-Euclidean geometry was made by Gerolamo Saccheri (1667–1733), although Saccheri himself believed he had fully vindicated Euclid's choice. Rather than try to prove the postulate directly as done in many of the earlier attempts, Saccheri assumed the negation of the postulate and sought to establish the truth of the parallel postulate by *reductio ad absurdum*. Specifically, he considered the following quadrilateral containing two right angles and two congruent sides.[19] It is straightforward to show within neutral geometry[20] that the angles α and β are congruent, leading to three possible cases: (1) angles α, β are right angles (which is equivalent to Euclid's Fifth); (2) angles α, β are obtuse; and (3) angles α, β are acute.

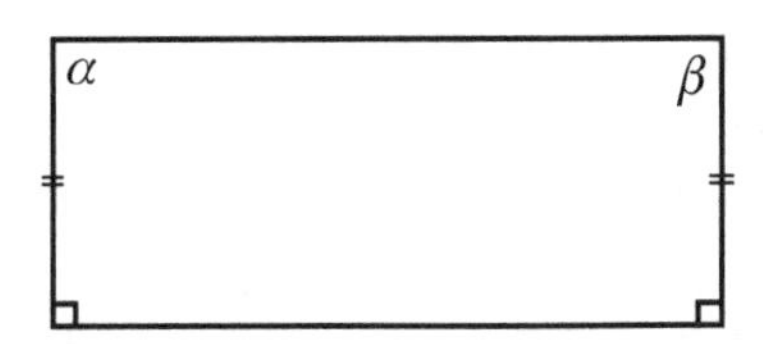

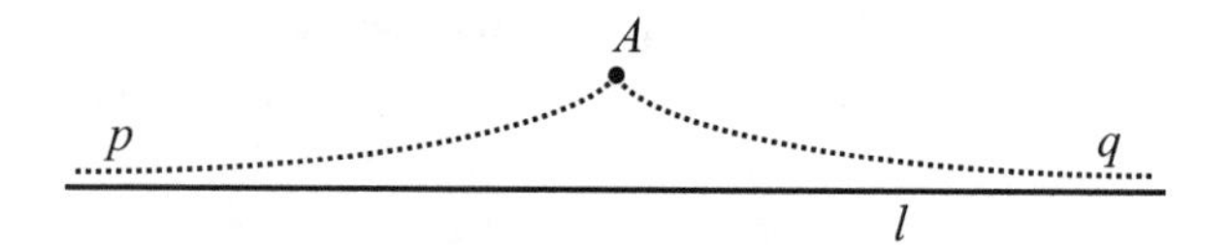

Saccheri believed he had succeeded in deriving a contradiction from the obtuse angle hypothesis, inadvertently assuming results of Euclidean geometry which do not hold in this case.[21] Unaware of his logical flaw, he turned to the acute angle hypothesis and succeeded in proving a sequence of theorems in what we would now call non-Euclidean geometry. In particular Saccheri considered the family of lines through a point A, where A does not lie on the line l, and showed that it contains two lines, p and q, which are themselves asymptotic to l and which separate those lines through A which intersect l from those lines through A which do not intersect l. He then argued that this result leads to the recognition of two straight lines [l and p] which "at one and the same point [at infinity] have in the same plane a common perpendicular."[22] Unable to derive the desired logical contradiction, Saccheri draws on his intuition of straight lines and concludes:

> The hypothesis of the acute angle is absolutely false, because it is repugnant to the nature of the straight line.[23]

Saccheri's inability to free himself of Euclidean intuitions was repeated by Lambert (1728–1777) and Legendre (1752–1853) whose efforts to improve on Saccheri's work ended with similar appeals to the 'correctness' of Euclid's geometry. A turning point in this development was the eighteenth century introduction of analytic techniques, which allowed the language of trigonometric formulas to be used in discussions of the problem. The work of Taurinus (1794–1874) on hyperbolic trigonometric formulas marks this turning point, although Taurinus himself was convinced that his work was not a description of any real geometry.[24] The critical breakthrough towards the acceptance of non-Euclidean geometry as real but paradoxical was finally taken by Lobachevsky (1793–1856) and Bolyai (1775–1856). Both men began their investigations by assuming a non-Euclidean geometry was possible and proceeded to follow this assumption to its logical and analytical conclusions. The formal validity of the non-Euclidean trigonometric formulas that they were able to derive provided convincing evidence that their original assumption was, indeed, a sound one.

With the publication of the works of Lobachevsky and Bolyai in 1829 (in Russian) and 1831 (in Latin) respectively, the unintuitive geometry of the acute angle hypothesis was publicly acknowledged as a logically rigorous alternative to Euclidean geometry for the first time. Despite the obvious break with accepted intuition, the work did not immediately draw wide attention. This may have been in part due to the fact that both men were isolated geographically and by language from the mainstream mathematical community, a fact which in turn may have contributed to their ability to break with the traditional view. This observation draws attention to the role of the individual in developing mathematical knowledge. Without *individual* reflections on and interpretations of shared problems and intuitions, the break with tradition accomplished by Lobachevsky and Bolyai could not have occurred.

The difficulty of breaking with tradition is indicated by Gauss's (1777–1855) reluctance to publish his own development of non-Euclidean geometry. Although the conclusions were logical, they were not intuitive, as he writes in a letter to Taurinus dated November 8, 1824:

> The assumption that the angle sum [of a triangle] is less than 180° leads to a curious geometry, quite different from ours but thoroughly consistent.... The theorems of this geometry appear to be paradoxical, and, to the uninitiated, absurd, but calm, steady reflection reveals that they contain nothing at all impossible.[25]

In a later letter to Bessel (dated 27 January 1829) Gauss admits that he "fears the howl of the Boeotians"[26] upon publication of these 'absurd' results. In fact, the truth of Euclidean geometry had been accepted implicitly for centuries, due both to the weight of Euclid's authority and to the fact that this geometry accords with our experience of the physical world. In the eighteenth century, Kant's doctrine of space as a pure intuition independent of empirical experience further emphasized Euclid's geometry as the one 'true' geometry.[27] Perhaps more surprising than the long delay in recognition of a non-Euclidean geometry is the fact that this recognition occurred at all. Had it not been for the posthumous publication of Gauss' correspondence in the 1860s, testifying again to the role played by pronouncements of an authority in building acceptance of novel concepts, this recognition might have been further delayed.

Two other nineteenth century developments, which aided the eventual acceptance of non-Euclidean geometry as legitimate, may also provide insight for the educator. The first of these was the development of new algebras possessing some, but not all, the properties of traditional algebra. These new algebras included a non-commutative matrix algebra, Boole's algebra of symbolic logic with its dual distributive properties, Hamilton's non-commutative quaternions, and the subsequent development of a vector algebra with two multiplicative operations. The acceptance of these various algebras shows that mathematics was increasingly being viewed as the (abstract) study of structures independent of physical reality and traditional 'rules'. It also illustrates how work in apparently unrelated areas can be of mutual value.

The second important nineteenth century development was Riemann's development of a general system of geometry in 1868, which in turn led to the creation of models of non-Euclidean geometry. These models enhanced the acceptability of the new geometries by providing a concrete interpretation for them. Additionally, because the models use a part of Euclidean space to interpret non-Euclidean space, any inconsistency within the non-Euclidean space corresponded to an inconsistency in the 'accepted' Euclidean model. What began as an effort to remove what Saccheri and others viewed as a 'blemish' in Euclid's work thereby ended with the parallel postulate being vindicated as the essential assumption which distinguishes Euclid's geometry from other, equally consistent geometries whose existence he could not possibly have guessed.

Infinity

The concept of infinity is another enigma which has intrigued scholars since the time of the Greeks, dating back at least to the appearance of Zeno's (in)famous paradoxes in the fifth century BCE. These paradoxes suggest that the assumption that space and time are infinitely divisible, as well as the assumption that space and time are composed of infinitely many indivisible atoms, will lead to contradictions concerning motion. 'Achilles', for example, asserts that "in a race the quickest runner can never overtake the slowest, since the pursuer must first reach the point whence the pursued started, so that the slower must always hold a lead".[28] Since this process continues *ad infinitum,* it would seem also that neither runner finishes the race, being unable to cover an infinite number of points in a finite amount of time. Under the alternative assumption that space is composed of indivisible atoms, Zeno argued that "the flying arrow is at rest"[29] since at each moment it occupies a space equal to itself.

Theories concerning Zeno's motivation in putting forward these arguments range from the desire to confound the sophists at their own rhetorical games to a more scholarly desire to add credence to his teacher Parmenides' tenet that "All change is illusory". Whatever his original objective, Zeno did succeed in drawing attention to problems concerning the infinitely large and small.[30] Aristotle was among those who took up the task of refuting Zeno's arguments about motion which "cause so much disquietude to those who try to solve the problems that they present".[31] Aristotle accepts the hypothesis that both space and time are infinitely divisible, and attempts to avoid Zeno's paradoxes by insisting that this 'infinite division' is only *potential* in nature. Although he rejects the notion of the 'actually infinite' in both a physical and mathematical sense as a result, Aristotle comments that:

> Our account does not rob the mathematicians of their science.... In point of fact they do not need the infinite and do not use it.[32]

This banishment of the actually infinite from mathematics remained intact for centuries, even after the distinction between number and magnitude had been removed. As late as 1831, Gauss writes to Schumacher:

> As to your proof, I must protest most vehemently against your use of the infinite as something consummated, as that is never permitted in mathematics... No contradictions will arise as long as Finite Man does not mistake the infinite for something fixed.[33]

A similar reference to 'Finite Man' was made earlier by Galileo, one of a small number of mathematicians who entertained the idea of an actually completed infinite set. On the First Day of his *Dialogues Concerning the Two New Sciences,* Galileo's character Salviati establishes a one-to-one correspondence between the set of natural numbers and the set of perfect squares, implicitly assuming the completion of both sets.[34] Rather than accept the notion that these two sets are the same size, however, Galileo uses this example as justification for rejecting the notion that cardinality can be applied to 'infinite quantities', asserting:

> This is one of the difficulties which arise when we attempt, with our finite minds, to discuss the infinite, assigning to it those properties which we give to the finite and limited; but this I think is wrong, for we cannot speak of infinite quantities as being the one greater or less than or equal to another ... the attributes 'equal', 'greater' and 'less' are not applicable to the infinite, but only to finite, quantities.[35]

The mathematician responsible for the critical breakthrough concerning infinity was Georg Cantor (1845–1918). Cantor's interest in infinite sets grew from his study of trigonometric series representation, which in turn led him to a study of the real number continuum.[36] His investigations led him to the observation that, although both sets are infinite, the set of real numbers is fundamentally different than the set of rational numbers, the former being dense, continuous and complete, while the latter is only dense. These differences suggested to Cantor that, despite the conviction of earlier mathematicians, there is a sense in which there really are fewer rational numbers than real numbers.[37] As a means of characterizing this inherent difference in richness, Cantor was led to the notion of one-to-one correspondences, proving in 1873 that such a correspondence is impossible between the set of real numbers and the set of natural numbers.[38] Having thus established the existence of at least two sizes of infinity, Cantor spent the next decade exploring these ideas, eventually developing an "unbounded ascending ladder"[39] of infinite (or transfinite) numbers which measure the differing sizes of actually infinite sets.

The principle that every 'consistent' set, whether finite or infinite, contains some definite number of elements and can therefore be assigned a cardinality marks Cantor's primary contribution. According to this stance, earlier confusion about the actually infinite stemmed from the belief that infinite sets must possess the same properties as finite sets in order to be considered coherent, a belief that is clearly suggested in the quote from Galileo. In essence, previous mathematicians had confused the concept of 'set' with their intuition of 'finite set'; their intuitions about finite sets were correct, but they had been overgeneralized.[40] In fact, an infinite set has the peculiar property that its elements can be placed in one-to-one correspondence with the elements of some of its proper subsets,[41] so that the whole is not necessarily bigger than its parts.[42] Rather than reject the notion of infinite sets due to this peculiarity, Cantor defined the concept of 'cardinality' using one-to-one correspondences, the very notion which led to earlier rejections of infinite sets.

This use of the paradoxical feature of a concept as the property by which it is formally defined is a twist of special interest to mathematics educators. Here we see that it was not the intuition *per se* that was incorrect—infinite sets *do* possess this peculiar property. Rejection of this intuition was shaped in part by prevailing cultural attitudes concerning the infinite, but also by the uncanny ability of mathematicians to avoid its use in their work for centuries.[43] Cantor was finally influenced to accept the intuition because of its usefulness in understanding the continuum,[44] where "each potential infinite, if it is rigorously applicable mathematically, presupposes an actual infinite"[45] so that even "the provision of a foundation for the theory of irrational numerical quantities [real number] cannot be effected without the use of the actual infinite in some form."[46]

Cantor's emphasis on the indispensability of the actually infinite did not, however, lead to the immediate acceptance of his work. In particular, objections were raised concerning the non-constructive nature of many of his proofs. Further concerns about his work arose when paradoxes involving the intuitive concept of 'set' appeared, raising questions concerning the logical consistency of the theory.[47] Again, the question of whether these paradoxes presented a real 'crisis' for the foundations of mathematics is debatable. For the 'working' mathematician of that time and today, the essential issue has again been the tremendous value of set theory as a tool in analysis and other areas of mathematics. Rather than lose the value and power of set theory, the majority of mathematicians choose to accept the formal axiomatization of set theory developed in response to these paradoxes. By 1903, we hear Russell declare:

> The Infinitesimal Calculus, though it cannot wholly dispense with infinity, ... contrives to hide it away before facing the world. Cantor has abandoned this cowardly policy, and has brought the skeleton out of its cupboard ... like many skeletons, it was wholly dependent on its cupboard, and vanished in the light of day.[48]

Pedagogical Implications: The Role of Intuition and Experience

The focus in the examples above has been on the relation between mathematical understanding and mathematical intuition. We adopt here the definition of *intuition* as "an accumulation of attitudes based on experience" given by Wilder [1967], a definition which captures the two essential features of intuition. First, it is the *intuition* of mathematicians that leads them to their definitions and theorems, not vice versa. Cantor did not begin with a definition of 'transfinite numbers', but was led to it on the basis of the intuition he had acquired at the end of extensive study and similarly for Lobachevsky and Bolyai. Such study constitutes the second key feature of intuition; namely, intuition is based on *experience*, and is not derived from logic.

With this definition in place, we may describe the process of mathematical development illustrated in the above examples as follows. An anomaly arises when experience conflicts with established intuitions. The anomaly is resolved when appropriate new intuitions based on further experience are established. Along the way, the 'paradoxes' become little more than curiosities for the uninitiated. Translating this process to the classroom, students can be expected to gain intuition, and thereby understanding, through the acquisition of experience, a view affirmed by research on the learning process.[49] Within this process, the teacher's role is to structure experiences in order to stimulate the desired intuition and to prepare the students to accept that intuition as 'correct'. Once developed, the teacher can then direct the students' intuition towards the appropriate definitions, theorems and processes.

Note here the implication that examples should *precede* definitions, in direct contrast to current practice in many mathematics classrooms. Consider, for instance, the usual introduction to group theory found in undergraduate texts: the definition of 'group' is presented, followed by a list of examples illustrating this definition, with perhaps an example or two of an operation that does not define a group. The assumption is that these examples are familiar to students from earlier studies, so that it will be clear (with minimal checking) that the necessary properties hold, although why these particular properties are necessary is likely to be less clear. The alternative approach suggested by the historical episodes is to allow students to analyze known examples and counterexamples in order to identify their important common characteristics *before* any mention of the term 'group' is made. Once these essential features have been extracted from the examples, the definition can be introduced. The anomalous lack of 'commutativity' in the definition is thereby removed as the students see that this property was not (algebraically) present in several of their examples.

A similar instructional strategy can be applied to many other concepts. The primary requirement for its success is prior student familiarity with concrete examples[50] which illustrate the desired intuition. Historical studies can be of particular value in identifying relevant examples for student exploration.[51] The curriculum which results from such analyses need not follow the original chronological order. For instance, although the development of non-Euclidean models followed the work of Lobachevsky and Bolyai, their role in gaining acceptance of these geometries suggests that these models should perhaps be explored in the classroom prior to the logical study of alternate parallel postulates. Folio [1985] has used such an approach with liberal arts students, reporting that it is "the experiencing of the *process* [of exploring the models] that has given meaning to the new ideas [of non-Euclidean geometries]". In other words, the students gain valuable experience in interpreting models and become better prepared (psychologically) for the possibility of non-Euclidean parallel postulates by this approach.

Pedagogical Implications: The Role of Anomalies and Confusion

Using history as a guide, we see that the process of preparing an individual to accept new intuitions can be lengthy. History also suggests that anomalies within students' understanding can facilitate this process, particularly in cases where old intuitions must be discarded along the way. One such concept is that of an 'irrational number', typically presented in the United States at grade 7 or 8 in terms of square roots of non-square numbers and their decimal expansions. Given the practical need to approximate most rational numbers, as well as the ease with which calculators provide approximations for rationals and irrationals alike, it is not surprising that students fail to grasp either the meaning or the relevance of this distinction. By posing the problem in terms of magnitudes,[52] however, one can play on the strong geometrical intuition which says that a small common measure must be possible. When faced with a pair of magnitudes for which no such measure can be found, both the terminology and import of the notion becomes more relevant to the student. Rosenfeld [1978] implements this idea by first looking at specific examples of commensurable segments a, b and their common unit u in order to translate this geometric relation to the algebraic statements $a = mu$, $b = nu$, $a/b = m/n$. Once the relation of commensurable magnitudes to rational numbers is established, the indirect proof of the irrationality of $\sqrt{2}$ is presented, starting from its geometric statement. The approach of Arcavi and Bruckheimer [1984] is more strictly geometric; after introducing the repeated subtraction procedure (*antenaresis*) as a means to obtain a common measure, the fact that "the process can be continued indefinitely"[53] in the case of the side and diagonal of a regular pentagon is used to demonstrate their incommensurability.

Anomalies can be used in a similar way to define relevant questions of study, much as the anomalous parallel postulate did in the development of non-Euclidean geometry. Sierpinksa [1994] remarks that a discrepancy in one's understanding is, in fact, *necessary* to produce questions which are both sensible and interesting to the student. 'Sensible and interesting questions', 'attention' and 'intention' are the three psychological conditions identified by Sierpinska as necessary for understanding. Unresolved anomalies can thus provide further instructional value by maintaining student attention to questions and by motivating them to understand and resolve these questions. Cordeiro [1988] who introduced paradoxes of infinity to a class of sixth grade students, some of whom were classified as remedial, describes an intriguing experiment of this type. Using Hilbert's Infinity Hotel,[54] these students were introduced to the use of one-to-one correspondences as a means of distinguishing between finite and infinite sets at a relatively early stage in their mathematical development. Although the models they then produced of Infinity Hotel and transportation devices for its guests showed some confusion about the infinite/finite distinction, this topic became the "stuff of idle conversation after school",[55] indicating a high degree of engagement in the study. The fact that the students themselves were able to point out some confusion in each other's work further indicates that their intuition concerning infinity was undergoing change as a result of their 'play'.

One concern expressed about using anomalies as a pedagogical device in this way is that the presence of anomalies may lead to confusion instead of learning. The historical examples suggest just the opposite: the presence of anomalies in mathematics seemed to *stimulate* the growth of mathematics, not suppress it. This growth was often accompanied by increased levels of abstraction and changes in accepted standards of rigor. In fact, it has been hypothesized that the need to resolve the contradictory mathematical knowledge of their Egyptian and Babylonian predecessors motivated the Greeks to turn towards deductive logic as a basis for mathematics. These observations are again consistent with current learning theory which hypothesizes that learning takes place when cognitive schemata are transformed through the processes of assimilation and accommodation.[56] Implicit in this theory is the suggestion that merely 'correcting' students who hold false intuitions is not effective instructional practice since errors are now viewed as indications of existing cognitive schemata or conflicts within them.[57] As such, errors are likely to systematically reoccur *until* the students' internal intuitions or schemata are modified in response to individual experiences and reflections. In conjunction with earlier observations about the role of the individual in developing mathematics, this suggests that classroom cooperative work alone is not sufficient to build individual understanding, despite the emphasis placed on communication with peers by constructivist theories. Effective opportunities for individual reflection are also needed, a suggestion which is being confirmed by educational research.[58] In this regard, historical materials, and especially original source material readings, can again be of value by encouraging such reflection in the context of book reviews, editorial reports, or essays based on historical problems.[59]

Another notable principle of constructivist learning theory is the self-perpetuating nature it ascribes to cognitive schemata. This characteristic of cognitive schemata leads the individual to resist cognitive modifications, much as mathematicians resisted changes in their shared schemata involving number, geometry and infinity. The Greek distinction between magnitude and number is evidence of how much resistance a culture can have in the case of accommodation, where significant restructuring of the cognitive schemata is required. In view of the amount of 'good' mathematics which was produced in conjunction with this 'false' intuition, we see that refraining from correcting certain student errors may not be as detrimental as is typically feared, provided the 'correct' intuition is eventually achieved.[60] For an individual, resistance to the accommodation process is so strong that cognitive disequilibration (i.e., an encounter with an anomaly) is viewed as necessary to motivate the change. Some educational research suggests that cognitive disequilibration may also benefit learning in situations requiring only the less extreme process of assimilation. In their research on the formation of rational number concepts, for instance, Behr *et al.* [1983] found that a 'good' manipulative aid is one that actually causes some confusion.

Thus, although the creation of confusion seems contrary to the role of a teacher, we see that both history and psychology suggest that confusion can be of benefit to students. We close with one final suggestion as to how instructors can draw on the role played by anomalies in shaping the historical development of mathematics. Namely, sharing stories of past paradoxes is an interesting way to show students that confusion is an inherent part of mathematical studies, and that the difficulty which such confusion represents "should be to us a guide post on the mazy paths to hidden truths, and ultimately a reminder of our pleasure in the successful solution."[61]

Acknowledgments

The author gratefully acknowledges the suggestions of the referee concerning various historical references, and thanks George Heine and Victor Katz for their valuable recommendations concerning the content and structure of the paper.

Bibliography

Arcavi, A. and Bruckheimer, B.: 1984, *The Irrational Numbers: A source-work collection for in-service and pre-service teacher courses,* Rehovot: The Weizmann Institute of Science. ERIC RIE Number ED251302.

Behr, M. and Harel, G.,: 1990, "Students' Errors, Misconceptions, and Cognitive Conflict in Application of Procedures," *Focus on Learning Problems in Mathematics* (12:3, 12:4). 75–84.

Behr, M., Lesh, R., Post, T. and Silver, E.: 1983, "Rational-Number Concepts," *Acquisition of Mathematics and Processes,* New York: Academic Press, pp. 91–126.

Berggren, J.: 1984, "History of Greek Mathematics," *Historia Mathematica* (11) 394–410.

Bolzano, B.: 1950, *Paradoxes of the Infinite,* Trans. D. H. Steel, London: Routledge & Kegan Paul. (Originally published as *Paradoxien des Unendlichen,* 1851.)

Bonola, R.: 1955, *Non-Euclidean Geometry: A Critical Study of Its Developments,* New York: Dover.

Borasi, R.: 1996, *Reconceiving Mathematics Instruction: A Focus on Errors,* New Jersey: Ablex Publishing Corporation.

Burton, D.: 1995, *Burton's History of Mathematics: An Introduction,* Dubuque: Wm. C. Brown Publishers.

Burton, D. & Van Osdol, D.: 1995, "Toward the Definition of an Abstract Ring," in *Learn from the Masters,* eds. F. Swetz *et al.,* Washington: The Mathematical Association of America, pp. 241–251.

Calinger, R.: 1982, *Classics in Mathematics,* New Jersey: Prentice Hall.

Cordeiro, P.: 1988, "Playing with Infinity in the Sixth Grade," *Language Arts* (65:6) 557–566.

Crowe, M.: 1975, "Ten 'laws' concerning patterns of change in the history of mathematics," in *Revolutions in Mathematics,* ed. D. Gillies, Oxford: Clarendon Press, pp. 1–20. Reprinted from *Historia Mathematica* (2), 161–166.

Dauben, J.: 1979, *Georg Cantor: His Mathematics and Philosophy of the Infinite,* Cambridge: Harvard University Press.

Daumas, D.: 1996, "Activités en classe sur l'irrationalité: de Pythagore à Théon de Smyrne," *Contributions à une approche historique de l'enseignement des mathématiques. Actes de la 6e Université d'été interdisciplinaire sur l'histoire des mathématiques,* Besançon: IREM, pp. 417–427.

Dedekind, R.: 1851, "Stetigkeit und irrationale Zahlen," in *Classics in Mathematics,* ed. R. Calinger, New Jersey: Prentice Hall, pp. 627–633. Reprinted from *Essays on the Theory of Numbers,* Lasalle: Open Court Publishing Co.

Falk, R.: 1994, "Infinity: A Cognitive Challenge," *Theory & Psychology* (4:1) 35–60.

Fauvel, J. and Gray, J., editors: 1987, *The History of Mathematics: A Reader,* London: The Open University.

Fishbein, E. Jehiam, R. & Cohen, D.: 1995, "The Concept of Irrational Numbers in High-School Students and Prospective Teachers," *Educational Studies in Mathematics* (29) 29–44.

Folio, C.: 1985, "Bringing Non-Euclidean Geometry Down to Earth," *Mathematics Teacher* (78) 430–431.

Fowler, D. H.: 1987, *The Mathematics of Plato's Academy,* Oxford: Clarendon Press.

——: 1994, "The Story of the Discovery of Incommensurability, Revisited," in *Trends in the Historiography of Science,* eds. K. Gavroglu *et al.,* Netherlands: Kluwer Academic Publishers, pp. 221–235.

Freudenthal, H.: 1966, "Y avait-il une crise des fondements des mathématiques dans l'antiquité?," *Bulletin de la Société Mathématique de Belgique* (18) 43–55.

Gillies, D. (editor): 1992, *Revolutions in Mathematics,* Oxford: Clarendon Press.

Gray, J.: 1979, *Ideas of Space,* Oxford: Clarendon Press.

Hallett, M.: 1984, *Cantorian set theory and limitation of size,* Oxford: Clarendon Press.

Heath, T. (translator): *Euclid's Elements,* The Great Books of the Western World, Pennsylvania: The Franklin Library.

Hutchins, R. M. (editor): 1952, "Galileo's Dialogues Concerning the Two New Sciences," Translated by H. Crew and A. de Salvio, *Great Books of the Western World* (28) pp. 129–260.

Hamilton, E. and Cairns, H., editors.: 1961, *The Collected Dialogues of Plato, Including the Letters,* New Jersey: Princeton University Press.

Hilbert, D.: 1900, "Mathematical Problems: Lecture Delivered Before the International Congress of Mathematicians at Paris in 1900," in *Classics in Mathematics,* ed. R. Calinger, New Jersey: Prentice Hall, pp. 698–718. Reprinted from *Bulletin of the American Mathematical Society,* vol. 8 (1902), pp. 437–479.

Jones, P.: 1956a, "Irrationals or Incommensurables I: Their Discovery and a 'Logical Scandal'," *From Five Fingers to Infinity: A Journey through the History of Mathematics,* ed. F. Swetz, pp. 172–175. Reprinted from Mathematics Teacher (49) 123–127.

——: 1956b, "Irrationals or Incommensurables III: The Greek Solution," in *From Five Fingers to Infinity: A Journey through the History of Mathematics,* ed. F. Swetz, pp. 176–179. Reprinted from *Mathematics Teacher* (49) 282–285.

Katz, V.: 1993, *A History of Mathematics: An Introduction,* New York: Harper Collins.

Kline, M.: 1972, *Mathematical Thought from Ancient to Modern Times,* New York: Oxford University Press.

Knorr, W.: 1975, *The Evolution of the Euclidean Elements,* Dordrecht: D. Reidel.

——: 1981, "On The Early History of Axiomatics: The Interaction of Mathematics and Philosophy in Greek Antiquity," in *Theory Change, Ancient Axiomatics, and Galileo's Methodology: Proceedings of the 1978 Pisa Conference on the History and Philosophy of Science (1),* Eds. J. Hintikka *et al.,* Dordrecht: D. Reidel Publishing Company, pp. 145–186.

——: 1986, *The Ancient Tradition of Geometric Problems,* Boston: Birkhäuser.

McKeon, R, editor: 1941, *The Basic Works of Aristotle,* New York: Random House.

Mehrtens, H.: 1976, "T. S. Kuhn's theories and mathematics: a discussion paper on the 'new historiography' of mathematics," in *Revolutions in Mathematics,* ed. D. Gillies, Oxford: Clarendon Press, pp. 21–41. Reprinted from *Historia Mathematica* (3) 197–320.

Moore, A. W.: 1990, *The Infinite,* London: Routledge, Chapman and Hall, Inc.

Rolwing, R. and Levine, M.: 1969, "The Parallel Postulate," in *From Five Fingers to Infinity: A Journey through the History of Mathematics,* ed. F. Swetz, pp. 592–596. Reprinted from *Mathematics Teacher* (62) 665–669.

Romberg, T.: 1992, "Mathematics Learning and Teaching," *Teaching Thinking,* New Jersey: Lawrence Erlbaum Associates, Inc., pp. 43– 64.

Rosenfeld, B. A.: 1988, *A History of Non-Euclidean Geometry: Evolution of the Concept of a Geometric Space,* Translated by A. Shenitzer, New York: Springer-Verlag.

Rosenfeld, B.: 1978, "An Often-Overlooked Motivation for Proving the Existence of Irrationals," *MATYC Journal* (12:1) 40–42.

Russell, B.: 1903, *The principles of mathematics,* London: George Allen and Unwin.

Shulman, B.: 1995, "Not Just a Spice: The Historical Perspective as an Essential Ingredient in Every Math Class," in *Third International History, Philosophy, and Science Teaching Conference, Proceedings* (2), eds. F. Finley *et al.*, Minneapolis: University of Minnesota, 1042–1049.

Sierpinska, A.: 1994, *Understanding in Mathematics,* London: Falmer Press.

Skemp, R.: 1987, *The Psychology of Learning Mathematics,* Hillsdale: Lawrence Erlbaum Associates.

Swetz, F. (editor): 1994, *From Five Fingers to Infinity: A Journey through the History of Mathematics,* Chicago: Open Court.

Swetz, F., Fauvel, J., Bekken, O., Johansson, B. & Katz, V. (editors): 1995, *Learn from the Masters,* Washington: The Mathematical Association of America.

Szabó, A.: 1964, "Transformation of mathematics into deductive science and the beginnings of its foundation on definitions and axioms," *Scripta Mathematica* (27) 27-48A, 113–139.

Tzanakis, C.: 1993, "Reversing the Customary Deductive Teaching of Mathematics Using Its History: The Case of Abstract Algebraic Concepts," *History and Epistemology in Mathematics Education: First European Summer University Proceedings,* Montpellier: IREM of Montpellier, pp. 271–273.

von Fritz, K.: 1945, "The Discovery of Incommensurability by Hippasus of Metapantum," *Annals of Mathematics* (46) 242–264.

Wilder, R.: 1967, "The Role of Intuition," *Science* (156). 605–610.

——: 1980, *Mathematics as a Cultural System,* Oxford: Pergamon Press.

Endnotes

[1] This is especially seen in laws 1, 2, 3 and 9, reproduced below:

(1) New mathematical concepts frequently come forth not at the bidding, but against the efforts, at times strenuous efforts, of the mathematicians who create them.

(2) Many new mathematical concepts, even though logically acceptable, meet forceful resistance after their appearance and achieve acceptance only after an extended period of time.

(3) Although the demands of logic, consistency, and rigour have at times urged the rejection of some concepts now accepted, the usefulness of these concepts has repeatedly forced mathematicians to accept and to tolerate them, even in the face of strong feelings of discomfort.

(9) Mathematicians have always possessed a vast repertoire of techniques for dissolving or avoiding the problems produced by apparent logical contradictions, and thereby preventing crises in mathematics.

[2] 399 BCE is the fictive date of the dialogue which was written in the year 368/367 BCE Although ancient tradition dates it earlier, Knorr [1975], pp. 36–49, argues convincingly for dating the discovery between 430 and 410 BCE.

[3] One suggestion as to why Theodorus stopped with $\sqrt{17}$ remarks that the diagram known as the Spiral of Theodorus circles back on itself once one reaches the square root of 19. An alternative theory is that Theodorus did not possess a general proof schema, but approached each case separately and encountered mathematical difficulties with the proof for $\sqrt{17}$. See Knorr [1975], pp. 83ff., concerning this latter theory.

[4] Aristotle, *Prior Analytics,* 41a 23 ff.

[5] Fowler [1987] remarks (pp. 296–297) that the word 'square' does not occur in the Greek text of Aristotle and offers two proofs (pp. 304–308) which fit Aristotle's remarks concerning an "odd-even" proof; the second of Fowler's proof techniques can be applied to any regular polygon.

[6] von Fritz [1945], p. 259.

[7] See also Jones [1956a] for details of this argument and Arcavi & Bruckheimer [1985] for a classroom activity based on this process.

[8] The veracity of this story is called into question by Fowler [1994], pp. 225–226.

[9] Aristotle, Metaphysics, 985b 26ff.

[10] Both Knorr [1975], pp. 36–42, and Freudenthal [1966] offer arguments against the 'foundational crisis' view of incommensurables. Fowler [1987, 1994] also questions the 'crisis' theory, as well as the traditional hypothesis that dealing with incommensurable magnitudes was the goal of the Eudoxean Theory of Proportion. Fowler's thesis is based on a non-traditional view of the role of antenaresis in Greek mathematics. See also Berggren [1984], pp. 398–402, for more on this debate.

[11] The Eudoxean Theory of Proportion replaced an earlier, but more cumbersome, proportion theory for magnitudes which was based on the process of antenaresis. See Knorr [1975], especially pp. 255–261.

[12] See Katz [1993] , pp. 72–77, for the mathematical details of both the pre-Eudoxean Proportion Theory based on antenaresis and the later Eudoxean Proportion Theory.

[13] Knorr [1986] argues that such geometric problem solving was the primary motivating factor for the development of much of Greek mathematics. See especially chapters 2 and 3 for an overview of the technical advances in geometry during this period.

[14] In fact, the techniques are interchangeable for commensurable magnitudes, as established in Euclid, Book X.

[15] Aristotle, Metaphysics, 983a 17 ff.

[16] Translation by Heath [1956].

[17] Gray [1979], pp. 34–36, remarks that attention to the problem posed by the parallel postulate predates Euclid since both the existence *and* uniqueness of parallel lines was required to allow parallels to be used in transporting angles in geometrical constructions. Although a proof eluded Greek geometers, the notion of *straight line* seemed to imply that the lines described in the fifth postulate would necessarily meet if they were indeed straight, so that "it grated that such an assumption had to be made" (p. 36). The aim of the fifth postulate and other early research on the uniqueness of parallels was thus "to graft onto geometry without parallels (an unexceptionable but dull subject) something unexceptionable but powerful enough to make the subject interesting." (Gray, p. 37)

[18] See Rosenfeld [1987], chapter 2, for a detailed survey of efforts to resolve the problem of the parallels prior to the work of Lobachevsky and Bolyai.

[19] The Islamic mathematicians Omar Khayyam and Nasir Eddin in al-Tusi also considered this quadrilateral in their attempts to prove the parallel postulate.

[20] Euclidean geometry without the parallel postulate.

[21] In particular, Saccheri used *Elements* I-17 which states that any two angles of a triangle are together less than two right angles, a result which depends on the ability to extend a straight line to any given length. Euclid explicitly used this latter assumption, which does not hold under the obtuse angle hypothesis.

[22] As quoted by Gray [1979], p. 61. See also Bonola [1955], pp. 22–44, for details of Saccheri's work.

[23] Fauvel and Gray [1987], p. 514.

[24] Taurinus obtained these formulas by "pure substitution" of imaginary quantities for the real quantities appearing in spherical trigonometric formulas, following a suggestion of Lambert's that the geometry of the acute angle hypothesis appears to hold on the surface of a sphere of imaginary radius.

[25] As quoted by Rolwing and Levine, [1994], p. 594.

[26] Fauvel and Gray [1987], p. 498.

[27] Gray notes, however, that Kant "was not committed to a geometry based upon the parallel postulate", and that he did not reject the logical possibility of a non-Euclidean geometry. See Gray [1979], p. 75.

[28] Aristotle, *Physics*, 239b, 14ff.

[29] Aristotle, *Physics*, 239b, 30ff.

[30] Szabó [1964] has suggested that Zeno's paradoxes were also critical in establishing the axiomatic method in mathematics. This view is critiqued by Knorr [1981] and others. See also Knorr [1986], pp. 86–88.

[31] Aristotle, *Physics*, 239b, 10ff.

[32] Aristotle, *Physics*, 207b 28ff.

[33] As quoted by Burton [1995], p. 593.

[34] See page 144 of the Great Books of the Western World translation of Galileo's *Dialogues Concerning the Two New Sciences*.

[35] *ibid*, p. 144.

[36] Building on Heine's 1870 result that a function can be uniquely represented by a trigonometric series of the form $f(x) = \frac{1}{2}a_0 + \sum a_n \sin(nx) + b_n \cos(nx)$ if that series is uniformly convergent in general, Cantor proved in that same year that a trigonometric series representation is unique provided only that the series *converges for every* x. By 1872, Cantor was able to weaken the assumption further, replacing *convergence for every* x *by convergence everywhere except on a set* P *for which some derived set* $P^{(n)}$ *is empty*, where n was considered to be finite and the sets $P^{(n)}$ be defined iteratively as sets of limit points. That is, given a set of real numbers P, the set P' is the collection of limit points of P and $P^{(n+1)} = \left[P^{(n)}\right]'$. Finally, Cantor set $P^{(\infty)} = \bigcap_{n=1}^{\infty} P^{(n)}$ and continued this first use of a set operation to a transfinite level. He later replaced the symbol ∞ by ω to emphasize that these were, in fact, actual numbers and not merely symbols. See Dauben [1979], pp. 30–46, for further detail.

[37] This suggestion is also made by Dedekind who states "the straight line L is infinitely richer in point-individuals than the domain R of rational numbers in number-individuals" in his discussion of completeness in Section III of his 1851 paper *Stetigkeit und irrationale Zahlen*.

[38] This proof appears in Cantor's 1874 paper *On a Property of the Collection of All Real Algebraic Numbers* in which he also shows that there are infinitely many transcendental numbers by establishing a one-to-one correspondence between the set of algebraic numbers and the set of natural numbers.

[39] Cantor, *Über Undendlichen, lineare Punktmannigfaltigkeiten*, as quoted by Hallett [1979], p. 39.

[40] Cantor himself wrote to Gustav Eneström: "All so-called proofs against the possibility of actually infinite numbers are faulty ...from the outset they expect or even impose all the properties of finite numbers upon the numbers in question, while on the other hand the infinite numbers, if they are to be considered in any form at all must (in their contrast to the finite numbers) constitute an entirely new kind of number." See Dauben [1979], p. 125.

[41] In his 1851 *Paradoxien des Undendlichen*, Bolzano also used this feature of infinite sets to formally define the actually infinite, although his work was not widely known by his contemporaries. In contrast to Cantor, Bolzano chose the part-whole conception as a mechanism for comparing the cardinality of sets, instead of the correspondence criteria. See Moreno & Waldegg [1991] for a historico-critical comparison of Bolzano's and Cantor's conceptions of the infinite and their relation to high school students' responses to the concept.

[42] Falk [1994] remarks that the notion that the part must be smaller than the whole is, in fact, learned in several developmental stages in early childhood, typically completed by age 7 or 8, so that it is not an *a priori* belief.

[43] We are reminded here of Crowe's ninth law; see footnote 1 above.

[44] The consistency of the mathematics of the transfinite was also an important factor in Cantor's philosophical justification of the work. See Hallett [1984], pp. 14–24, and Dauben [1984], pp. 128–132.

[45] Cantor , *Über die verschiedenen Ansichten in Bezug auf die actualunendlichen Zahlen* [1886], as quoted by Hallett [1984] , p. 25.

[46] *ibid*, p. 26.

[47] Cantor drew additional criticism by appealing to theological arguments to address these paradoxes and other objections. See Hallett [1984], pp. 1–48, for an analysis of Cantor's line of arguments in this respect.

[48] Russell [1903], p. 304.

[49] See Romberg [1992], pp. 53–59, for an overview of the basic principles underlying current learning theories in mathematics.

[50] The term 'concrete' does not refer necessarily to actual physical materials, but to examples and experiences which aim at the same level of abstraction at which the student in currently operating.

[51] See, for example, Tzanakis in this volume for a historical study of examples important in the development of group theory; Burton and Van Osdol [1995] for a study in ring theory, and Tzanakis [1993] on complex numbers.

[52] As these presentations typically require significant algebraic or geometric prerequisites on the part of the student, it is unlikely that they could be used at the middle school level where irrationals are typically introduced. Fishbein *et al.* [1995] have found that most Israeli high school students still lacked a clear concept of irrational numbers (as did many preservice teachers in their study), and recommend (page 43) that students be led to "*live the difficulty* of accepting that for two segments one may *not* find any common unit, no matter how small—that is, not to disregard the difficulty but to confront it!" (Emphasis in original)

[53] Arcavi and Bruckheimer [1985], p. 10 of the answer sheet to "The Pythagoreans" worksheet. This source-work collection of activities designed for in-service and pre-service teachers at the junior high level combines original and secondary source readings with activity worksheets tracing the historical development of irrational numbers from the Pythagoreans through Dedekind's definition in the 20th century. Similar activities used at the secondary level are briefly described in Daumas [1996].

[54] See Gamow [1947], p. 28, for a description of this paradox.

[55] Cordeiro [1988], p. 560.

[56] *Assimilation* is the process by which new experiences are integrated into existing cognitive schemata, whereas *accommodation* is the process of re-structuring existing schemata in order to make sense of recent experiences.

[57] See, for example, Behr and Harel [1990], for a classification of some cognitive conflict types which result in systematic student errors.

[58] See, for example, Sierpinska [1994], pp. 66–68.

[59] See, for example, Shulman [1995], who has employed such assignments based on Archimedes' *Sand Reckoner* with undergraduate students and secondary teachers.

[60] Borasi [1996] proposes that errors should, in fact, be integrated into an inquiry instructional style, and not viewed as impediments to student progress.

[61] From Hilbert's 1900 Lecture to the International Congress of Mathematicians. See Calinger, p. 699.

The Historicity of the Notion of what is Obvious in Geometry

Evelyne Barbin
IREM, University Paris 7

Something that is evidently true, or obvious, is something that is "immediately grasped by our intelligence", it is therefore something that is essentially personal. A mathematical argument or assertion cannot be "obvious" independently of the person who makes such a judgment, who says, or writes, "that's obvious". The task of the historian then, is not a matter of identifying the "eternal" obvious, but of trying to determine those aspects of reasoning that were generally said to be, or accepted, as obvious. Taking this point of view, we shall proceed here not by looking at mathematical reasoning of past times with the benefit of our own familiar mathematics, but by reading declarations of the kinds of things that were held to be obvious at the time, and by examining the mathematical practices that were based on those convictions. We shall explore readings in Greek geometry of Proclus and Descartes, in analytic geometry of Lamy and Poncelet, and in descriptive geometry of Gergonne and Chasles. The comparisons between the readings of disciples and detractors have their own historicity. Opposing views generated controversies, which could be strident, and so we can certainly claim that the nature of what may be taken to be clearly evident, or obvious, is a matter of concern to the mathematician at a deep and significant level.

A reasoned discourse on images

Euclid proposes the following problem as the first Proposition of Book I of the *Elements*: to construct an equilateral triangle on a given finite straight line AB. The construction is presented in an impersonal way, the geometer effacing himself before what he contemplates:[1] "With centre A and distance AB let the circle BCD be described; again, with centre B and distance BA let the circle ACE be described; and from the point C, in which the circles cut one another, to the points A, B let the straight lines CA, CB be joined."[2] (See Figure 1.)

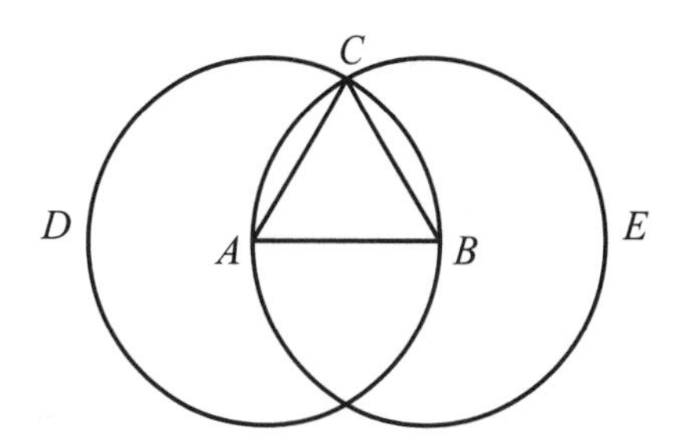

FIGURE 1

This construction depends on two Postulates stated at the beginning of Book I: we can draw a circle with any center and any radius, and we can draw a straight line from any given point to any other point. The reasoning by which we can conclude that the triangle ABC is equilateral is a reasoned discourse, consisting of a set of deductions which rely on one of the Common Notions, also stated at the beginning of Book I, namely that two magnitudes both equal to another magnitude are equal to each other. In his Commentary on the First Book of Euclid's *Elements,* written in the fifth century CE, Proclus distinguishes between the Postulates and the Common Notions, or Axioms: the first are things it is useful to allow so to be, the second are things which are obvious in themselves. But Proclus says that "both must have the character of being simple and readily grasped."[3]

In his commentaries on Proposition 1, Proclus relates Zeno's objection to this construction: no axiom asserts that the two lines CA and CB could not have a common part CE, and so the triangle AEB would not be equilateral (Figure 2). This objection is a matter of logic, it is offered as a challenge to the principles of geometry, since Euclid relies on an assertion that was not made explicitly in a Postulate or an Axiom. But Proclus raises no objection concerning the point C; he believes there is no need to prove that the two circles do in fact intersect. The visual evidence that the two circles do intersect is obvious enough. For Proclus, this Proposition has the virtue of "showing us the nature of things as images."[4] He writes that the equilateral triangle is the most beautiful of triangles, and it is very similar to the circle. The image allows us to perceive the close relationships between these two perfect figures.

But this does not mean that the geometer has only to rely on visual evidence. Thus, Proclus regards Proposition 20 of Book I of Euclid's *Elements,* which states that any two sides of a triangle are greater than the remaining side, legitimate as a proposition. He replied to the Epicureans, who derided the theorem because "it is evident to an ass and needs no proof," that "To this it should be replied that, granting the theorem is evident to sense-perception, it is still not clear for scientific thought."[5] Any assertion, obvious to the senses, but not having the status of an axiom, has to be demonstrated. What is obvious still needs to be proved by the geometer using logical reasoning.

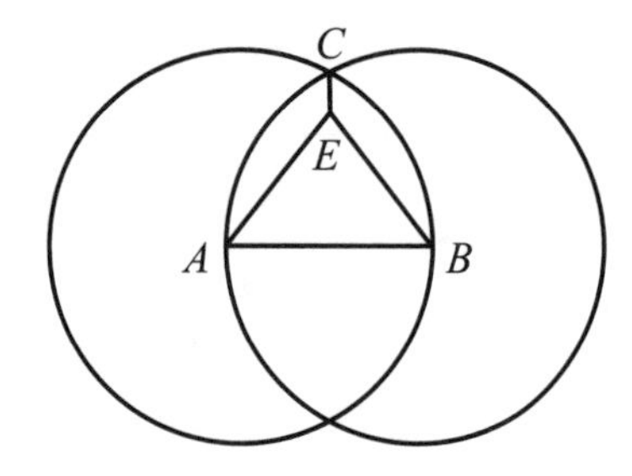

FIGURE 2

Proposition 20 is, in fact, a diorism[6] which allows us to solve the problem posed in Proposition 22, namely to construct a triangle having its sides equal to three given straight lines. Euclid attaches to the problem in Proposition 22 a condition: two of the straight lines taken together must be greater than the third. The construction, as in Proposition 1, depends on drawing two circles, which Euclid assumes will intersect. In his commentary, Proclus asks, why can he assume this? He points out that two circles can be disjoint, can touch each other, or can intersect, and he shows that the two first assumptions lead to contradictions. In the case of two circles touching, he only considers the case in which the circles are exterior to each other. Then he concludes that Euclid "was right in assuming that the circles intersect, since he had also posited that of three straight lines two of them together in any way are greater than the other one."[7] So Proclus's commentary here seems to be designed more to justify the need for the condition than to demonstrate the existence of a point of intersection. In the seventeenth century, Rene Descartes posed the question of how this could be done, but in a wider context.

Obviousness for Descartes

Descartes, in Rule IV of the *Rules for Direction of the Mind,* expresses his dissatisfaction with proofs in arithmetic and geometry in the following way: "But in neither subject did I come across [ancient] writers who fully satisfied me. I read much about numbers which I found to be true once I had gone over the calculations for myself; the writers displayed many geometrical truths before my very eyes, as it were, and derived them by means of logical arguments. But they did not seem to make it sufficiently clear to my mind why these things should be so and how they were discovered."[8] The works of those writers did not show why a mathematician would propose to prove some result, nor how he arrived at the proof. For Descartes, the form of logical reasoning allowed the truth of results to be incontestably asserted, but it did not provide a way for solving new problems.

In order better to understand Descartes' dissatisfaction, let us look at his view of Proposition 11 of Book 2 of Euclid's *Elements.* This requires a given straight line AB to be cut at a point H in such a way that the rectangle contained by the sides HB and BD, where BD is equal

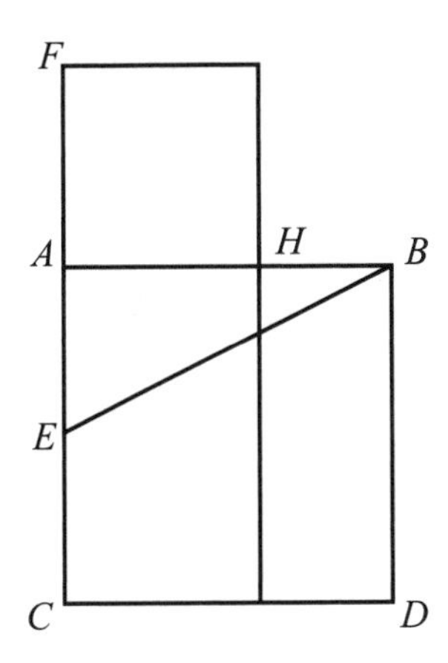

FIGURE 3

to AB, should be equal to the square on the side AH. Euclid sets out his construction before our eyes: first let the square of side AB be constructed; let E be taken as the mid-point of AC; let E be joined to B; let EF be made equal to EB; and let the square FH be drawn on AF (Figure 3).

Then, says Euclid: "I say that AB has been cut at H ..." as was required. He then goes on to prove that the rectangle is indeed equal to the square by a sequence of rigorous deductions that flow from the Common Notions or from earlier proved propositions. Descartes is obliged to admit that the result presented before his eyes is true, but he has no way of knowing how Euclid discovered the construction of the point H. He wants to know the process that was used to determine the point H, and he challenges the obviousness of the process: "I say that."

The obviousness that Descartes was interested in was not the obviousness of an image, nor that of logical discourse, but that of the geometer who works with objects that are obvious, or simple. To do this, he reduces all geometrical objects to straight lines, which can be related to each other with the aid of a unit straight line. In this way, to solve a problem is to establish relationships among (finite) straight lines. Descartes stated in his *Geometry* of 1637 his method for solving any problem in geometry thus: "If then, we wish to solve any problem, we first suppose the solution already effected, and give names to all the lines that seem needful for its construction, to those that are unknown as well as to those that are known. Then making no distinction between known and unknown lines, we must unravel the difficulty in any way that shows most naturally the relations between these lines, until we find it possible to express a single quantity in two ways. This will constitute an equation ...."[9] The known and unknown line segments are to be labeled with letters, and the solution is found by solving an equation. This is an analysis, in the sense that it goes from the unknown to the known; but here the known is not one or more propositions considered as obvious or known to be true. The known consists of given line segments or, in the case of a given curve, of relations between line segments.

When a result has been obtained by algebraic calculation, can it be considered to be a proof? Descartes' reply is yes. He writes that there are two ways of proving, one by analysis, the other by synthesis: "Analysis shows the true way by means of which the thing was discovered methodically ...; so that if the reader is willing to follow it ... he will understand it as perfectly as if he had discovered it for himself. Synthesis, by contrast, employs a directly opposite way ... it employs a long series of definitions, postulates, axioms, theorems and problems ... the reader is compelled to give his assent. However, this method is not as satisfying as the method of analysis, nor does it engage the minds of those who area eager to learn, since it does not show how the thing in question was discovered."[10] The "satisfaction of the mind" comes from the systematic procedure of the geometer, who decomposes the figure into its simple parts and then manipulates the algebraic symbols that represent those parts. The figure has not only been contemplated, it has been dissected through calculation. What is visually obvious leads to evidence provided by calculation. For Descartes, this evidence is a guarantee of the truth of the results that have been obtained, and this legitimises the calculation as the proof. We will see later, with Lamy, an example of the Cartesian method to solve problems.

That something is obvious, according to Descartes, does not depend on logical deductions derived from axioms, but is based on the manipulation of simple objects represented by symbols. The metaphor of contemplation or visual appearance is replaced by one of "touching". Thus, Descartes, in a letter to Mersenne in May 1630, distinguishes between comprehension and knowledge: "to comprehend something is to embrace it by thought; whereas to know something, it suffices to touch it by thought."[11]

Descartes' radical break with "ordinary mathematics" was sufficiently deep to make *The Geometry* incomprehensible to most of his contemporaries. But Descartes was to have his disciples, like Arnauld and Lamy. In *La logique ou l'art de penser* (*Logic or the Art of Thinking*), co-written with Nicole in 1662, Arnauld reproaches geometers like Euclid for having "more concern for certainty than for obviousness, and for convincing the mind rather than for enlightening it." In particular, he reproaches Euclid for having proved Proposition 20 of Book I, since it has no need of a proof, and of having proved Proposition 1, since it is a special case of Proposition 22 (see the statements of the propositions above). Arnauld was also unconvinced of the logical order of Euclid's *Elements* and he recommended instead a "natural order", more related to things being obvious than to the logical order of propositions.[12]

Lamy devotes a chapter of his *Elements of Geometry* to the Cartesian method. Using this method, he solves a number of propositions of the type given in Euclid's *Elements* Book II. For example, for the proposition mentioned above, he assumes the problem already solved and labels the known and unknown segments: here $AB = a$ and $AH = x$. He then works through the problem in order to establish relations between the known and unknown lengths, here $x^2 = a(a - x)$. He then solves the equation

$$x = -\frac{1}{2}a + \frac{\sqrt{5}}{2}a \quad \text{and} \quad x = -\frac{1}{2}a - \frac{\sqrt{5}}{2}a.$$

The first solution makes sense, because the required unknown is a line segment. The second is also legitimate from an algebraic point of view. What, however, is its status from a geometrical point of view? As we will see, the question of the geometrical status of negative or imaginary algebraic solutions was going to occur frequently in the nineteenth century.

Lamy also solved other problems, in the sort of "blind" way that algebraic calculations allow. For example, find the point F on the side CB of a given square such that the segment FE should be equal to a given segment (Figure 4).[13] Lamy lets the given segment FE have length b and sets $AB = a$, $AF = x$, and $CF = z$. He then works through the problem, writing relations derived from similar figures and from the Pythagorean theorem, until he finally obtains the quartic equation (1) in x:

$$x^2(b + x)^2 = a^2\left[(b + x)^2 + x^2\right]. \qquad (1)$$

Further progress is impeded by the complexity of the equation, so Lamy attacks the problem by using a supplementary

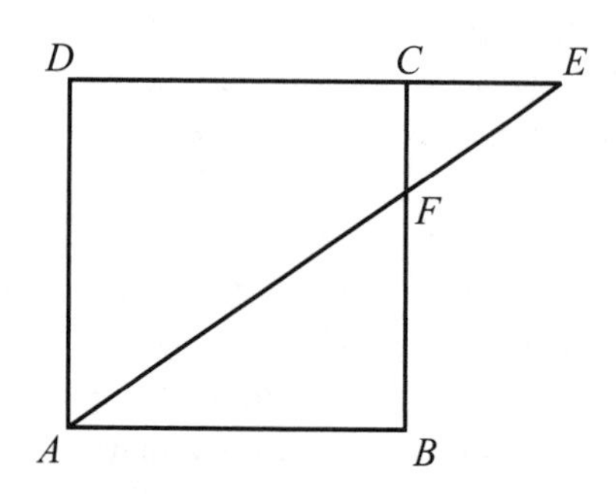

FIGURE 4

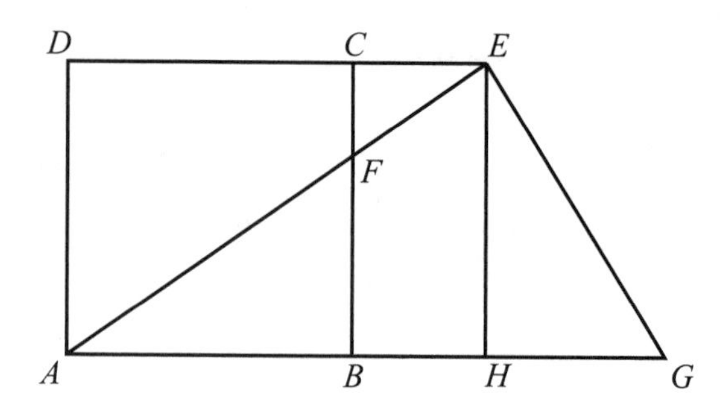

FIGURE 5

construction (Figure 5). He draws two lines from the point E, one parallel to AD and one perpendicular to AE, which cut AB produced at H and G respectively, and shows that $EG = x$. He then lets $BG = y$, and works through the problem again, this time obtaining the quadratic equation:

$$y^2 = a^2 + b^2. \qquad (2)$$

Before proceeding with the problem we should note that Descartes' criticism of the Ancients applies just as well here: how did Lamy discover his construction? Was it as a geometer contemplating the figure, or was it as an algebraist guided by the calculations? Lamy does not tell us. But now back to his equation (1). Treating this as a purely algebraic problem, if we take a new variable t, with

$$t^2 = (b + x)^2 + x^2, \quad \text{or} \quad t^2 = 2x(x + b) + b^2,$$

then equation (1) becomes: $x(b + x) = at$. Hence t has to satisfy the equation:

$$t^2 = 2at + b^2,$$

which, with another change of variable $y = t - a$, brings us back to Lamy's equation (2). But what is the geometrical significance of t and of y? We can see that t is the hypotenuse of a right-angled triangle with sides $(b + x)$ and x, like AG in the triangle AEG in Lamy's construction, and y is the difference between AG and AB, like BG.

This example, given by an admirer of Descartes, is a good illustration of the difficulties that a geometer may meet when using algebraic methods, even though the master himself had devoted a whole Book of his *Geometry* to working with equations. More than a century later, Gergonne was to propose problems in geometry for which geometric reasoning from the figures turned out to be vastly simpler than analytical methods.

Analytic Methods and Pure Geometry: a shared view of what could be taken as obvious

The works of such mathematicians as Desargues, Pascal or La Hire on perspective geometry went almost unnoticed in the eighteenth century, while the Cartesian method, extended to analytical geometry and infinitesimal calculus, became all powerful. However, towards the end of the century, Monge's lessons reawakened interest in purely geometric methods, and gave birth to what would later be called modern geometry. Moreover, Gergonne, as editor of the *Annales de mathématiques pures et appliquées,* promoted new ideas which challenged the established methods of the analysts. In volume I of the journal, which appeared in 1810–1811, he offered his readers a number of problems about minima. One of these was to determine in a plane the point for which the sum of its distances from three given points, or more generally from any number of points, should be a minimum.[14]

A correspondent at the Institute of Paris, named Tédenat, presented two methods of solution.[15] The first uses analytic methods, in which the given points are referred to two rectangular coordinate axes: the given points are $m, m', m'', \ldots$ with coordinates (a, b), (a', b'), $(a'', b''), \ldots$ and the required point M is (x, y). Hence the sum S of the distances $z, z', z'', \ldots$ of the given points from M can be expressed algebraically, and the condition for S to be a minimum is given by putting the derivatives of S with respect to x and y equal to zero. Tédenat obtains the two equations

$$\frac{x-a}{\sqrt{(x-a)^2+(y-b)^2}}+\frac{x-a'}{\sqrt{(x-a')^2+(y-b')^2}}+\frac{x-a''}{\sqrt{(x-a'')^2+(y-b'')^2}}+\cdots=0,$$

$$\frac{y-b}{\sqrt{(x-a)^2+(y-b)^2}}+\frac{y-b'}{\sqrt{(x-a')^2+(y-b')^2}}+\frac{y-b''}{\sqrt{(x-a'')^2+(y-b'')^2}}+\cdots=0,$$

which, he says, "theoretically speaking, are sufficient for the determination of this point; but unfortunately these equations, by their extreme complexity, cannot be, in the greater number of cases, a great help for the complete solution of the problem." He then lets $a, a', a'', \ldots$ stand for the slopes of the lines $mM, m'M, m''M, \ldots$ and notes that the equations now take on the very simple form

$$\cos a + \cos a' + \cos a'' + \cdots = 0$$

and

$$\sin a + \sin a' + \sin a'' + \cdots = 0.$$

It should be noted that when a straight line intersects a regular polygon then the angles the line makes with the sides of the polygon (or the sides extended) satisfy these equations. The rest of the argument is then purely geometric: the lines mM, $m'M$, $m''M$ must be parallel to the sides of such a polygon, and so for three points, the angles mMm' and $m'Mm''$ must both be equal to $120°$.

In the second solution, Tédenat lets $a, b, c, \ldots$ stand for the distances $mM, mM, m''M, \ldots$, A, B, C for the angles $mMm', m'Mm'', m''Mm''', \ldots$, and takes P as the area of the polygon whose vertices are the given points $m, m', m'', \ldots$. From this he obtains the three conditions:

$a+b+c+\cdots$ is a minimum,

$A+B+C+\cdots = 360°$,

$ab \sin A + bc \sin B + cd \sin C + \cdots = 2P.$

By considering $a, b, c, \ldots$ as "absolutely independent variables", he derives three equations with respect to $a, b, c, A, B, C, \ldots$ and obtains a system of differential equations which he solves in three pages of calculations—without any diagrams. However, it is clear that the calculations, in particular the changes of variables, are dictated by the form of the expressions: in other words, vision guides manipulation.

Gergonne offers, anonymously, a geometric solution of the above problem that is "extremely simple", of which he writes that it will be "easy, for any intelligent reader, to supply what he has intentionally left out." His solution rests on two geometric lemmas, which can be used for the solution of numerous problems about minima.[16]

The first lemma states: the point M on a straight line AB, such that the sum of its distances from two fixed points P and Q, both situated on the same side of AB, is a minimum, is the point such that the angles AMP and QMB are equal (Figure 6). In the figure, let P' and C be points such that PP' is perpendicular to AB and

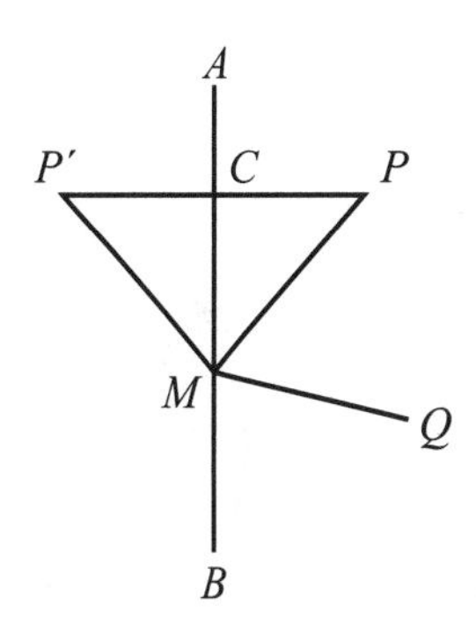

FIGURE 6

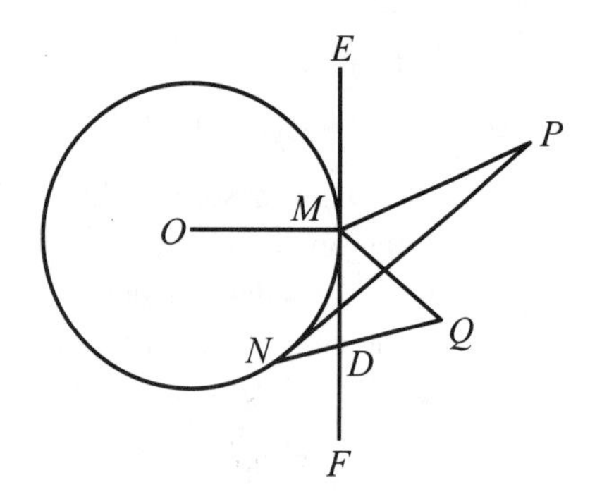

FIGURE 7

$CP = CP'$. Then if $MP + MQ$ is a minimum, so is $MP' + MQ$, and M, P', Q must lie on a straight line.

The second lemma states: the point M on the circumference of a given circle with centre O, such that the sum of its distances from two given points P and Q for which MP and MQ do not cut the circumference, is a minimum, is the point such that the angles OMP and OMQ are equal (Figure 7). In the figure, if another point N is taken on the circumference, and if NQ intersects the tangent drawn to the circle at M in the point D, then

$$MP + MQ < DP + DQ,$$

from the first lemma, and

$$DP+DQ < NP+NQ, \quad \text{so} \quad MP+MQ < NP+NQ.$$

Gergonne's reasoning can be seen in the figures, for example, the construction of the point M whose distances from three given points A, B, and C is a minimum (Figure 8). He supposes that the distance MC is given, and takes M as a point on the circumference of a circle with centre C, such that $MA + MB$ is a minimum. From the second lemma, the angles CMA and CMB are equal. Hence all three angles CMA, AMB and BMC must be equal (exchange C by B and then by A), and thus each is equal to 120°.

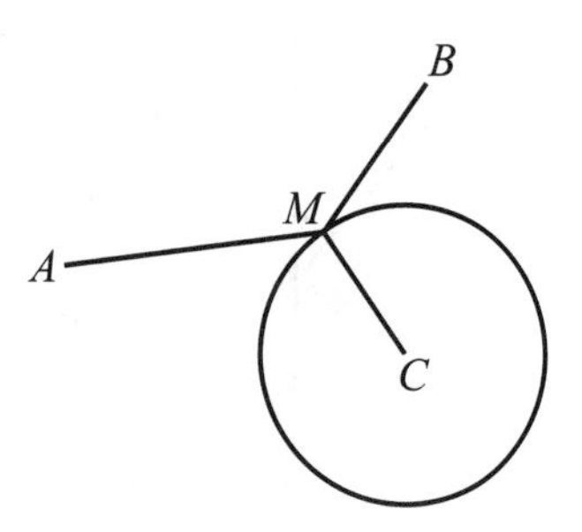

FIGURE 8

In his journal, Gergonne promoted the works of Monge, the theory of transversals of Carnot who wrote in 1806 that he wished to "free geometry from the hieroglyphics of algebra", and the theory of poles and polars developed by Brianchon in 1806. Gergonne also published, in 1817–1818, one of the first memoirs by Poncelet, a pupil of Monge. In it, Poncelet recounts that when he was a student at the Ecole Polytechnique, from 1808 to 1810, he wanted to find, by coordinate analysis, the curve whose radius of curvature at any point was twice the normal to the abscissa axis, and he was extremely surprised to obtain the equation of a parabola, instead of the expected cycloid. He realized that this result was due to "an influence of the sign +", and set about looking for a purely geometric proof of the principal properties of the parabola, "a proof unencumbered by any of the trappings of algebra."[17] While languishing in Russian prisons, with only the memories of his student days at the Ecole Polytechnique, where he had shown a keen interest in the works of Monge, Carnot and Brianchon, Poncelet wrote several notebooks devoted to the new synthetic geometry.

The third of his notebooks concerns descriptive properties of conics and the first principles of projection. According to the fourth principle, if we are given a conic and a straight line in the same plane, there exist an infinite number of ways of projecting the figures onto another plane in such a way that the projection of the conic is a circle and that of the straight line is the line at infinity. In examining the cases where a geometric proof of this principle comes up against impossibilities, Poncelet notes however that coordinate geometry does allow the result to be established and writes:

> It is this great generality of [coordinate] analysis, that we are able to bring to proofs in geometry in the same circumstances, that has justified this expression of the power of [coordinate] analysis.... In the course of a calculation, it often happens that certain expressions which arise implicitly are null, infinite, imaginary, or take a quite different form: we continue the calculation unworried.... This is not the same with geometry, as we are accustomed to regard it; since all the arguments, and all the consequences, cannot be perceived, grasped by the mind, unless they are portrayed to the imagination by perceptible objects, as soon as these objects are missing, reasoning ceases.[18]

Thus, according to Poncelet, coordinate analysis and geometry differ in the notions that are accepted by the mathematician as obvious. In the former, these are based on the

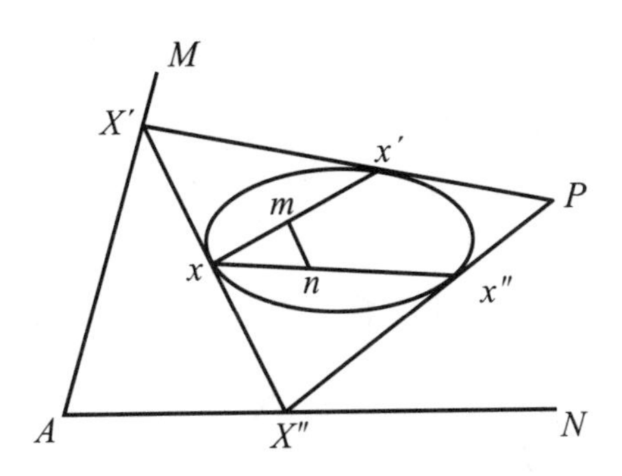

FIGURE 9

"blind" manipulation of symbols, and in the latter, on visual imagination. If coordinate geometry appears as all powerful mathematically, pure geometry is superior from the point of view of simplicity and elegance, and can be made equal to it if new principles are adopted. Poncelet goes on to examine a problem in which pure geometry and coordinate geometry are again compared: given two straight lines AM and AN, a conic in the same plane, and a tangent $X'X''$ to the conic with X' on AM and X'' on AN, find the locus of the point P, which is the intersection of two other tangents PX' and PX'', as the tangent $X'X''$ varies (Figure 9).

Poncelet begins with a "purely geometric solution." As $X'X''$, varies the chord xx' has to pass through a fixed point m (the pole of AM), and the chord xx'' has to pass through a fixed point n (the pole of AN). By Poncelet's fourth principle, there exists a projection such that the conic goes to a circle and the line to infinity. Hence, the above reduces to the following problem: given a circle, a point x on it and two chords xx' and xx'' parallel to given directions, to find the locus of the point P, which is the intersection of the tangents Px' and Px'' as x moves on the circumference of the circle (Figure 10). "It is clear", says Poncelet, that the angle $x'xx''$ is constant, so the chord $x'x''$ is constant and therefore the point P lies at a constant distance from the centre C of the circle. Thus P describes a circle

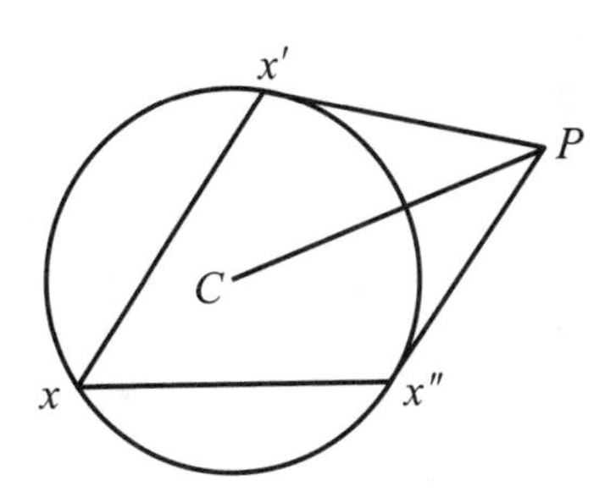

FIGURE 10

concentric to the given circle, and thus the point P in the original problem describes a conic. Poncelet comments on his solution: "If one wished to deal with the same question in all its generality by coordinate analysis, one would be thrown into calculations and eliminations of a disheartening length."[19] He does go on, however, to offer three long solutions by analytical geometry, "not to give a greater certitude to the first proof", but in order to determine the locus precisely by equations, and to show how solutions foreign to the question intrude into the calculations.

In a note concerning the determination of this locus, Poncelet again contrasts the resources of coordinate geometry with the reasoning in pure geometry, "where one abandons oneself to the personal, one might say intuitive, contemplation of the conditions and the givens of each problem."[20] This type of solution, he writes, "alone merits the epithets of elegance, speed and ingenuity." The visual evidence, which he invites the mathematician to share, is something that is obvious in space, and not in the plane. With plane geometry, the geometer may contemplate a figure as an image before him, but, with projective geometry, he himself is in a space in which the figure lies. To put it another way, in space, the image and the figure no longer coincide: the visual evidence in the plane and in space are certainly different. Take for example, Desargues' theorem, which states that if two triangles are such that the lines joining the corresponding vertices pass through the same point O, then the corresponding sides intersect in three collinear points M, N, P. It is not an obvious theorem of plane geometry, but the geometer can see this theorem as obvious if he embeds the two triangles in two planes in space whose intersection is the straight line MNP (Figure 11).

Chasles reports that in his lectures Monge never drew any figures, but he "knew how to make all the most complicated forms of space appear in space ... with no other aid than his hands, whose movements admirably supplemented his words."[21] In space, what is visually obvious requires the help of movement and talking, just as the new projective geometry needed new principles to extend the power of coordinate geometry. In the 1820s, Poncelet and Gergonne went on to propose two fundamental principles, the principle of continuity and the principle of duality, respectively. As we will see later, this was to cause a dispute between the two mathematicians, but beyond simple polemics and disputes of priority, their differences were profound. In fact, in a way, the principle of continuity derives from the evidence provided by motion and the principle of duality from the evidence of language.

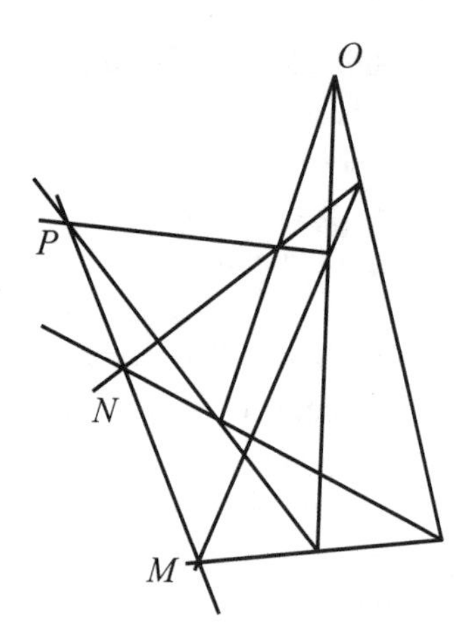

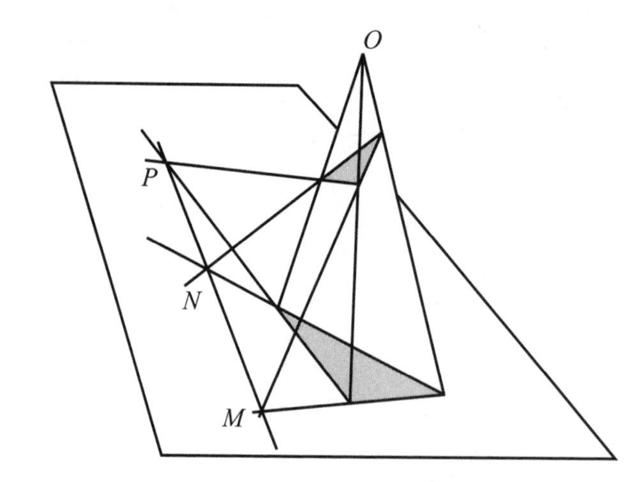

FIGURE 11

Questions of Principles: a dispute about what could be accepted as obvious

In his *Traité des propriétés projectives des figures* of 1822, Poncelet set out to make descriptive geometry independent of algebraic analysis. Wherein lies the extensive power of coordinate geometry? Why is the geometry of the Ancients lacking and what means can be found for it to enjoy success? Poncelet explains that algebra represents magnitudes by "abstract signs" which provide these magnitudes with the greatest possible lack of determination. Reasoning is implicit in the algebra and is an abstraction from the figure, whereas the geometry of the Ancients never draws conclusions which cannot be "portrayed to the imagination or to sight". But were it possible to use implicit reasoning, through abstraction from the figure, ancient geometry could become the rival of coordinate geometry. Poncelet goes on to state his principle of continuity. Consider some figure, in a general and to a certain extent undetermined position, and suppose we have found either "metric" or "descriptive" relations or properties of the figure by ordinary explicit reasoning. Then, says Poncelet, "is it not evident that if, keeping the same given things, one could vary the primitive figure by insensible degrees by imposing on certain parts of the figure a continuous but otherwise arbitrary movement, then the properties and relations found for the first system remain applicable to successive states of the system, provided always that one has regard for certain particular modifications that may intervene, ... modifications that will always be easy to recognise a priori and by infallible rules?"[22]

For Poncelet, the principle of continuity is a sort of axiom whose "obviousness is manifest, incontestable and does not need to be proved". For him, if it is not admitted as a method of proof, it can at least be admitted as a means of discovery or invention. It seemed to him unreasonable to reject in geometry ideas that were generally accepted in algebra. The obviousness of the principle of continuity did not, however, convince all of Poncelet's contemporaries. Poncelet's geometry was even described by some of them as "romantic geometry." In a communication to the Royal Academy of Sciences in January 1826, Cauchy wrote that the principle of continuity was "only a strong induction."[23] Poncelet replied to this later, writing that Cauchy only possessed an imperfect feeling for true geometry. In another communication to the Royal Academy, commenting on a memoir of Gergonne's about the theory of reciprocal polars, the writer explains that to put the result "beyond doubt, it appears necessary for them to substitute an analytic proof for M. Poncelet's geometric proof."[24] In 1827 Poncelet denounced "the habit generally acquired of according to algebra an almost undefined rigour." He accused correspondents of the Academy of having cast a sort of disfavour on his principles, and he regretted that the reproach of lack of rigour had been restated by several geometers, including Gergonne.[25] This latter had, the previous year, stated his principle of duality, and the two mathematicians were now in dispute about it.

In his *Considérations philosophiques sur les éléments de la science de l'étendue* (*Philosophical considerations on the elements of the science of space*)[26] of 1826, Gergonne separates the theories of the science of space into two categories: those concerning metric relations, and established by calculation, and those relating to the position of geometric objects, which can be deduced by calculation but can also be "completely freed" from calculation. He draws support from the works of Monge, which allowed one to conclude that the division of geometry into plane geometry and spatial geometry was not as natural as "twenty centuries of habit have been able to persuade us." He notes that in plane geometry, to each theorem there corresponds another, which can be derived by a simple interchange of the words "points" and "lines", while in the spatial geometry it is the words "points" and "planes" that have to be interchanged. "This sort of duality of the theorems constitutes a geometry of position." Gergonne sets out theorems in two columns since, he says, it is "superfluous to attach to this memoir figures, often more embarrassing than useful, in the geometry of space.... We only have here, in effect, logical deductions, always easy to follow, when the notation is chosen in a convenient way."[27]

The duality is obvious in the statement of Desargues' theorem and its converse, and setting it out in columns shows this clearly:

Desargues' Theorem	Dual of Desargues' Theorem
If two triangles are such that the *lines* joining the corresponding vertices pass through the same *point* O, then the corresponding sides intersect in three collinear *points*.	If two triangles are such that the *points* of intersection of corresponding sides lie on the same *line* O, then the corresponding vertices are joined by three concurrent *lines*.

We could equally show the duality of Pascal's theorem and Brianchon's theorem. The correspondent to Gergonne's memoir in the *Bulletin des sciences de Férussac* notes that the properties of poles and polars can also illustrate the property of duality.[28] But Gergonne extends the notion of duality in a much more general way to the whole of the geometry of space. His duality consists of a true work of translation, work done on language, as is shown in the first example of his 1826 memoir, which shows the duality of propositions for the plane and for space:

Two *points*, distinct from each other, given in space, determine an infinite line which, when these two *points* are designated A and B, can itself be designated by AB.	Two *planes*, non parallel, given in space, determine an infinite line which, when these two *planes* are designated by A and B, can itself be designated by AB.

In a note[29] of March 1827 about a memoir by Poncelet on the theory of reciprocal polars, Gergonne writes that certain statements of theorems are indeed susceptible of a "sort of translation which is the subject of Poncelet's meditations, without their authors having the air of doubting that this translation is possible."[30] He reproaches Poncelet for hastening a revolution, while he himself tries to make the truth accessible by senses. For him, the obstacle facing an easy propagation of the doctrines, which both he and Poncelet wished to popularise, lay in the obligation to speak in a language that had been created for a more restrained geometry, which is the plane geometry of the Ancients where the figures were not embedded in space. For, when "the language of a science is well made, logical deductions come from it with such ease, that the mind moves as it were of itself before the new verities."[31] It is therefore through work on language that the new doctrines can be made obvious: to present the new theory in the most advantageous way, it is necessary to first create a "language in keeping with it."

The above note is the origin of the dispute between the two mathematicians. Poncelet was furious to read in the *Bulletin des sciences de Férussac* that he had taken his research from Gergonne's duality. Poncelet was obliged to show that his idea of reciprocal polars was not one of duality, and to explain that he was even opposed to the latter idea, in particular to setting out propositions in columns. He reproached Gergonne with having established false propositions as general principles, and of having "tortured the meaning of words" in order to make the consequences correct.[32] The dispute was also to draw in Plücker, who announced that he "had discovered in a purely analytic way the secret of duality."

The intensity of Poncelet's remarks shows the depth of the disagreements: here were three conceptions of what could be taken as obvious, rooted in different practices and habits. The dispute about what could be taken as obvious remained lively until the end of the nineteenth century. For example, Hadamard was described by Hermite as a traitor because, in a memoir of 1898, the former had used geometric intuition while the latter despised geometry. Hermite protested vehemently against Hadamard, crying: "You have betrayed analysis for geometry."[33]

Logically obvious, visually obvious, and obvious by manipulation

In Hilbert's *Fundamentals of Geometry,* which appeared in 1899, the objects of geometry are defined in a purely grammatical and formal way; points, lines and planes are only words which are related together by sentences. This is close to Gergonne's idea of interchanging words in sentences. Thus, the first axioms state that:

1. There exists a *line* associated with two given *points* A and B to which these two points belong.
2. There is no other line to which the two points A and B belong.

3. On a line, there are at least two points; there exist at least three noncollinear points.
4. There exists a *plane* associated with three noncollinear *points* A, B, C to which these three points A, B, C belong.

This geometry is axiomatic, like that of the ancients, but here no visual evidence is supposed. The combination of logic and vision that is found in Euclidean geometry was blown apart by the invention of non-Euclidean geometries. In 1733, when faced with a choice between logical discourse and visual evidence, Saccheri wrote that the nature of the straight line was repugnant to the non-Euclidean hypothesis and so he chose visual evidence. The inventors of non-Euclidean geometry made the opposite choice. Poincaré, however, proposed a visual model of hyperbolic geometry, in accordance with a geometric vision.

Poincaré's reading of Hilbert's book can serve as a fitting epilogue to our story. In *Science et méthode,* which appeared in 1909, Poincaré contrasts two types of mathematical definitions: "those which seek to make an image become alive, and those where we are restricted to combine empty forms, perfectly intelligible, but purely intelligible forms, that abstraction has deprived of all matter."[34] For him, these latter definitions are those of Hilbertian geometry. For a geometer of the nineteenth century, like Poncelet, they were also those of coordinate geometry, where, for example, a circle is defined as an equation. Poincaré remarked that the definitions which are best understood by one were not those best suited to the other.

The historical stages to which we have referred convince us of the justice of this remark of Poincaré's. At each stage, new arguments or practices declared as obvious by a mathematician come up against the acquired habits of his contemporaries regarding what is understood to be obvious. To put it another way, and paradoxically, obviousness may not be obvious. The feeling, however intimate it may be, that something is obvious, must be understood more as a *habitus,*[35] rather than as something inborn.

Endnotes

[1] Plato said that "geometry obliges one to contemplate the essence of things" (*The Republic,* 521–527).

[2] Euclid, *The thirteen books of Euclid's elements,* tr. T.L. Heath, vol. I, University Press, Cambridge, 1908, p. 243.

[3] Proclus, *A commentary on the first book of Euclid's Elements,* tr. Glenn R. Morrow, Princeton University Press, Princeton, 1970, p.140. Aristotle, in his *Posterior Analytics,* writes that demonstrative science must start out from premises that are "true, first and immediate, more known than the conclusion, anterior to it, and which are the causes of it."

[4] Proclus, *op. cit.,* p. 183.

[5] Proclus, *op. cit.,* pp. 275–276.

[6] A necessary condition for the statement of a problem.

[7] Proclus, *op. cit.,* pp. 259–260.

[8] Descartes, *Règles pour la direction de l'esprit* (*Rules for the Direction of the Mind*). English tr. in *The Philosophical Writings of Descartes,* tr. J. Cottingham, R. Stoothoff, D. Murdoch, Cambridge: Cambridge University Press, 1985, vol. I, pp. 17–18.

[9] Descartes, *The Geometry,* tr. D. E. Smith & M. L. Latham, New York: Dover, pp. 6–9.

[10] Descartes, *Œuvres,* ed. Adam & Tannery, Paris, 1897–1913, vol. IX, pp. 121–122. English tr. in Cottingham et al., *The Philosophical Writings of Descartes,* vol. II, pp. 110–111.

[11] Descartes, *Œuvres,* ed. Adam & Tannery, Paris, 1897–1913, vol. I, p. 152.

[12] Arnauld & Nicole, *La logique ou l'art de penser,* PUF, pp. 326–331.

[13] Lamy, *Eléments de géométrie,* 2nd ed., 1695, pp. 351–353.

[14] *Annales de mathématiques pures et appliquées,* vol. 1 (1810–1811), p. 196.

[15] *ibid,* pp. 285–291.

[16] *ibid,* pp. 375–384.

[17] Poncelet, *Applications d'analyse et de géométrie,* ed. 1862, pp. 462–468.

[18] *ibid.,* pp. 124–125.

[19] *ibid.,* p. 145.

[20] *ibid.,* p. 151.

[21] Chasles, *Aperçu historique des méthodes en géométrie,* Gauthier-Villars, Paris, 1889, p. 209.

[22] Poncelet, *Traité des propriétés projectives des figures,* vol. I, Intr., p. xiii.

[23] *ibid.,* vol. 2, Intr., p. 356.

[24] *ibid.,* p. 361.

[25] Gergonne, note in vol. 7 of *Bulletin de Férussac,* 1827, pp. 109–117.

[26] Gergonne, "Considérations philosophiques sur les éléments de la science de l'étendue," *Annales de mathématiques pures et appliquées,* vol. XVI, 6 (Jan. 1826), pp. 209–231.

[27] *ibid.,* p. 212.

[28] Poncelet, *op. cit.,* p. 382.

[29] *ibid.,* pp. 390–393.

[30] *ibid.,* p. 392.

[31] *ibid.,* p. 392.

[32] *ibid.,* p. 374.

[33] Quoted by Mandelbrot in *Cahiers du séminaire d'histoire des mathématiques,* IHP, 6 (1985), pp. 1–46.

[34] Poincaré, *Science et méthode,* 1909, p. 127.

[35] Bourdieu defined a habitus as "a structured structure which functions like a structural structure", in *Le sens pratique,* Editions de minuit, 1980, p. 88.

Use of History in a Research Work on the Teaching of Linear Algebra

Jean-Luc Dorier
Laboratoire Leibniz
Grenoble, France

Linear algebra is universally recognized as a very important subject not only within mathematics but also with regards to its applications to physics, chemistry, economics, and so on. In the modern organization of mathematical subjects, a vector space is one of the simplest algebraic structures. Nevertheless students, who often have great difficulty at the start of a linear algebra course in the first year of university, do not easily recognize this simplicity. This paper will focus on the French context of this teaching, but several research works have found similar difficulties in various countries around the world.[1]

In France, the teaching of linear algebra was entirely remodeled with the "reform of modern mathematics" in the sixties. At that time, the influence of Bourbaki and a few others led to the idea—which was based on a very democratic concern—that geometry could be more easily accessible to students if it were founded on the axioms of the structure of affine spaces. Therefore the axiomatic theory of finite-dimensional vector spaces was taught in the first year of secondary school (age 15). The fate of this reform and the reaction it aroused are well known. Therefore, from the beginning of the eighties, the reform of the teaching of mathematics in French secondary schools gradually led to the removal of any subject related to modern algebra. Moreover, the teaching of geometry focused on the study of transformations on elementary figures, and analytical geometry is now barely taught at secondary level. On the other hand, formal theories became unpopular and students entering university nowadays have very little practice of any formal mathematical subject.

This situation created a total change in the background of students, for whom the teaching of linear algebra represented the first contact with such a "modern" approach. Of course the teaching in the first year of university changed and became less theoretical. In many universities,[2] it was decided to prepare the students for the teaching of linear algebra by a preparatory course in Cartesian geometry and/or by a course in logic and set theory. Yet, in secondary school, students still learn the basics of vector geometry and the solving of systems of linear equations by Gaussian elimination. Therefore, they have some knowledge on which the teaching of linear algebra can be based. For the moment, the idea of teaching students the axiomatic elementary theory of vector spaces within the first two years of a science university has not been questioned seriously, and the teaching of linear algebra, in France, remains quite formal.

In this context, A. Robert and J. Robinet (1989) showed that the main criticisms made by students toward linear algebra concern the use of formalism, including the overwhelming number of new definitions and the lack of connection with what they already know in mathematics. It is quite clear that many students have the feeling of landing on a new planet and are not able to find their way in this new world. On the other hand, teachers usually complain of their students' erratic use of the basic tools of logic or set theory. They also complain that students have no skills in elementary Cartesian geometry and consequently cannot use intuition to build geometrical representations of the basic concepts of the theory of vector spaces. These complaints correspond to a certain reality, but the few attempts of remediation—with previous teaching in Cartesian geometry and/or logic and set theory—did not seem to improve the situation substantially. Indeed, many works have shown that difficulties with logic or set theory cannot be interpreted without taking into account the specific context in which these tools are used.

In my first work (1990a), I tested with statistical tools the correlation between the difficulties with the use of the formal definition of linear independence and the difficulties with the use of mathematical implication in other contexts. Although these two types of difficulties seemed at first closely connected, the results showed clearly that no systematic correlation could be made. This means that students' difficulties with the formal aspect of the theory of vector spaces are not just a general problem with formalism. They are mostly a difficulty of understanding the specific use of formalism in the theory of vector spaces and the interpretation of the formal concepts relative to more intuitive contexts like geometry or systems of linear equations, in which they historically emerged. We will analyze this point in more detail in this paper.

A. Robert, J. Robinet, M. Rogalski and I have developed a research program on the learning and teaching of linear algebra in the first year of a French science university. This work, which started some ten years ago, includes not only the elaboration and evaluation of experimental teaching but also epistemological reflections. These are built in a dialectical process on a historical analysis of the genesis of the concepts of linear algebra and a pedagogical analysis of the teaching and the difficulties of the students.[3] In this paper, I will try to summarize the main issues of our research, focusing on a restricted set of concepts: linear dependence and independence, generators, basis, dimension and rank.

I. Logic versus Meaning : the Role of Formalism

Linear dependence and independence, generators, basis, dimension and rank are the elementary concepts which constitute the foundations of the theory of vector spaces. For any mathematician, they seem very simple, clearly interrelated notions. Indeed, in the formal language of modern algebra they correspond to easily expressible definitions. Moreover, the logic of a deductive presentation induces a "natural" order among them (more or less the order given above) which reflects their intrinsic network of relations.

In his doctoral dissertation, A. Behaj (1999) has interviewed several students from the second to the fourth year of French and Moroccan universities. He asked them, in pairs, to build the skeleton of a course, to be addressed to first year students, on these basic notions of linear algebra. It is quite surprising to see that nearly all of them have the same initial reaction: they present the notions in their supposedly natural logical order. They do it consciously and they give justifications on the ground of logic and simplicity. Yet, most of them, when they are asked to give applications and exercises in relation to the course, gradually abandon (more or less explicitly) this "natural" order and reorganize the network of relations according to a less conscious construction, after using them in several contexts. Similar phenomena have been revealed when interviewing teachers. Not that they change their presentation, but they clearly—although it may not be really conscious—show two contradictory concerns when structuring their course: a concern for the logic of a deductive presentation and a concern for the applicability of their course to exercises and problems.

These two concerns induce different organizations of the notions. Therefore, most of the time, the teacher's course is apparently organized according to standards of logic, proper to the rigor of a mathematical text. However, the choices of examples, remarks, and so on create a secondary level of organization in relation to the solving of exercises and problems and also according to what teachers know of the difficulties of their students. For instance, the understanding of the notion of generators is very rarely evaluated in exercises. Indeed, teachers know that once the dimension of a subspace is known to be n, then n independent vectors generate the subspace; they also know that proving that vectors are independent is easier than proving that they generate a subspace. Thus the notion of generator holds a very different position within their formal presentation of the theory from the one it has within the solving of exercises.

Such observations may certainly be made about several different mathematical notions. Yet, this group of notions is particularly interesting, because each of them is somehow elementary, not in the sense that it is simple, but in the sense that it is an element which will be part of a more complex set of notions. In this sense, it is true that the notions of linear independence and generators are more elementary than the notion of basis, because a basis is usually defined as a set of linearly independent generators. Yet, a basis can be defined as a minimal set of generators or as a set of units, of which each element of the vector space is a unique linear combination. In these two approaches, there is no reason to say that linear independence is more elementary than the notion of basis, which can be defined independently of the notion of linear independence. On the other hand, a finite-dimensional vector space may be defined as having a finite maximal set of independent vectors. Then a basis can be defined as any maximal set of independent vectors, without using the notion of generators, which has then no reason for being more elementary than the notion of basis. Although this last alternative is somehow "unnatural", the first two are sometimes chosen in textbooks or actual teaching. However all three are logically consistent.

I draw two conclusions from these remarks :

- Any logical construction is partly arbitrary and cannot be qualified as natural, without further epistemological investigations.
- The nature and meaning of concepts is to be found beyond their logical interrelations.

Although these may seem obvious, they are very important as far as a theory such as that of vector spaces is concerned. Indeed, in its modern axiomatic version, this theory has been so highly formalized that it is tempting to reduce its content to the logical network of relations between formal concepts. On the other hand, as a reaction to this extreme position, one may want to give only "practical" knowledge with reference to "practical" contexts like geometry, linear equations, differential equations, and so on. But this last option reduces the meaning and general nature of linear algebra. Indeed, the formal aspect of the theory of vector spaces is the result of its general nature and a condition for its simplicity. Therefore one cannot spare the students the difficulty of formalism if the theory is to be understood with all the meaning it has now acquired. Moreover, we put forward the hypothesis that the necessity for formalism has to be understood very early in the learning of the theory. Our historical analysis (Dorier 1995a, 1997 and 2000 (first part)) has been, to a great extent, conducted in order to clarify and support the preceding statement.

Our epistemological reflection has led us to understand more clearly the different stages of unification and generalization in the genesis of linear algebra but also the role played by the different contexts of origin (in geometry, linear systems and determinants, algebra, and functional analysis). The axiomatic approach of the theory only prevailed around 1930 and took a long time to be accepted by mathematicians, even in the early history of functional analysis where infinite determinants were preferred by many until Banach's 1932 *Théorie des opérations linéaire* (Dorier 1996). Yet, once accepted, the axiomatic approach was quick to replace all previous analytical approaches, and its power of generalization and unification was universally recognized. In teaching, however, formalism should not be introduced too early and imposed without care. A formal concept has to be introduced with reference to students' conceptions previously acquired in intuitively based contexts, as a means for generalization and simplification.

Let us take as an example the notions of linear independence and dependence.

II. The Case of Linear Dependence and Independence

In the language of the modern theory these two notions are extremely simple. They can be defined as two logically opposite notions in the language of basic set theory. Yet, even if students can be easily trained to solve standard questions like "is this set of vectors independent or not?" in various contexts, the use of these notions in less straightforward situations may be much less easy. On the other hand, the historical elaboration of the formal definition of linear independence was not as easy as one could imagine. I will start by developing these two aspects: the students' difficulties and the historical evolution. Then, in a third part, I will draw some conclusions on the basis of an epistemological synthesis of the first two parts.

1) The difficulties of the students.

Anyone who has taught a basic course in linear algebra knows how difficult it may be for a student to understand the formal definition of linear independence and to apply it in various contexts. Moreover, once students have proved their ability to check whether a set of n-tuples, equations, polynomials or functions are independent,[4] they may not be able to use the concept of linear independence in more formal contexts.

Let us take a few examples of exercises[5] given to beginning students to obtain a better idea of this kind of difficulty.

1. *Let U, V, and W be three vectors in $\mathbf{R}^3$. If they are pairwise non-collinear, are they independent?* Many students think that this proposition is trivial, and they are sure that it is true. If they want to prove it by using the formal definition, they are led to say more or less any kind

of nonsense just for the sake of giving a formal proof (as required by their teacher) to a statement of which they are convinced.[6] Nevertheless, this mistake is one aspect of a more general difficulty with the global aspect of the concept of linear (in)dependence. Indeed, many students are inclined to treat the question of linear (in)dependence by successive approximations starting with two vectors, and then introducing the others one by one. We will say that they have a local approach to a global question. Indeed, in many cases, at least if it is well controlled, this approach may be correct and actually quite efficient, yet, it is a source of mistakes in several situations. The students have built themselves what G. Vergnaud (1990) calls *théorèmes-en-acte* (i.e., rules of action or theorems which are valid in some restricted situations but create mistakes when abusively generalized to more general cases). Here is a non-exhaustive list of *théorèmes-en-acte* connected with the local approach of linear (in)dependence, that we have noticed in students' activities:

- if U and V are independent of W, then U, V, and W are globally independent;
- if U_1 is not a linear combination of $U_2, U_3, \ldots, U_k$, then $U_1, U_2, \ldots, U_k$ are independent;
- if U_1, V_1, and V_2 are independent and if U_2, V_1, and V_2 are independent, then U_1, U_2, V_1, and V_2 are independent.[7]

2.1. *Let U, V, and W be three vectors in $\mathbf{R}^3$, and f a linear operator in $\mathbf{R}^3$. If U, V, and W are independent, are $f(U)$, $f(V)$, and $f(W)$ independent?*

2.2. *Let U, V, and W be three vectors in $\mathbf{R}^3$, and f a linear operator in $\mathbf{R}^3$. If $f(U)$, $f(V)$, and $f(W)$ are independent, are U, V, and W independent?* To answer these questions, beginners usually try to use the formal definition without first building concrete examples that would help them to obtain an idea of the result. Then they try different combinations with the hypotheses and the conclusions, and finally answer "yes" to the first question and "no" to the second one, despite coming close to writing the correct proof for the correct answers. Here is a reconstructed proof that reflects the difficulties of the students:

> If $\alpha U + \beta V + \gamma W = 0$ then $f(\alpha U + \beta V + \gamma W) = 0$ so f being a linear operator, $\alpha f(U) + \beta f(V) + \gamma f(W) = 0$; now as U, V, and W are independent, then $\alpha = \beta = \gamma = 0$; so $f(U)$, $f(V)$, and $f(W)$ are independent.

In their initial analysis, A. Robert and J. Robinet concluded that this type of answer was an incorrect use of mathematical implication characterized by the confusion between hypothesis and conclusion. This is indeed a serious difficulty in the use of the formal definition of linear independence. As mentioned above, I tested the validity of this hypothesis with different students. Before the teaching of linear algebra, I set up a test to evaluate the students' ability in elementary logic and particularly in the use of mathematical implication (Dorier 1990a and b), and after the teaching of linear algebra, I gave the above exercise to the students. The results showed that the correlation was insignificant; in some cases it was even negative. Yet, on the whole (both tests included many questions), there was quite a good correlation between the two tests. This shows that if a certain level of ability in logic is necessary to understand the formalism of the theory of vector spaces, general knowledge, rather than specific competence is needed. Furthermore, if some difficulties in linear algebra are due to formalism, they are specific to linear algebra and have to be overcome essentially in this context.

On the other hand, some teachers may argue that, in general, students have many difficulties with proof and rigor. Several experiments that we have made with students showed that if they have connected the formal concepts to more intuitive conceptions, they are in fact able to build very rigorous proofs. In the case of the preceding exercise, for instance, if you ask the students after the test to illustrate the result with an example individually, let us say in geometry, they usually realize very quickly that there is something wrong. It does not mean that they are able to correct their wrong statement, but they know it is not correct. Therefore, one main issue in the teaching of linear algebra is to give our students better ways of connecting the formal objects of the theory to their previous conceptions, in order to have a better intuitively based learning. This implies not only giving examples but also showing how all these examples are connected and what the role of the formal concepts is with regard to the mathematical activity involved.

2) Historical background.

The concept of linear (in)dependence emerged historically at first in the context of linear equations (Dorier 1993, 1995, 1997 and 2000). Euler's 1750 text entitled *Sur une contradiction apparente dans la doctrine des lignes courbes* is the first one in which a question of dependence between equations was discussed. Euler's concern was to solve Cramer's paradox. This paradox first drawn out by Cramer and MacLaurin was based on two propositions commonly admitted in the beginning of the eighteenth century:

1. *An algebraic curve of order n is uniquely determined by $n(n+3)/2$ of its points.* This was clear from

elementary combinatorics by counting the coefficients of the equation of such a curve.

2. *Two algebraic curves of orders n and m intersect in nm points.* It was known that some points may be multiple, infinite or imaginary, but one knew cases when all these mn points were finite, real, and distinct.

Therefore for $n > 3$, it appears that n^2 points common to two algebraic curves of order n would not be sufficient to determine uniquely an algebraic curve, while the first proposition states that $n(n+3)/2$ ($< n^2$) should determine one and only one such curve. This is Cramer's paradox. Euler discussed the validity of both propositions and of their consequences and concluded that the first one, based on the fact that n linear equations determine exactly n unknown values, should be restricted. At that time, the general idea that n equations determine n unknowns was so strong that nobody had taken the pains to discuss the exceptional case, until Euler pointed out this problem.

He starts by an example with two equations :

> Let us just look at these two equations $3x–2y=5$ and $4y=6x–10$; one will see immediately that it is not possible to determine the two unknowns x and y, because if one eliminates x then the other unknown y disappears by itself and one gets an identical equation, from which it is not possible to determine anything. The reason for this accident is quite obvious as the second equation can be changed into $6x - 4y = 10$, which being simply the first one doubled, is thus not different.[8]

It is clear—especially by reading the end of this quotation—that Euler does not intend to fool his reader, even though he artificially hides the similarity of the two equations. Yet, it is also clear that it is not the fact that the two equations are similar that determines the dependence of the equations, but the fact that something unusual—an accident—happens in the final step of the solving process. This accident reveals the dependence of the equations, because, although there are two of them, these equations do not determine two unknowns. Mathematically speaking, the two statements are logically connected; a linear dependence between n equations in n unknowns is equivalent to the fact that the system will not have a unique solution. However, these two properties correspond to two different conceptions of dependence. To be able to distinguish these two conceptions, I will call Euler's conception, *inclusive dependence*. I wish to insist on the fact that this conception is natural in the context in which Euler and all the mathematicians of his time were working, that is to say with regard to the solving of equations and not the study of equations as objects on their own.

To make my statement clearer, let us see what Euler says in the case of four equations. He gives the following example:

$$5x + 7y - 4z + 3v - 24 = 0,$$
$$2x - 3y + 5z - 6v - 20 = 0,$$
$$x + 13y - 14z + 15v + 16 = 0,$$
$$3x + 10y - 9z + 9v - 4 = 0.$$

> These would be worth only two, since having extracted from the third the value of
>
> $$x = -13y + 14z - 15v - 16$$
>
> and having substitued this value in the second to get:
>
> $$y = (33z - 36v - 52)/29 \text{ and } x = (-23z + 33v + 212)/29,$$
>
> these two values of x and y being substituted in the first and the fourth equations will lead to identical equations,[9] so that the quantities z and v will remain undetermined.[10]

Here again the proof is based on the solving of the system of four equations by substitution that leads to two undetermined quantities. Euler does not even mention any linear relations between the equations, although they may seem rather obvious, like: $(1) - (2) = (4)$ and $(1) - 2x(2) = (3)$, for instance. Therefore, the property expressed by Euler is not the modern concept of linear dependence. It is, however, a property of the equations that makes the network of constraints they impose on the unknowns equivalent to two constraints and not four. This is what we propose to call inclusive dependence. Another aspect of Euler's work is worth special attention, the passage from two to three equations. Indeed, for three equations, Euler says:

> The first one, being not different from the third one, does not contribute at all in the determination of the three unknowns. But there is also the case when one of the three equations is contained in the two others.... So when it is said that to determine three unknowns, it is sufficient to have three equations, it is necessary to add the restriction that these three equations are so different that none of them is already comprised in the others.[11]

Euler's use of terms such as *comprised* or *contained* refers to the conception of inclusive dependence as we explained above. It does not mean that Euler was not aware of the logical equivalence with the fact that there exist linear relations between the equations, but, within his practice with linear equations, the conception of inclusive dependence is more consistent and efficient. Yet, there is a difficulty for further development; indeed, the conception of inclusive dependence is limited to the context of equations and cannot be applied to other objects, like n-tuples for instance. Therefore inclusive dependence is context-dependent although linear dependence is a general concept that applies to any linear structure.

It is important to notice also that Euler separates the case when two equations are equal from the case when the three equations are globally dependent. This points out an intrinsic difficulty of the concept of dependence, which has to take all the equations in a whole into account, not only the relations in pairs. We have seen that students have real difficulties with this point.

Moreover, in this text, Euler was able to bring out issues that can be considered in many respects as the first consistent ideas on the concept of rank. For instance the quotation above, about four equations, is an illustration of the relation:

$$\text{(number of unknowns)} - \text{(rank of the system)} = \text{(dimension of the set of solutions)}.$$

Indeed, there are four unknowns; the equations "are worth only two" (their rank is two); and "two unknown quantities will remain undetermined" (the dimension of the set of solutions is two). Of course this result remains implicit and is not formalized. Euler shows, however, through a few numerical examples never exceeding five unknowns and five equations, that he had an intuitive yet accurate and consistent conception of the relation between the size of the set of solutions and the number of relations of dependence between the equations of the system. In this sense, the specific context of Cramer's paradox helped him to elaborate such a reflection as shown in the following statement:

> When two lines of fourth order[12] intersect in 16 points, as 14 points, when they lead to equations all different among themselves,[13] are sufficient to determine a line of this order, these 16 points will always be such that three or even more of the equations they produce are already comprised in the others. So that these 16 points do not determine anything more than if they were 13 or 12 or even fewer and in order to determine the line completely we will have to add to these 16 points one or two points.[14]

We will see that it took more than a century for the concept of rank to come to its maturity.

The year 1750 is also the year when Cramer published the treatise that introduced the use of determinants, which were to dominate the study of linear equations until the first quarter of the twentieth century. In this context, dependence of equations but also of n-tuples was characterized by the vanishing of the determinant. However, determinants produced technical tools that were not always appropriate for easily approaching questions like the relation between the size of the set of solutions and the number of relations of dependence between the equations, as Euler had done in a very intuitive approach. On the other hand, the conception of inclusive dependence, which was still dominant, prevented equations and n-tuples being treated as identical objects with regards to linearity. Thus, this was an obstacle to the use of duality reasoning.

But the concept of rank is intrinsically connected to duality, since it is not only an invariant of a subspace but also of its orthogonal complement (i.e. its representation by equations). In the context of equations, to make all the aspects of rank explicit, it is necessary to consider all the equivalent systems to a given system of homogeneous linear equations and to show that its set of solutions cannot be represented by fewer than a certain number (the rank of the system) of equations. This process uses duality (see Dorier 1993). Therefore abandoning inclusive dependence for linear dependence was a necessary and decisive step toward the determination of the concept of rank.

In the second half of the nineteenth century many mathematicians made great progress toward this goal. But they used very elaborate technical tools within the theory of determinants and never really made anything explicit. All the works that we have analyzed in this period aimed to give a rule for solving any system of linear equations in the most concise manner. The idea was to choose one of the nonzero minors of maximal order r. The r unknowns and r equations involved in this minor were called the main equations and the main unknowns.[15] One can apply Cramer's rule to the r main equations, the remaining unknowns being put on the other side of the equality with the constant terms. Therefore, the value of the r main unknowns are given as functions of the other unknowns. By substitution of these values in the other equations, one gets the conditions of consistency of the system. This practical rule for solving a system made the value of r central in the relation between the size of the set of solutions and the number of relations of dependence among the equations. Yet the specific choice of the minor had to be discussed with regard to the invariance of the result. In order to abstract the concept of rank from this technique, one had to take a more general point of view and take up the dual problem of considering the set of solutions in connection with all systems having this set of solutions.

Frobenius made the decisive improvement in 1875. In his article about the problem of Pfaff, there is an important section about linear systems. He considers a general system[16] of homogeneous linear equations:

$$a_1^{(\mu)}u_1 + a_2^{(\mu)}u_2 + \cdots + a_n^{(\mu)}u_n = 0, \quad (\mu = 1, \ldots, m)$$

Immediately, he gives the definition of linear independence of k solutions of this system of linear homogeneous equations:

> Several particular solutions
>
> $$A_1^{(\chi)}, A_2^{(\chi)}, \ldots, A_n^{(\chi)} \quad (\chi = 1, 2, \ldots, k)^{17}$$
>
> will be said to be *independent* or *different,* if $c_1 A_\alpha^{(1)} + c_2 A_\alpha^{(2)} + \cdots + c_k A_\alpha^{(k)}$ cannot be all zero for $\alpha = 1, 2, \ldots, n$, without $c_1, c_2, \ldots, c_k$ being all zero, in other words if the k linear forms $A_1^{(\chi)}u_1 + A_2^{(\chi)}u_2 + \cdots + A_n^{(\chi)}u_n$ $(\chi = 1, \ldots, k)$ are independent.[18]

Not only is this a definition quite similar to the modern definition of linear independence but it also explicitly shows the similarity between n-tuples of solutions and equations with regard to their linearity. This simple idea allows Frobenius to give in a couple of pages a full overview of the properties of the rank of a system (still defined implicitly[19] as the maximal order of a nonzero minor). The main idea is to use the concept of an associated (*zugeordnet* or *adjungirt*) system, which is, in modern terms, the representation by equations of the orthogonal subspace.

Let us consider the system[20]

$$\begin{aligned} a_{11}x_1 + \cdots + a_{1n}x_n &= 0 \\ \cdots & \\ a_{p1}x_1 + \cdots + a_{pn}x_n &= 0. \end{aligned} \tag{I}$$

If $(A_1^{(\chi)}, A_2^{(\chi)}, \ldots, A_n^{(\chi)})$ $(\chi = 1, 2, \ldots, n - r)$, with r being the maximal order of nonzero minors, is a basis of solutions of (I), the associated system is:

$$\begin{aligned} A_1^{(1)}x_1 + \cdots + A_n^{(1)}x_n &= 0 \\ \cdots & \\ A_1^{(n-r)}x_1 + \cdots + A_n^{(n-r)}x_n &= 0. \end{aligned} \tag{I*}$$

Again if $(B_1^{(\nu)}, B_2^{(\nu)}, \ldots, B_n^{(\nu)})$ $(\nu = 1, 2, \ldots, q)$ is a basis of solutions of (I*), then the associated system is:

$$\begin{aligned} B_1^{(1)}x_1 + \cdots + B_n^{(1)}x_n &= 0 \\ \cdots & \\ B_1^{(q)}x_1 + \cdots + B_n^{(q)}x_n &= 0 \end{aligned} \tag{I**}$$

Frobenius proves that, whatever the choice of basis, (I**) is equivalent to (I) and $q = r$.

This first result on duality in finite-dimensional vector spaces shows the double level of invariance connected to rank both for the system and for the set of solutions. Moreover, Frobenius' approach allows a system to be seen as part of a class of equivalent systems having the same set of solutions, a fundamental step toward the representation of subspaces by equations.

This short analysis[21] shows how adopting a formal definition (here of linear dependence and independence) may be a fundamental step in the construction of a theory, and is therefore an essential intrinsic constituent of this theory and not only a change of style. It also points out several epistemological difficulties attached to the concepts of linear dependence and independence.

3) Epistemological synthesis.

In his work, R. Ousman (1996) gave a test to students in their final year of *lycée* (just before entering university). Through this test, he wanted to analyze the students' conception of dependence in the context of linear equations and in geometry before the teaching of the theory of vector spaces. He gave several examples of systems of linear equations and asked the students whether the equations were independent or not. The answers showed that the students justified their answers through the solving of the system and very rarely by mentioning linear relations between the equations. In other words they very rarely give a justification in terms of linear combinations but most of the time in terms of equations vanishing or unknowns remaining undetermined. Their conception of (in)dependence is, like Euler's, that of inclusive dependence and not linear dependence. Yet, this is not surprising, because these students, like Euler and the mathematicians of his time, are only concerned with the solving of the system. Therefore, inclusive dependence is more natural and more relevant for them.

This established fact and our previous historical analysis lead us to a pedagogical issue. When entering university, students already have ideas about concepts like linear (in)dependence in several contexts;[22] when they learn the formal concept, they have to understand the connection with their previous conceptions. If not, they may have two ideas of the same concept and yet not know clearly that they refer to the same concept. Moreover, making the connection helps in giving the formal concept a better intuitive foundation. Yet, the students must also understand the role of the formal concept and have an idea of the improvement it brings. In the case of linear algebra, and more specifically of linear (in)dependence, the formal concept is the only means to comprehend all the different types of "vectors" in the same manner with regard to their linearity. In other words, students must be aware of the unifying and generalizing nature of the formal concept. Therefore we build teaching situations leading students to reflect on the epistemological nature of the concepts with explicit reference to their previous knowledge (Dorier 1992, 1995b and 1997 and Dorier *et al.* 1994a and b). In this approach, the historical analysis is a source of inspiration as well as a means of control. Yet, these activities must not be only words from the teacher, nor a reconstruction of the historical development; they must reconstruct an epistemologically controlled genesis taking into account the specific constraints of the teaching context.

For instance, with regard to linear (in)dependence, French students entering university normally use Gaussian elimination for solving systems of linear equations. It is therefore possible in the beginning of the teaching of linear algebra to make them reflect on this technique not only as a tool but also as a means to investigate the properties of systems of linear equations. This does not conform to the historical development, as the study of linear equations was historically mostly within the theory of determinants. Yet, Gaussian elimination is a much less technical tool and a better way for showing the connection between inclusive dependence and linear dependence, because identical equations (in the case when the equations are dependent) are obtained by successive linear combinations of the initial equations. Moreover, this is a context in which such question as "what is the relation between the size of the set of solutions of a homogeneous system and the number of relations of dependence between the equations?" can be investigated with the students as a first intuitive approach to the concept of rank.[23] M. Rogalski has experimented with teaching sequences illustrating these ideas (Rogalski 1991, Dorier *et al.* 1994a and b and Dorier 1992, 1997 and 2000).

Here is for instance an exercise given to the students to illustrate this idea:

> A magic square of order 4 and of sum zero is a square matrix of order 4 with real coefficients such that the sum along each column, each row and both diagonals is zero. Without any calculation, give an evaluation as precisely as possible of the number of entries that you can freely chose in any magic square of order 4 and of sum zero.

The number that the students have to evaluate is of course the dimension of the space of magic squares of order 4 and sum zero. It is less than 16 (the number of entries in a square of side 4) and more than $16 - 10 = 6$, because the coefficients are solutions of a system of 10 equations. To be more precise one must know the number of dependences, i.e., the rank, of the equations. Without calculation, it is easy to see that at least three equations are independent (for instance the three equations expressing that the sum along each of the first three rows is zero), so that the rank is at least 3 and the dimension is therefore less than $16-3 = 13$. It would not be very difficult to be even more precise.

This exercise is interesting because it operates with large dimensions, yet the equations are quite simple. Moreover its concrete framework makes the question more accessible to the students, even if they do not know the formal concepts of dimension and rank. It is important though to prevent tedious calculations and to emphasize intuitive reasoning even if the outcome is less accurate.

On another level, the historical and pedagogical analyses confirm the fact that there is an epistemological difficulty in treating the concept of linear (in)dependence as a global property (remember the distinctions made by Euler). It follows that special care must be taken in the teaching regarding this point. For instance, exercises such as the first one analyzed above can be discussed with the students. Moreover, the teacher, knowing the type of *théorèmes-en-acte* that students may have built, must help them in understanding their mistakes and therefore in correcting them more efficiently.

Finally we will give the outline of a teaching experiment that we have used for the final step when introducing the formal theory after having made as many connections as possible with previous knowledge and conceptions in order to build better intuitive foundations.

After the definition of a vector space and subspace and linear combination, the notion of generator is defined. Because a set of generators gathers all the information we have on the subspace, it is therefore interesting to reduce it to the minimum. Therefore, the question is to know when it is possible to take away one generator, the remaining vectors still being generators for the whole subspace. The students easily find that the necessary and sufficient condition is that the vector to be taken away must be a linear combination of the others. This provides the definition of linear dependence: "a vector is linearly dependent on others if and only if it is a linear combination of them". This definition is very intuitive, yet it is not completely formal, and it needs to be specified for sets of one vector. It provides without difficulty the definition of a set of independent vectors as a set of which no vector is a linear combination of the others.

To feel the need for a more formal definition, one just has to apply this definition. Indeed, students must answer the question: "are these vectors independent or not?". With the definition above, they need to check that each vector, one after the other, is a linear combination of the others. After a few examples, with at least three vectors, it is easy to explain to the students that it would be better to have a definition in which all the vectors play the same role. (It is also interesting to insist on the fact that this is a general statement in mathematics.) One is now ready to transform the definition of linear dependence into "vectors are linearly dependent if and only if there exists a zero linear combination of them, whose coefficients are not all zero." The definition of linear independence being the negation of this, it is therefore a pure problem of logic to reach the formal definition of linear independence. A pure problem of logic, but in a precise context, where the concepts have made sense to the students with an intuitive background.

This approach has been proven to be efficient with regard to the students' ability to use the definitions of linear dependence and independence, even in formal contexts such as in the three exercises discussed in section II.1. Moreover, it is quite a discovery for the student to realize that a formal definition may be more practical than an "intuitive" one. In Behaj's work, quoted above, it was clearly shown that this fact is not clear for many students and even for some of their teachers. Most of them keep seeing the fact that a vector is a linear combination of the others as a consequence of the definition of linear dependence. Therefore they believe that this consequence is the practical way of proving that vectors are or are not independent, even if that goes contrary to their use of these definitions. Yet, imagine that one has to check whether three vectors u, v, and w are independent. If one proves that u is not a linear combination of v and w, there is still a chance that v and w would be collinear in which case the three vectors are dependent. Many students would conclude from the first step that the three vectors are independent. In doing so they would not be easily contradicted, because in most of the cases the result is true.

One can easily imagine that starting by proving that one vector is not a linear combination of the others will get even more dangerous with more than three vectors. In theory, if one applies this definition, one has to check that each vector is not a linear combination of the others. There are shortcuts (like in our example, if you check that v and w are not collinear) but they require students to have a good control of the meaning of linear dependence. Therefore the formal definition is a good means to prevent false justifications. On the other hand, it is useful when one knows that some vectors are dependent to use it through the fact that one vector is a linear combination of the others because this is what is meaningful.

4) Conclusions.

Formalism is what students themselves confess to fear most in the theory of vector spaces. One pedagogical solution is to avoid formalism as far as possible, or at least to make it appear as a final stage gradually. Because we think that formalism is essential in this theory (our historical analysis has confirmed this epistemological fact), we give a different answer: formalism must be presented in relation to intuitive approaches as the means of understanding the fundamental role of unification and generalization of the theory. This has to be an explicit goal of the teaching. This is not incompatible with a gradual approach toward formalism, but it induces a different way of thinking out the previous stages. Formalism is not only the final stage in a gradual process in which objects become more and more general; it must appear as the only means of comprehending different aspects within the same language. The difficulty here is to give a functional aspect to formalism while approaching it more intuitively.

Linear dependence is a formal notion that unifies different types of dependence which interact with various previous intuitive conceptions of the students. It has been shown above how in the historical development of linear algebra the understanding of this fact was essential for the construction of the concepts of rank and duality. In teaching, this questioning has to be made explicit, if we do not want misunderstanding to persist. Therefore, even at the lowest levels of the theory, the question of formalism has to be raised in connection with various contexts where the students have built previous intuitive conceptions. Formalism has to be introduced as the answer to a problem that students are able to understand and to make their own, in relation to their previous knowledge, in fields where linear algebra is relevant. These include at least geometry and linear equations but may also include polynomials or functions, although in those fields one may encounter more difficulties.

III. Conclusion

From the example of linear (in)dependence, we can now draw some conclusions about the epistemological reflection we have conducted on the bases of our didactical and historical research. In the experimental teaching, we did not use historical texts directly with the students, even though we refer sometimes to historical facts, for instance when introducing new concepts. Moreover, in the case of linear algebra, our analysis shows that teaching may gain from taking some distance from the historical order of development. Indeed, the unifying and generalizing aspects of the theory of vector spaces is not only a fundamental character of the theory but is also a very recent one—it only began to be used in the 1930s. Moreover, the final developments were linked essentially with functional analysis and brought out issues which are beyond the mathematical background of first year university students.

All the subjects we can model with vector spaces at this level of teaching have historically been solved with other tools, mostly the theory of determinants. Yet, for reasons we have explained above, we think that Gaussian elimination is a much better adapted tool than determinants for studying linear equations and introducing basic notions of linear algebra. Thus the artificial genesis that our work led us to build for the teaching of linear algebra differs in many ways from the historical genesis. It is generally admitted that teaching should not and cannot reproduce all the historical aspects of the development of a mathematical discipline. Sometimes the constraints are only due to the limitations of time and cognitive or institutional aspects of the teaching situation; in the case of linear algebra there is a more complex epistemological constraint. Nevertheless, although artificial, the genesis induced by the teaching has to take the historical development into account. It cannot be based only on logical constraints, as it may have been at the beginning of modern mathematics. It must also take into account the whole history of linear algebra as far as possible, even in its latest transformations. This implies that the formal aspect of the theory must appear as a final stage of maturity, in a context where it makes sense. We have tried to show how this can be done concerning the concept of linear (in)dependence. The historical analysis is then a fundamental tool at least on two levels :

(1) It provides a source of inspiration and an epistemological control for the building of an artificial genesis.
(2) It helps in understanding and analyzing the mistakes of the students. An erratic use of formal tools (using logic and set theory) can then be interpreted as a missing connection in the ontological process. This is fundamental, as an erratic mistake is usually corrected by the teacher without further comment, and is therefore likely to reappear. But, if the teacher can locate the missing connection in the ontological process leading to the mistake, a much better remediation is possible.

Moreover, the historical analysis interacts with the pedagogical analysis on a more global level. For instance, in the case of linear algebra, it shows the necessity of interactions between different frameworks and registers of representation. In our experimental teaching, we have tried to make the student more aware of this possibility and its importance for a better understanding of formal concepts. In her doctoral dissertation, M. Dias (1998)—see also Alves Dias and Artigue(1995)—has shown the lack of use of changes of framework, register of representation and point of view in traditional teaching. She is now building and evaluating the effects of teaching sequences in order to encourage *cognitive flexibility* in the students. The analysis of the historical role of the different uses of changes of framework, register of representation, point of view or style (Granger 1995) may be very interesting work to support this type of pedagogical analysis. We have already analyzed the role of the geometrical framework in the emergence of vector spaces of functions (Dorier 1996), and are planning to enlarge this type of analysis to different aspects of the genesis of linear algebra.

We think that there are many means of interaction between historical and pedagogical analyses. Both of them provide epistemological reflections which are complementary, and both for historical and pedagogical research one benefits by trying to emphasize the similarity and complementarity of approach. We hope that this paper has shown the relevance of this statement in a very specific context.

References

Alves Dias, M., 1998: *Problèmes d'articulation entre points de vue "cartesian" et "paramétrique" dans l'ensiegnement de l'algèbre linéaire,* Doctoral dissertation, Paris 7 University.

Alves Dias, M. and Artigue, M., 1995: Articulation problems between different systems of symbolic representations in linear algebra, in *The proceedings of the 19th annual meeting of the international group for the Psychology of Mathematics Education,* Recife (Brazil): Universidade Federal de Pernambuco, 3 vol., 2: 34–41.

Behaj, A., 1999: *Eléments de structurations à propos de l'enseignement et l'apprentissage à long terme de l'algèbre linéaire,* Doctoral dissertation, Dhar El Mehraz University, Fès, Morocco.

Behaj, A. and Arsac, G., 1998: "La conception d'un cours d'algèbre linéaire," *Recherches en Didactique des Mathématiques* 18 (3) 333–370.

Cramer, G., 1750: *Introduction á l'analyse des courbes algébriques,* Genève: Cramer et Philibert.

Dorier, J.-L., 1990a: *Contribution à l'étude de l'enseignement à l'université des premiers concepts d'algèbre linéaire. approches historique et didactique,* Thèse de Doctorat de l'Université J. Fourier-Grenoble 1.

——, 1990b: *Analyse dans le suivi de productions d'étudiants de DEUG A en algèbre linéaire,* Cahier DIDIREM n°6, IREM de Paris VII.

——, 1990c: *Analyse historique de l'émergence des concepts élémentaires d'algèbre linéaire,* Cahier DIDIREM n°7, IREM de Paris VII.

——, 1991: Sur l'enseignement des concepts élémentaires d'algèbre linéaire à l'université, *Recherches en Didactique des Mathématiques* 11(2/3), 325–364.

——, 1992: *Illustrer l'Aspect Unificateur et Généralisateur de l'Algèbre Linéaire,* Cahier DIDIREM n°14, IREM de Paris VII.

——, 1993: L'émergence du concept de rang dans l'étude des systèmes d'équations linéaires, *Cahiers du séminaire d'histoire des mathématiques* (2e série) 3 (1993), 159–190.

Dorier, J.-L., Robert A., Robinet J. and Rogalski M. (1994a): L'enseignement de l'algèbre linéaire en DEUG première année, essai d'évaluation d'une ingènierie longue et questions, in *Vingt ans de Didactique des Mathématiques en France,* Artigue M. *et al.* (eds), Grenoble : La Pensée Sauvage, 328–342.

Dorier, J.-L, Robert, A., Robinet, J. and Rogalski, M., 1994b: The teaching of linear algebra in first year of French science university, in *Proceedings of the 18th conference of the international group for the Psychology of Mathematics Education,* Lisbon, 4 vol., 4: 137–144.

Dorier, J.-L, 1995a: A general outline of the genesis of vector space theory, *Historia Mathematica* 22(3), 227–261.

——, 1995b: Meta lever in the teaching of unifying and generalizing concepts in mathematics, *Educational Studies in Mathematics* 29(2), 175– 197.

——, 1996a: Basis and Dimension: From Grassmann to van der Waerden, in G. Schubring (ed.), *Hermann Günther Grassmann (1809–1877): Visionary Mathematican, Scientist and Neohumanist Scholar,* Boston Studies in the Philosophy of Science 187, Dordrecht: Kluwer Academic Publisher, 175–196.

——, 1996b: Genèse des premiers espaces vectoriels de fonctions, *Revue d'Histoire des Mathématiques* 2, 265–307.

—— (ed.) 1997: *L'enseignement de l'algèbre linéaire en question,* Grenoble: La Pensée Sauvage.

—— (ed), 2000: *The Teaching of Linear Algebra in Question,* Dordrecht: Kluwer Academics Publishers.

Euler, L., 1750: Sur une contradiction apparente dans la doctrine des lignes courbes, *Mémoires de l'Académie des Sciences de Berlin* 4, 219– 223, or, *Opera omnia,* Lausanne: Teubner-Orell Füssli-Turicini, 1911–76, 26:33–45.

Frobenius, G. F., 1875: Über das Pfaffsche Problem, *Journal für die reine und angewandte Mathematik* 82, 230–315.

——, 1879: Über homogene totale Diffrentialgleichungen, *Journal für die reine und angewandte Mathematik* 86, 1–19.

Granger, G. G., 1988: *Essai d'une philosophie du style,* Paris : Editions Odile Jacob.

Ousman, R., 1996: *Contribution à l'enseignement de l'algèbre linéaire en première année d'université,* Thèse de doctorat de l'université de Rennes I.

Robert, A. and Robinet, J., 1989: *Quelques résultats sur l'apprentissage de l'algèbre linéaire en première année de DEUG,* Cahier de Didactique des Mathématiques 53, IREM de Paris VII.

Rogalski, M., 1990a: Pourquoi un tel échec de l'enseignement de l'algèbre linéaire?, in *Enseigner autrement les mathématiques en DEUG Première Année,* Commission inter-IREM université, 279–291.

——, 1991: *Un enseignement de l'algèbre linéaire en DEUG A première année,* Cahier de Didactique des Mathématiques 53, IREM de Paris VII.

——, 1994: L'enseignement de l'algèbre linéaire en première année de DEUG A, *La Gazette des Mathématiciens* 60, 39–62.

Vergnaud, G., 1990: La théorie des champs conceptuels, *Recherches en Didactiques des Mathématiques* 10(2/3), 133–170.

Endnotes

[1] See, e.g., the special issue on linear algebra of *The College Mathematics Journal* 24 (1) (1993).

[2] Unlike secondary schools, for which a national program is imposed by the ministery of education, the curricula of universities are decided locally, even though differences from one university to the other are usually superficial.

[3] The results of this work are now gathered in a book (Dorier 1997) with the presentation of other works (in France, Canada, USA and Morocco). (Dorier 2000) is a revised English version of this book.

[4] I will not say more in this paper about the difficulties students may have in carrying out these standard tasks, not because there are none but because I will focus on another level of difficulty, less technical and more conceptual.

[5] The three examples given here have been tested by A. Robert and J. Robinet (1989). They are not original exercises, but they reveal important recurrent mistakes of the students.

[6] This is what French didacticans will call "un effet de contrat".

[7] For instance when asked what is the intersection of the two subspaces generated by U_1, U_2, and V_1, V_2, students prove that neither U_1 nor U_2, are a linear combination of V_1 and V_2, and conclude that the intersection is reduced to 0.

[8] "On n'a qu'à regarder ces deux équations: $3x - 2y = 5$ et $4y = 6x - 10$ et on verra d'abord qu'il n'est pas possible d'en déterminer les deux inconnues x et y, puisqu'en éliminant l'une x, l'autre s'en va d'elle-même et on obtient une équation identique, dont on est en état de ne déterminer rien. La raison de cet accident saute d'abord aux yeux puisque la seconde équation se change en $6x - 4y = 10$, qui n'étant que la première doublée, n'en diffère point." [Euler 1750, 226]

[9] Euler does not mean here that the two equations are the same but that each of them is an identity, like $0 = 0$, i.e., always true.

[10]
$$"5x + 7y - 4z + 3v - 24 = 0,$$

$$2x - 3y + 5z - 6v - 20 = 0,$$

$$x + 13y - 14z + 15v + 16 = 0,$$

$$3x + 10y - 9z + 9v - 4 = 0,$$

elles ne vaudroient que deux. car ayant tiré de la troisième la valeur de

$$x = -13y + 14z - 15v - 16$$

et l'ayant substituée dans la seconde pour avoir:

$$y = (33z - 36v - 52)/29 \text{ et } x = (-23z + 33v + 212)/29,$$

ces deux valeurs de x et de y étant substituées dans la première et la quatrième équation conduiront à des équations identiques, de sorte que les quantités z et v resteront indéterminés. [*ibid.*, 227]

[11] " La première ne différant pas de la troisiéme, ne contribue en rien à la détermination des trois inconnues. Mais il y a aussi le cas, où une des trois équations est contenue dans les deux autres conjointement ... Ainsi quand on dit que pour déterminer trois inconnues, il suffit d'avoir trois équations, il y faut rajouter cette restriction, que ces trois équations diffèrent tellement entr'elles, qu'aucune ne soit déjà contenue dasn les deux autres." [*ibid.*, 226]

[12] This is the usual term of the time to designate what is today known as an algebraic curve of order 4.

[13] Euler uses this ambiguous formulation several times to express the fact that the equations are independent. It is another proof of the difficulty of the global aspect of dependence as well as of the difference between linear and inclusive dependence.

[14] "Quand deux lignes du quatrième ordre s'entrecoupent en 16 points, puisque 14 points, lorsqu'ils conduisent à des équations toutes différentes entr'elles, sont suffisants pour déterminer une ligne de cet ordre, ces 16 points seront toujours tels que trois ou plusieurs des équations qui en résultent sont déjà comprises dans les autres. De sorte que ces 16 points ne déterminent plus que s'il n'y en avoit que 13 ou 12 ou encore moins et partant pour déterminer la courbe entièrement on pourra encore à ces 16 points ajouter un ou deux points." [*ibid.*, 233]

[15] The terms "*équations principales*" and "*inconnues principales*" used by Rouché and Fontenay is very popular in France. Yet it seems that no English equivalent has been used systematically.

[16] In fact he starts with a system of independent equations and then generalizes to any kind of system.

[17] These are k solutions. We would maybe note today $A^{(\chi)} = (A_1^{(\chi)}, A_2^{(\chi)}, \ldots, A_n^{(\chi)})$ each of them. Therefore: $a_1^{(\mu)} A_1^{(\chi)} + a_2^{(\mu)} A_2^{(\chi)} + \cdots + a_n^{(\mu)} A_n^{(\chi)} = 0$, for each $\mu = 1, \ldots, m$ and each $\chi = 1, 2, \ldots, k$.

[18] "Mehrere particuläre Lösungen

$$A_1^{(\chi)}, A_2^{(\chi)}, \ldots, A_n^{(\chi)} \quad (\chi = 1, 2, \ldots, k)$$

sollen daher *unhabhängig* oder *verschieden* heissen, wenn $c_1 A_\alpha^{(1)} + c_2 A_\alpha^{(2)} + \cdots + c_k A_\alpha^{(k)}$ nicht für $\alpha = 1, 2, \ldots n$, verschwinden kann, ohne dass $c_1, c_2, \ldots, c_k$ sämmtlich gleich Null sind, mit andern worten wenn die k linearen Formen $A_1^{(\chi)} u_1 + A_2^{(\chi)} u_2 + \cdots + A_n^{(\chi)} u_n$ $(\chi = 1, \ldots, k)$ unhanbhängig sind." [Frobenius 1875, 255]

[19] The term of rank (*Rang*) will be introduced for the first time in [Frobenius 1879, 1] .

[20] This is a summary of Frobenius' ideas with adapted notations and vocabulary.

[21] For more detail see [Dorier 1993 and 1997].

[22] We only mentioned the context of equations here, but it is clear that they have conceptions in other contexts. In geometry, if they know the vectors (as it is the case in France) their conception will be close to the formal concept, yet, they had a conception prior to vectorial geometry, connected to the ideas of alignment and coplanarity.

[23] As far as we know Gauss himself never came near such an investigation.

Presenting the Relation between Mathematics and Physics on the Basis of their History: a Genetic Approach[1]

Constantinos Tzanakis
University of Crete
Rethymnon, Greece

1. Introduction

There are quite divergent opinions about the role the history of mathematics could play in the presentation[2] of mathematics itself. A very common attitude is simply to ignore it, arguing that a deductive approach is better suited for this purpose, since in this way all concepts, theorems and proofs can be introduced in a clearcut way. On the other extreme, a rather naive attitude is to follow the historical development of a mathematical discipline as closely as possible, presumably using original books, papers, and so on. It is clear that both methods have serious defects.

In a strictly deductive approach, the motivation for the introduction of new concepts, theories or proofs, is hidden, hence a deeper understanding is not easily acquired. Moreover, in such an approach the emphasis is more on the results and less on the questions and problems that led to them. "From a logical point of view, only the answers are needed, but from a psychological point of view, learning the answers without knowing the questions is so difficult that it is almost impossible" ([14] p. vii). Finally, after maturity has been reached for a mathematical discipline, its deductive—or even strictly axiomatic—presentation is most suited to reveal its logical structure and completeness and it can be useful to those knowing the subject, or at least having enough acquaintance with it or with other related subjects.

On the other hand, a strictly historical approach is not didactically appropriate,[3] since, contrary to what is sometimes naively assumed, the historical evolution of a scientific domain is almost never straightforward and cumulative. It involves periods of stagnation and confusion, and new concepts or proofs are not introduced in the simplest and most transparent way. Actually, a historical presentation is primarily concerned with "an accurate record of the main ideas and events which played a part in the evolution of the subject", examining problems and theories only to the extent that they are essential to an understanding of these events ([14] p. vi). We shall adopt an approach between these two extremes, in which knowledge of the basic steps of the historical development of a mathematical subject plays an essential role in the presentation ([28], [52], [53]). We may call such an approach genetic, since it is neither strictly deductive, nor strictly historical, but its fundamental thesis is that a subject is presented only after one has been motivated enough to do so, in the sense that questions and problems which this presentation may answer have been sufficiently appreciated. Moreover, in such an approach, emphasis is less on how to use theories, methods and concepts and more on why these theories, methods and concepts provide an answer to specific mathematical problems and questions.

The term genetic is here used in a sense very close to that used by Toeplitz already since 1926, and it appears systematically in [51] and in a more or less similar way in [14], [37], [38], [48]. More specifically, in a genetic approach it is often helpful to follow the historical evolution of the subject in the sense that

(i) the teacher or author has a basic knowledge of the subject's history;
(ii) the crucial steps of the historical development have been thus appreciated;
(iii) key ideas and problems that stimulated this evolution are reconstructed in a modern context so that they become didactically appropriate for the introduction of new concepts, methods or theories. This usually implies that the historical evolution is not respected in any strict sense (see particularly the example in section 3.2);
(iv) many details of this reconstruction are given as exercises, which, in this way, become essential for a full understanding of the subject (see below).

In this way it is possible

(a) to recognize conceptual, epistemological or even philosophical obstacles that appeared in the historical process and which may reappear in the teaching process and to decide whether and to what extent a subject can be presented at a particular level of instruction (see [5] p. 178 and section 3.1).

(b) to use examples that served as prototypes in the historical process, thus giving the student the opportunity to understand the motivation behind the introduction of a new concept, theory, method, or proof and grasp its content in a more profound way.[4] In addition, the student is thus encouraged to formulate conjectures and to determine why his conjectures, or other similar ones that have been put forward in the past, do or do not supply satisfactory answers to the already existing problems (see section 3.4).

(c) to interrelate domains which at first glance appear completely different, thus making it possible to appreciate the fact that fruitful research activity in a mathematical domain never stands in isolation to similar activities in other domains, but, on the contrary, is often motivated by questions and problems coming from apparently unrelated disciplines (see section 3.3).

(d) to make problem solving an essential ingredient of the presentation, indispensable for a complete understanding of the subject; many details of the historically relevant examples can be restated in sequences of exercises of an increasing level of difficulty, so that each one presupposes (some of) its predecessors. In this way the student has the possibility of arriving at presumably nontrivial results, starting from easy corollaries of the main subject; at the same time the size of a textbook or the teaching time is kept at a reasonable level (see section 3.2). This is perhaps the most natural way to acquire the "technical" knowledge of how to calculate correctly and quickly ([54] p. 281). In addition, by avoiding in this way the explicit treatment of all the details, the frequently raised objection against this presentation, namely, that it presupposes a good deal of time on the part of the teacher or author, is thus refuted. However, one should be careful not to abuse this point, for instance, by presenting fundamental concepts, theorems or proofs, central to the subject, through a sequence of exercises just to save teaching time or printed pages of a book!

Although the above mentioned four points give a brief outline of what can be achieved, it should be emphasized that, in a genetic approach, there is no uniquely specified way of presentation of a given subject. Therefore, it is not a method, in the strict sense of an algorithm. Rather, it is a general attitude towards the presentation of scientific subjects in which the desire prevails to explain the motivation behind the introduction of new concepts, theories, or key ideas of proofs on the basis of the historical evolution of the subject.

From what has already been said, it is clear that a genetic approach is not restricted to mathematics only, but that it can be applied to any scientific domain capable of deductive presentation, in particular to physics. Thus, in what follows we will illustrate this approach by considering in some detail examples taken from the historical development of mathematics and physics, which at the same time emphasize the close relationship of these two sciences. Consequently, in the next section a brief account of how this relationship appears historically is given, whereas in section 3 we pass to specific, historically important, scientifically relevant, and didactically appropriate examples.

2. On the relation between mathematics and physics

It must be admitted that, at any level of education, there is often a strict separation between mathematics and physics. This fact reflects a corresponding separation of these sciences at the research level, since often physicists do not accept the way mathematicians think. They argue that the latter always stay in a universe of ideal logical rigor, having nothing to do with the real world. On the other hand, mathematicians are suspicious of physicists, characterizing them as simple-minded users of mathematics who do not pay sufficient attention to logical completeness and rigor. However, such ideas are not compatible with the fact that, throughout their history, the two sciences have always had a close relationship clearly revealed in the work of great mathematicians, whose contributions to physics rival their purely mathematical works.[5] Evidently we do not mean that the mathematical activity is or should be motivated only by its applications, for instance to physics; we simply want to stress the fact that their close relationship indicates that any difference is not so much in method or problems, but mainly in their purpose, [19] ch. II. Historically this relationship between mathematics and physics appears in three different ways:

(a) There is a parallel development of physical theories and the appropriate mathematical framework, often due to the same persons. Typical examples are (i) the foundation of infinitesimal calculus and classical mechanics in the seventeenth century, mainly by Newton and Leibniz (ii) the development of vector analysis in the second half of the nineteenth century, by Maxwell, Gibbs, Heaviside and others, in connection with Maxwell's electromagnetic theory, [7].

(b) New mathematical theories, concepts or methods are formulated in order to solve already existing physical problems, or to provide a solid foundation to methods and concepts of physics. We may think here of classical Fourier analysis, emerging from the partial differential equation of heat conduction; or, more recently (i) Dirac's delta function in quantum mechanics, [11] ch. 3, and its later clarification in the context of L. Schwarz's theory of generalized functions [41] (ii) the foundation of ergodic theory in the 20's and 30's mainly by G. Birkhoff, J. von Neumann and E. Hopf ([17] ch. 3, [25] ch. 3), motivated by the problems posed by the introduction in the second half of the nineteenth century of Boltzmann's ergodic hypothesis in classical statistical mechanics ([15] ch. 2, [2] p. 10–12, part II section 32).

(c) The formulation of a mathematical theory precedes its physical applications. Its use is often made after the corresponding physical problems naturally indicate the necessity of an appropriate mathematical framework. Typical examples here are (i) Einstein's work on the foundations of the general theory of relativity in the period 1907–1916, on the basis of Riemannian geometry and tensor analysis developed in the nineteenth century by Riemann, Christoffel, Ricci, Levi-Civita and others ([35] ch. 12, [50] p. 167–168, [47] ch. 4) (ii) the invention of Heisenberg's matrix mechanics in 1925, in which he realized on the basis of purely empirical (spectroscopic) facts that atomic magnitudes have the algebraic structure of (infinite dimensional) complex matrices ([59], [22] Appendix, [34] ch. III).

The above examples are indicative of the intimate connection between mathematics and physics. Therefore, in the light of the discussion in section 1, while presenting them, their relationship cannot be ignored, but, on the contrary, it should be unfolded as much as possible. Their relationship may be used in both directions, either by motivating the introduction of abstract general mathematical concepts, methods or theories on the basis of physical problems to which the new mathematics provides an answer (section 3.4), or by using mathematical arguments to clarify the meaning of new physical ideas or theories (sections 3.2, 3.3).

As an illustration of how this can be done, certain examples are analyzed in the next section.

3. Specific examples

As already mentioned in section 1, points (a)–(d) constitute general statements about what can be achieved by using history in teaching mathematics. In what follows, these four points will be illustrated by outlining a genetic presentation of particular examples emphasizing the intimate relationship between mathematics and physics; specifically, in connection with

(a) the recognition of conceptual difficulties and deciding whether a subject can be taught at a particular level of instruction: The teaching of abstract algebraic concepts to last-year high school students;[6]

(b) the motivation for the introduction of new concepts, methods or theories: The introduction of concepts of functional analysis to undergraduate mathematics students, on the basis of quantum mechanical problems;

(c) the possibility to interrelate *a priori* different domains and concepts and motivate the invention of new mathematical methods and physical theories: Hamilton's unified treatement of geometrical optics and classical mechanics, which influenced the discovery of quantum mechanics, as it can be incorporated in an undergraduate course on partial differential equations (PDE) and the calculus of variations, classical or quantum mechanics, or optics;

(d) the role of problem solving: Presenting to last year high school students the foundations of the Special Theory of Relativity as an application of matrix algebra.

3.1 Recognizing conceptual difficulties: The teaching of abstract algebraic concepts. Here we consider the question of whether it is possible to teach algebraic concepts like group, ring, field and vector space in their full generality to last-year high school students, as was done in many countries for several years, under the influence of the "modern mathematics" current in mathematics education. We study this problem on the basis of the historical development of the group concept.

We first notice the following facts, confirmed by the author's experience of teaching the subject for three years in Greek schools:

(a) In all textbooks currently used in Greek schools, these abstract concepts are presented in their full generality while most examples are trivial 'algebraizations' of well-known algebraic properties of numbers accepted as self-evident (e.g., algebraic structures induced by ordinary point operations on real valued functions; cf. the scepticism expressed in [16] ch. 28). As a consequence of this, most teachers are forced to follow this approach.

(b) Students have serious difficulties solving exercises that are simple applications of the definitions, e.g., solve the exercise: "Show that the interval $(-1, 1]$ equipped with the operation

$$x * y = \frac{x+y}{1+xy}$$

is an abelian group" (cf. however, section 3.2 where this apparently artificial exercise finds its right place).

(c) Students cannot understand an algebraic structure in its abstract form; for instance, that in the same set different operations define different structures, e.g., $\mathbf{R}^2$ as a real vector space and as the field of complex numbers.

(d) Students do not understand the isomorphism concept, e.g., $\mathbf{R}^2$ with the structure of the field of complex numbers and $\mathbf{R}\cup\{i\}$ with the ordinary operations extended linearly.

The above is of more or less general validity; Greek mathematics teachers agree that these facts have been confirmed through personal experience ([56]). As a consequence, the official mathematics curriculum has recently been modified and abstract algebraic concepts play a less prominent role.

The main reason for these difficulties seems to be that such concepts are introduced in an abstract way and in full generality, at a moment when the students have not had experience with many concrete examples. On the other hand, a very rough knowledge of their historical development shows that algebraic structures are introduced and established usually when (i) many concrete problems have appeared, to which a particular structure offers a (not easily substituted) solution (e.g., the group concept and the solution by radicals of the nth degree polynomial equation) (ii) it becomes evident that there are sets with the same structure but having elements of a totally different nature (e.g., vectors in analytic geometry and solutions of linear ordinary differential equations).

As an example supporting (i), (ii) above, we give a very schematic and necessarily incomplete account of concrete and mathematically important problems and ideas which played a crucial role in the development and establishment of the group concept:

(1) The problem of solving by radicals the nth degree polynomial equation (Galois 1832—see, e.g., [4] p. 638ff, [31] sections 719–720; for more details see [10] vol. I, ch. II section III.B)

(2) Problems and ideas concerning transformations in geometry and physics.

(a) The geometric representation of complex numbers (Wessel 1797, Argand 1806, Gauss 1831), their representation as ordered pairs of real numbers (Hamilton 1833–1837) and their relation to plane rotations. These motivated the search for generalized complex numbers with a similar relation to space rotations. As is well known, this led to the first nontrivial noncommutative algebraic structure, namely quaternions [53]; and it is a fact that the significance of the properties of operations like commutativity, associativity, and so on, was appreciated only when structures were invented which do not obey them ([44] p. 678).

(b) The importance of the concept of a group of transformations in geometry and physics. Specifically

- Klein's Erlangen program, in which geometry is defined as a theory of invariants of a set under a group of transformations on it ([26] and the preface by Dieudonné and the historical analysis of its birth by P.F. Russo; see also [48] ch. 18, particularly section 18.5, [4] p. 591–592)
- Lie groups (of transformations) applied to geometry, differential equations and mechanics, as developed from about 1870 onwards, especially by Lie (see, e.g., [4] p. 591, [24] p. 157, [31] section 721; for a short account of the history of the theory of invariants in this context see [33] section 3.3; for a detailed treatment, [3] p. 309–323).
- Galileo, Lorentz and arbitrary (differentiable) coordinate transformations in relativity theory (Poincaré 1904–1905, Minkowski 1908, Einstein 1909–1915; see Einstein in [46] ch. VII especially section 3,[7] Minkowski in [46] ch. V especially sections I, II, [35] ch. 12 especially sections $12(c, d)$ and pp. 129–130).
- Isometric mappings of surfaces and more generally, of Riemannian manifolds, as they implicitly appear for instance in Gauss' Theorema Egregium (1827) and Riemann's generalization (1854) (see Riemann's original

lecture in [47], particularly pp. 144– 145 and 176, and the account given in [10] Vol. II, pp. 197–200)

(c) The development of matrix algebra by Cayley (1855, 1858; see, e.g., [4] section XXVI.6; for his predecessors see [10] vol. I, p. 95–97) and the associated linear representations of groups (Frobenius, Schur and Burnside, 1896–1910) stimulated by ideas of Cayley (1854), Weber (1884) and Dedekind (1896)—see [10] vol. I, pp. 117–119, [3] pp. 154–155.

In view of the above outline of the historical evolution of the group concept, it is clear that, in its full generality, it can only be appreciated when its significance in the context of many concrete and mathematically relevant examples has been revealed.[8] It is also evident that at the high school level this can be achieved only partially, especially by presenting examples of groups of transformations, and therefore teaching just the most basic and elementary facts on abstract groups is useless or even harmful! Actually the following remarks of F. Klein on abstract group theory have a more or less general validity in connection with teaching algebraic concepts in their abstract form right from the beginning: "[The] abstract formulation is excellent, but it is not at all directed to the discovery of new ideas and methods; rather it represents the conclusion of a preceding development. Hence it greatly facilitates instruction insofar as one can use it to give complete and simple proofs of known theorems. On the other hand, it makes the subject much more difficult for the learner, for he is faced with a closed system, not knowing how these definitions were arrived at and absolutely nothing is presented to his imagination . . . the method has the drawback that it does not stimulate thought" ([27] p. 316).

The main conclusions to be drawn from the above discussion are: At the high school level the student cannot have experience of many important concrete examples of such abstract algebraic concepts, an absolutely necessary requirement for the significance of these concepts to be really appreciated.[9] Therefore such concepts cannot and should not be taught in full generality at the high school level, but should be presented only implicitly through concrete and "natural" examples, like groups of invertible functions under composition, linear transformation groups in geometry and physics (cf. the next subsection), the group of nth roots of unity, etc.

3.2 The role of problem solving: Special Relativity and Matrix Algebra. As already mentioned in the previous subsection, in many countries, matrix algebra and elements of (finite dimensional) vector space theory are introduced in high school's last-year mathematics. Given that the aim and power of algebraic methods and concepts lies in the unified approach to distinct, quite different concrete problems through abstraction, it is to be expected that high school students, having a lack of mathematical experience amd mathematical maturity, meet severe difficulties in the study of abstract algebraic concepts, like vector spaces of matrices, linear transformations, group structures and so on. Therefore, if such concepts can be taught at all at this level, this ought to be done by giving as many concrete examples as possible (cf. [52]). For instance, the matrix concept and matrix operations can be introduced via their relation to geometric linear transformations (cf. [53] sections 2, 3, 5). Below we give an outline of an elementary, but nevertheless fairly complete account of the foundations of the special theory of relativity, much in the spirit of Minkowski's original ideas about spacetime, using simple matrix algebra ([46] ch. V). Of course Minkowski's lecture is very compact and makes no use of matrix algebra, a custom in mathematical physics that became established several years after the appearance of quantum mechanics. This is an example of the way the presentation of a subject is inspired by a general knowledge of the historical development, but does not respect it in a strict sense (cf. section 1). The present approach is based on the presentation of this subject to last-year Greek high school students, with a mathematics and science orientation.

Specifically we will show how the use of the algebra of 2×2 real matrices may lead to the Lorentz transformation in two dimensions (one spatial and one temporal) in a very simple way and in close analogy to a similar treatment of plane rotations of elementary analytic geometry.

(i) We introduce plane rotations geometrically and show that any such rotation by an angle θ is described by a 2×2 orthogonal matrix A_θ of determinant 1 and vice versa, and prove that such matrices satisfy the composition law (an example of an abelian transformation group)

$$A_\theta A_{\theta'} = A_{\theta+\theta'} \tag{1}$$

(ii) Subsequently we may show that A_θ conserves the euclidean norm x^2+y^2 and that it satisfies

$$AA^t = I \tag{2}$$

A^t being the transpose of A and I the identity matrix.

(iii) After this preparation, the two postulates of special relativity can be introduced, namely

(a) the existence of inertial coordinate systems, moving with respect to each other with constant velocity and for which Newton's law of inertia holds

(b) the speed of light in a vacuum is constant, equal to c say, in all inertial systems.

The aim is to specify the form of the transformation from one inertial system to another. Since by (a) straight lines are mapped to straight lines, we get that such transformations are linear.[10] Then by (b) the light cone given by

$$x^2 - c^2t^2 = 0 \tag{3}$$

is conserved, where (x, ct) are cartesian coordinates in the spacetime plane and it can be shown by elementary, though somewhat lengthy, calculations, that in general the spacetime "distance" $x^2 - c^2t^2$ is constant in all inertial systems.

(iv) In close analogy to plane rotations, we look for linear transformations of the plane conserving this spacetime "distance" and we easily find that the corresponding matrices, A say, satisfy (cf. (2))

$$A\eta A^t = \eta \tag{4}$$

where η is a 2×2 diagonal matrix with diagonal elements $\eta_{11} = -\eta_{22} = 1$. Then simple algebra leads to the following expression for the elements a_{ij} of A

$$a_{11} = a_{22} = \gamma, a_{12} = a_{21} = \alpha, \gamma^2 - \alpha^2 = 1 \tag{5}$$

where we have required that A is reducible continuously to the identity.

(v) If a certain inertial system moves with velocity v relative to another one, then by considering a point at rest with respect to the latter, it is easily derived that

$$\gamma = (1 - \beta^2)^{-1/2}, \quad \beta = v/c \tag{6}$$

so that we have obtained the familiar form of the Lorentz transformation.

(vi) It is a computational exercise to show that under matrix multiplication, the Lorentz transformations A_β (cf. (6)) satisfy the composition law (another example of an abelian transformation group)

$$A_\beta A_{\beta'} = A_{(\beta+\beta')/(1+\beta\beta')} \tag{7}$$

which readily implies the "relativistic law of velocity addition"

$$\beta'' = \beta * \beta' = (\beta + \beta')/(1 + \beta\beta'),\ \beta \in (-1, 1) \tag{8}$$

giving a nice, nonartificial example of a group structure defined on a subset of real numbers (cf. section 3.1).

(vii) Equation (6) gives an interesting opportunity to introduce hyperbolic functions, since it is thus possible to make more explicit the analogy between (7) and (1), through the parametrization $\beta = \tanh\phi$. This implies that (8) is equivalent to

$$\tanh\phi'' = \tanh(\phi + \phi')$$

hence (7) is equivalent to

$$A_{\phi+\phi'} = A_\phi A_{\phi'}$$

Many details of the above steps can be given as exercises, but lack of space does not permit a complete account of what should be given in the main presentation and what should be left as an exercise. More about this procedure and further examples are given in [58].

This is but one example of how the use of simple mathematics helps high school students to grasp the essentials of important physical theories, which are supposed to be "difficult" and unintelligible, at the same time appreciating the power and beauty of abstract mathematical methods and concepts. It is also evident that further development of special relativity can be given along these lines without difficulty, by deriving time dilation, length contraction and so on, depending on the more general structure of the mathematics and physics curriculum in which the present example is incorporated.

3.3 Interrelating different domains: Optics, Mechanics and Differential Equations. In a high school calculus course it is possible to give a unified treatment of the laws of reflection and refraction, using Fermat's principle of Least Time as an elementary application of differential calculus to extremum problems (e.g., [43] pp. 25–27). This can be compared to more elementary geometrical proofs, like that of Hero ([21] ch. 22); or may be used as a first step for the introduction to the brachistochrone problem as a prototype subject that finally leads, at the university level, to the calculus of variations ([43] section 1.6, [32]).

On the other hand, at an undergraduate university level, the basic idea that lies behind this innocent-looking treatment may have far reaching consequences, helping one to introduce naturally both new mathematical methods and fundamental physical ideas and concepts, as in fact it happened historically. This is a good example exhibiting the intimate connection between mathematical theories and methods and physical ideas and concepts. Its presentation may be based on an appropriate reconstruction of the following historical facts:

(a) As a mathematical theory, geometrical optics can be founded essentially only on Fermat's principle of Least Time as a variational principle. Then, using basic methods of the calculus of variations, as Hamilton did in the period 1824–1832 (e.g., [27] ch. V), the so-called Hamilton's eikonal PDE of geometrical optics for the optical length of the light rays is obtained (see, e.g., [27] ch. V, [36] ch. I [13] part IV, section VI.1). On the other hand, the dynamical evolution of a classical physical system with a finite number of degrees of freedom and constant total (mechanical) energy follows by similar methods using Maupertuis' principle of Least Action, as formulated by Hamilton (1834–5) and Jacobi (from 1837 onwards). This leads to the Hamilton-Jacobi PDE for Hamilton's characteristic function. This formal similarity of these a priori totally distinct physical theories, was a main motivation for Hamilton in developing a unified mathematical theory based on such variational principles ([27] ch. V, [36] ch. I, [13] part IV, sections VI. 2–5).

(b) From a mathematical point of view, it is possible in this way to arrive naturally at the conclusion that the solution of a first order PDE (more precisely, the determination of a complete integral of it), is equivalent to the solution of a system of ordinary differential equations attached to it, the so-called Hamilton's canonical equations. Depending on whether the subject is included in a mathematics or physics curriculum, the emphasis may be on the theoretical background of the method (Jacobi's method—e.g., [45] sections II.13–14, [6] sections II.7–8) or on the solution of mechanical problems (Hamilton-Jacobi theory—e.g., [20] sections 10.1, 10.3, [30] ch. VIII).

(c) From a physical point of view on the other hand, one may notice that the formal similarity mentioned in (a) above between geometrical optics and classical mechanics, has far-reaching consequences: geometrical optics (and in particular the eikonal equation) can be obtained as an approximation of wave optics when the wave length of light is small compared with the characteristic dimensions of the system under investigation. In the same way, classical mechanics (and in particular the Hamilton-Jacobi equation) may be thought of as an approximation to some (more accurate) wave mechanics. This in fact was the idea put forward by de Broglie in his work on the extension of the wave-particle duality from radiation to matter. It was explicitly elaborated by Schrödinger, thus leading him to the equation bearing his name and his wave mechanics ([8], [39], see also [23] section 5.3, [54]).

Actually this historical fact can be reconstructed in a rather interesting way by taking the mathematical similarity between the laws and magnitudes of geometrical optics and classical mechanics in the strict sense of proportionality, i.e., each magnitude of the one theory being proportional to a corresponding one of the other, in principle with a different proportionality factor in each case. Then simple calculations show that all proportionality factors can be determined in terms of one constant that remains unspecified. Moreover, this and the use of the wave equation for the wave function of optics finally yield in mechanics an equation mathematically identical to Schrödinger's equation, with the above unspecified constant playing the role of Planck's constant.[11] Evidently, such an approach has important advantages while teaching undergraduate quantum mechanics, for several reasons that were discussed in [54]. At the same time, it provides an interesting example of using analogy as a type of reasoning distinct from deductive or inductive reasoning, which often played a crucial role in applications of mathematics to physics (for further details see [55]).

3.4 Motivating the introduction of new concepts: Functional Analysis and the foundations of Quantum Mechanics. As a final example, we consider the introduction of abstract mathematical concepts at the level of an undergraduate mathematics course; specifically, we consider the interesting possibility (though not the only one!) of introducing some basic concepts of functional analysis motivated by quantum mechanical problems.

As a theory of infinite dimensional vector spaces, functional analysis appears already at the beginning of this century in the works of Fredholm, Hilbert, Volterra, Riesz and others, in the study of integral equations ([9], Introduction). However, it is well known that there was a great impulse in its development just after the invention of quantum mechanics (QM), which called for new mathematical concepts and methods. As a typical example we may consider von Neumann's contributions to the mathematical foundations of QM, [61]. What is perhaps less known is that in his works he introduces for the first time concepts and methods which often appear today *a priori* and in full generality in a functional analysis course, whereas this generality was motivated by the mathematical questions QM posed. There are many examples of abstract concepts whose introduction could be motivated on the basis of quantum mechanical problems. I will explain one example in some detail and mention briefly three others:

(a) Undergraduate courses on functional analysis often start with the definition of a Hilbert space as a linear space equipped with a scalar product, so that the induced distance makes it a complete metric space. Then a separable Hilbert space is defined by the rather mysterious condition that there exists a countable dense subset. On the basis of this, one proves the theorem characterizing such spaces as those having either an orthonormal (ON) countable subset spanning the space, or an ON countable subset not perpendicular to any other element, or satisfying a generalized Parseval identity with respect to a countable ON subset, and so on. Thus the question of the motivation behind these definitions is naturally raised. Therefore, inspired by the historical development of the subject, the following steps may be followed:

(i) Describe in detail important examples of (infinite dimensional) separable Hilbert spaces like the space $L^2(\mathbf{R})$ of Lebesgue quadratically integrable complex valued functions, or the space l^2 of infinite complex sequences $\alpha = (\alpha_1, \alpha_2, \ldots)$ such that $\sum_k |\alpha_k|^2 < +\infty$, a straightforward generalization of the familiar n-dimensional Euclidean space.

(ii) It is possible to present in a simple form the basic mathematical problem of matrix mechanics invented by Heisenberg, Born, and Jordan in 1925 and of Schrödinger's 1926 wave mechanics; namely, the diagonalization of the hamiltonian matrix H in the space l^2 and the solution of Schrödinger's equation in $L^2(\mathbf{R})$ respectively. The crucial point here is that these physically and mathematically a priori totally different theories of atomic phenomena yield identical correct experimental results. Consequently, as far as agreement with experiment is concerned, both theories should be accepted and therefore the question of their relation naturally arises.

(iii) One can repeat the essentially heuristic and nonrigorous arguments of Schrödinger, to show formally that Schrödinger's equation reduces to a matrix eigenvalue problem, once an ON basis of $L^2(\mathbf{R})$ has been chosen ([39] paper no 4).

(iv) Motivated by this approach, one can show rigorously that l^2 and $L^2(\mathbf{R})$ are isometric Hilbert spaces, a fact already apparent in the works of Riesz and Fisher in 1907 ([3] p. 267, [12]).

(v) Based on the heuristic approach in (iii), one is justified in reversing the argument and considering linear spaces with a scalar product spanned by a countable ON set, thus arriving at the isomorphism of all such spaces and their characterization mentioned previously. In fact this is essentially the approach followed by von Neumann in [61] (see also [9] p. 172 and [49] p. 2).

It is clear that many of the details in the above steps can be given as exercises, thus keeping the size of the presentation at a reasonable level.

(b) The spectrum of a bounded operator and the corresponding spectral resolution appear already in the work of Hilbert (1906) on integral equations and the work of Riesz on compact (hence bounded) operators (1916–18). However, many physically relevant quantum mechanical operators are not bounded (e.g., differential operators)! This led von Neumann to introduce the concept of a closed operator, to the study of such operators which are densely defined, to the distinction between self-adjoint and hermitian operators, and so on. ([9] ch. VII section 4, [23] section 6.3 pp. 318–319).

(c) The central problem of matrix mechanics was the diagonalization of the Hamiltonian operator by an appropriate similarity transformation. The physical requirement that quantum mechanical probabilities of energy states should be conserved under such transformations gives the latter in terms of unitary operators. This was a main motivation for the clarification of the spectral analysis of normal operators, a special case of which are self-adjoint and unitary operators.

(d) Originally, quantum mechanical operators were represented by infinite dimensional matrices, the diagonal elements of which gave quantum probabilities. This was the main motivation for the study of hermitian extensions of a hermitian operator, leading to criteria for the extension of a closed hermitian operator to a self adjoint operator, and consequently to the result that all hermitian extensions have the same matrix in a given complete orthonormal basis. Therefore, the matrix representation of Hilbert space operators is ambiguous ([9] ch. VII section 4). This led to the conclusion, that the mathematically appropriate formulation of the statistical interpretation of QM is given via the eigenprojections appearing in the spectral analysis of an operator ([23] ch. 6).

The examples presented in this section give a very brief outline of the way the historical development of a mathematical discipline can motivate its genetic presentation in the sense described in section 1.

Acknowledgement. I would like to thank Dr. Y. Thomaidis for his critical comments on an earlier version of this paper and the referee for suggesting several improvements of it.

References

1. Arnold, V.I.: 1978, *Mathematical methods of classical mechanics,* Springer.
2. Boltzmann, L.: 1964, *Lectures on gas theory,* California University Press.
3. Bourbaki, N.: 1974, *Éléments d' histoire des Mathématiques,* Hermann.
4. Boyer, C.B.: 1968, *A history of Mathematics,* Princeton University Press.

5. Brousseau, G.: 1983, "Les obstacles épistémologiques et les problèmes en mathématiques," *Recherche en didactique des Mathématiques,* **4** 165–198.
6. Courant, R. and Hilbert D.: 1962, *Methods of Mathematical Physics,* vol. II, J. Wiley.
7. Crowe, M.J.: 1967, *A history of vector analysis,* Dover.
8. de Broglie, L.: 1927, "Recherches sure la Théorie des Quanta," *Ann. de Physique* **3** 22–128.
9. Dieudonné, J.: 1981, *History of Functional Analysis,* North Holland.
10. ——: 1978, *Abregé d' Histoire des Mathématiques 1700–1900,* Hermann.
11. Dirac, P.A.M.: 1958, *The Principles of Quantum Mechanics,* Oxford University Press.
12. Dorier, J-L.: 1996,"First vector spaces on functions" in M.J. Lagarto, A. Vieira, and E. Veloso (eds.), *Proceedings of the 2nd European Summer University on History and Epistemology in Mathematics Education,* vol. II, Braga, Portugal, p. 238.
13. Dugas, R.: 1988, *A History of Mechanics,* Dover.
14. Edwards, H.M.: 1977, *Fermat's Last Theorem: A Genetic Introduction to Algebraic Number Theory,* Springer.
15. Ehrenfest, P. and Ehrenfest, T.: 1990, *The Conceptual Foundations of the Statistical Approach to Mechanics,* Dover.
16. Eves, H.: 1983, *Great Moments in Mathematics (after 1650),* The Mathematical Association of America.
17. Farquhar, I.E.: 1964, *Ergodic Theory in Statistical Mechanics,* Wiley.
18. Fermi, E.: 1961, *Notes on Quantum Mechanics,* University of Chicago Press.
19. Flato, M.: 1990, *Le pouvoir des Mathématiques,* Hachette.
20. Goldstein, H.: 1980, *Classical Mechanics,* Addison Wesley.
21. *Greek Mathematical Works, vol. II,* The Loeb Classical Library, Harvard University Press/Heinemann, 1941.
22. Heisenberg, W.: 1949, *The Physical Principles of the Quantum Theory,* Dover.
23. Jammer, M.: 1966, *The Conceptual Development of Quantum Mechanics,* McGraw Hill.
24. ——: 1969, *Concepts of Space,* Harvard University Press.
25. Khinchin, A.I.: 1948, *Mathematical Foundations of Statistical Mechanics,* Dover.
26. Klein, F.: 1974, *Le programme de Erlangen,* Gauthier-Villars.
27. ——: 1979, *Development of Mathematics in the 19th century,* Mathematics and Science Press.
28. Kline, M.: 1990, *Why Johnny can't add? The Failure of the new Mathematics,* (Greek edition) ch. 4.
29. Kolmogorov, A.N.: 1957, "The general theory of dynamical systems and Classical Mechanics," in *Prodeedings of the 1954 International Congress of Mathematicians,* North-Holland. Reprinted in *Foundations of Mechanics* by R. Abraham, J.E. Marsden, Benjamin/Cammings Publishing Co., 1978, Appendix D.
30. Lanczos, C.: 1970, *The Variational Principles of Mechanics,* University of Toronto Press.
31. Loria, G.: 1950, *Storia delle Matematich,* U. Hoepli. plus .25pt
32. Martin, C.M.: 1996, "Géométrie et mouvement: un example de autoformation," in M.J. Lagarto, A. Vieira, and E. Velodo (eds.), *Proceedings of the 2nd European Summer University on History and Epistemology in Mathematics Education,* vol. II, Braga, Portugal, p. 73–79.
33. Mehra, J.: 1973, "Einstein, Hilbert and the Theory of Gravitation", in J. Mehra (ed.), *The Physicist's Concept of Nature,* D. Reidel.
34. Mehra, J. and Rechenberg, H.: 1982, *The Historical Development of Quantum Theory,* vol. 3, Springer.
35. Pais, A.: 1982, *Subtle is the Lord: The Science and Life of A. Einstein,* Oxford University Press.
36. Pauli, W.: 1973, *Optics and the Theory of Electrons,* MIT Press.
37. Pólya, G. 1954, *Induction and Analogy in Mathematics,* Princeton University Press.
38. Rademacher, H. and Toeplitz, O.: 1990, *The enjoyment of mathematics,* Dover.
39. Schrödinger, E.: 1982, *Wave Mechanics,* Chelsea, p. 13ff.
40. Schubring, G.: 1977, "Die historisch-genetische Orientierung in der Mathematik-Didaktik," *Zentralblatt für Didaktik der Mathematik,* **9** 209–213.
41. Schwarz, L.: 1950, *Théorie des distributions,* Herman.
42. Sierpinska, A.: 1991, "Quelques idées sur la méthodologie de la recherche en didactique des mathématiques, liées à la notion de l'obstacle épistémologique," *Cahiers de didactique des Mathématiques,* **7** 11–28.
43. Simmons, G.F. 1974, *Differential Equations with Applications and Historical Notes,* McGraw Hill.
44. Smith, D.E.: 1959, *A Source Book in Mathematics,* Dover.
45. Sneddon, I.: 1957, *Elements of Partial Differential Equations,* Addison Wesley.
46. Sommerfeld, A. (ed.): 1952, *The Principle of Relativity,* Dover.
47. Spivak, M.: 1970, *Differential Geometry Vol. II,* Publish or Perish.
48. Stilwell, J.: 1989, *Mathematics and Its History,* Springer.
49. Stone, M. H.: 1932, *Linear Transformations in Hilbert Space,* American Mathematical Society.
50. Struik, D.J.: 1948, *A Concise History of Mathematics,* Dover.
51. Toeplitz, O.: 1963, *The Calculus : A Genetic Approach,* University of Chicago.
52. Tzanakis, C.: 1995, "Reversing the customary deductive teaching of mathematics by using its history: the case of abstract algebraic concepts," in *Proceedings of the First European Summer University on History and Epistemology in Mathematics Education,* IREM de Montpellier, p. 271–273.
53. ——: 1995, "Rotations, Complex Numbers and Quaternions," *Int. J. Math. Education Sci. Techn.,* **26** 45–60.
54. —— and Coutsomitros C.: 1988, "A genetic approach to the presentation of physics: the case of quantum theory," *Europ. J. Phys.* **9** 276–282.
55. ——: 1997, "The quest of beauty in research and teaching of Mathematics and Physics: An historical approach," *Nonlinear Analysis, Theory, Methods and Applications,* **30** 2097–2105.
56. ——: 1991, "Is it Possible to Teach Abstract Algebraic Concepts in High School? A historical approach," *Euclides* **7** 24–34, (in Greek).
57. ——: 1998, "Discovering by Analogy: the case of Schrödinger's Equation," *European J. of Physics* **19** 69–75.

58. ——: 1999, "Unfolding interrelations between mathematics and physics motivated by history: Two examples," *Int. J. Math. Education Sci. Tech.*, 30, 103–118.
59. van der Waerden, B.L. (ed.): 1967, *Sources of Quantum Mechanics,* Dover, pp. 19–36, 261–321.
60. van der Waerden, B. L.: 1985, *A History of Algebra: From al-Khwarizmi to Emmy Noether,* Springer, p. 137, 153–154.
61. von Neumann, J.: 1932, *Mathematische Grundlagen der Quantenmechanik,* Springer.
62. Weyl, H.: 1950, *The Theory of Groups and Quantum Mechanics,* Dover.

Endnotes

[1] This paper is based on the talk given and a 3-hour workshop organized, at the meeting on 'History and Education of Mathematics', held in Braga, Portugal from 24 to 30 July 1996 (see the Proceedings of this conference, Braga 1996, vol. II p. 96 and vol. I p. 305).

[2] By presentation we mean, either teaching in a classroom or writing a textbook, survey article etc.

[3] cf. [5] p. 193: "...En aucun cas, il ne saurait suffire..., d'appliquer sans modification l'étude historique à l'étude didactique", see also [42] p. 13, [40] p. 211.

[4] "...the meaning of a concept is not completely determined by its modern definition, but it is the resultant of its development in the past as well as in the present" ([42] p. 13, my translation).

[5] For examples in this century, think for instance of Hilbert, e.g., in connection with the foundation of general relativity, [33] ch. 7, Minkowski's geometrization of special relativity, [46] ch. 5, von Neumann's "Grundlagen der Quantentheorie," [61], and Kolmogorov's revitalization of classical mechanics and dynamical system theory, [29], [1] Appendix 8.

[6] In this example the relation between mathematics and physics plays a secondary role, but it is more suited for our purposes than other better-known and better-analyzed examples, like the concept of velocity and its relation to the derivative of a function.

[7] That geometrically speaking, relativity theory was very close to Klein's Erlangen program, has been stressed by Klein himself, quoted in [33] p. 113.

[8] Similar comments, which however will not be given here, hold for the concepts of a ring, field and vector space, which evolved gradually in the nineteenth century as a consequence of important developments in algebra, number theory and geometry (see, e.g., [10] vol. I, chs. II, III [3] chs. 3–7, [27] chs. II, IV, VII).

[9] This seems to be historically supported as well. For instance (i) in its abstract form, the vector space concept was defined by Peano in 1888 ([10] vol. I, p. 94, [12]), but his work passed largely unoticed, presumably because it was rather premature at that time. On the contrary, there is a tendency to believe that the modern definition appeared first in 1930 in Weyl's book on group theory and quantum mechanics, [62] p. 1–2, simply because it proved rather fruitful in that context. (ii) Although Cayley had already defined an abstract finite group in 1854, it was not until the 1890's that the modern general definition was given by Weber ([10], vol. II, p. 116, [60]).

[10] This step uses the continuity of a function of two variables and therefore it is more advanced in this sense. Nevertheless, it is intuitively clear to the students and therefore it presents no serious difficulties in its presentation.

[11] This approach is presented in detail elsewhere ([57]). A similar reconstruction has been given in [54], [18].

Part IV

The Use of History in Teacher Training

A Historical Approach to Developing the Cultural Significance of Mathematics among First Year Preservice Primary School Teachers

Ian Isaacs, V. Mohan Ram, Ann Richards
Northern Territory University, Australia

A unit of mathematics, *The Cultural Origins of Mathematics,* was introduced to first year preservice primary mathematics teachers with the goal of modifying their world view of mathematics. By emphasizing the social and cultural factors that influenced the historical development of elementary mathematics, it was hoped that the students would develop some appreciation of the cultural significance of mathematics.

Introduction

Most of the students who enter the three-year course for the preparation of primary teachers at the Northern Territory University have a weak background in the content of secondary school mathematics and a mechanistic and utilitarian perception of the subject. About 40% are mature age students and the remainder are recent graduates of Northern Territory secondary schools. Generally they hold beliefs about mathematics which portray this subject as (i) mainly concerned with the facilitating of the buying and selling of goods; (ii) one in which there is one best method to getting right answers; and (iii) including the branches of algebra and geometry which are useless for everyday living.

In 1995 the new three year Bachelor of Teaching course, consisting of twenty-four units, provided the three mathematics education lecturers at the NTU with the opportunity to introduce a unit, spanning two semesters, which we entitled *The Cultural Origins of Mathematics.* We hoped that this unit would give the students a broader perspective on the place of mathematics in the cultures of different societies over the last 5000 or so years and that it would allow them to recognize the role of mathematics as a response to the physical and intellectual problems of those cultures which eventually led to the development of the subject as we know it today.

The Unit—*The Cultural Origins of Mathematics*

In this unit we aimed to modify the belief systems and perceptions of these preservice teachers regarding the nature of mathematics and the purpose of school mathematics. To set the scene for the geometrical ideas we planned to develop in the first semester, we focused on how the various societies from China, India, Egypt and Greece dealt with geometrical concepts and notions in their practical and intellectual life. The students explored:

(i) geometry as a practical science used to solve real problems,

(ii) geometry as a constructive and aesthetic medium where patterns, transformations and geometrical relationships predominate,
(iii) the ritualistic requirements for accurate constructions,
(iv) measurement as an introduction to numbers which are not rational, and
(v) logical justifications in geometry.

By examining the attempts of the ancient Egyptian, Greek, Hindu, and Chinese mathematicians to solve problems related to finding lengths and areas and constructing polygons of specified dimensions and shapes, we hoped the students would begin to appreciate geometry as a tool for solving spatial problems. By working to solve problems similar to those posed by mathematicians of the day, in surveying situations or the design of shapes appropriate for ritualistic functions, the students were challenged to undergo similar thought processes to those of the ancient mathematicians.

Episodes from the mathematics of ancient China, India, Egypt and Greece were used to illustrate and inform the students of the quantitative problems faced by those mathematicians. However, the problems were modified to make them readily accessible to the students. The students were allowed the use of modern aids and nomenclature. The lecturers bore in mind that problems had to be framed in such a way that the majority of the students would successfully solve them after a little thought. [The underlying assumption on which the unit was based was: Success breeds confidence and further success.]

For example, rather than solve the problem posed by Socrates in Plato's *Meno* which was constructing a square of area twice that of 4 square units, the problem was modified to finding the area of a square whose area was half of 4 cm^2 [see Figure 1]. For most of the students, it was then a much easier task to construct a square of area 8 cm^2.

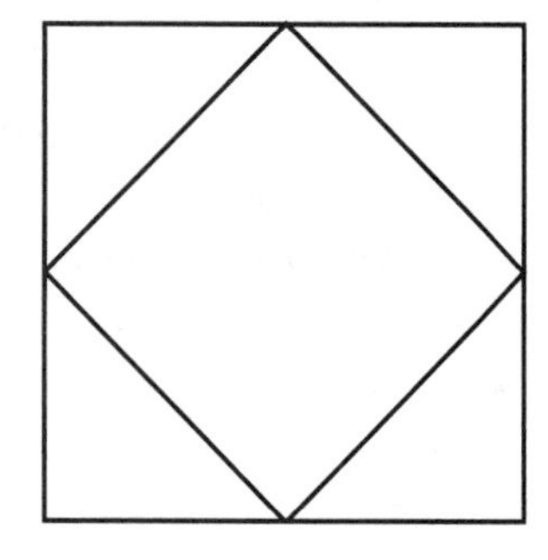

FIGURE 1
Constructing a square of area 2 cm^2 given a square of area 4 cm^2

ACTIVITY 1

Your task is to carry out a survey of a region which has at least four straight boundaries and is not a rectangle or other simple geometric shape, e.g.,

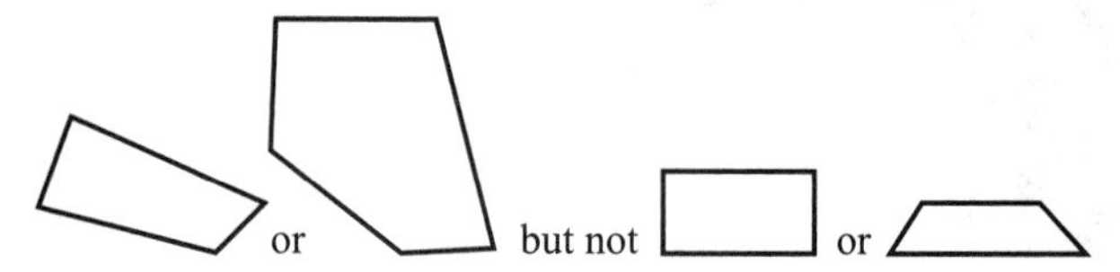

Your initial unit of measurement is to be paces, and these need to be maintained at an even length to ensure accuracy. You need to choose a recorder and a pacer and preferably check all your measurements using a second pacer from your group.
In your group you will need to:
(1) Identify your region. This should not be too small—aim for all distances to measure to be in the range of 30–60 paces.
(2) Draw a rough sketch of your area labeling the vertices.
(3) Measure all distances by pacing.
**Remember to triangulate—i.e., measure certain diagonals within your polygonal region so that this can be marked as composed of triangles whose dimensions have been measured.
(4) Record all measurements appropriately on your sketch.
(5) Have your pacer walk the marked 20 meter length to find how many of his/her paces are equivalent to 20 meters.
Back inside
(6) Convert all your measurements to meters and record.
(7) Using the grid provided, decide on a scale which will allow you to convert your meter measurements to scaled values to provide a plan of your area which will occupy most of the page.
(8) Use the available equipment to construct an accurate plan of your region on paper or in your book. Note: The calculation of your area will be set at a later date.

Extension—for Week 3/4
Use your plan to calculate the area of your region. Your final value should be actual area in square meters (m^2).
Possible Methods:
(1) Consider the region as composed of triangles and find the area of each triangle (accurately) and total these areas.
(2) Using the scale diagram of your polygonal area, construct a triangle with the same area using the method of reducing the number of sides (week 3).
Find the area of the resulting single triangle.

FIGURE 2
Task to estimate the area of a quadrilateral/pentagon

Methodology

This section contains a brief description of the five themes that the students were asked to explore during the first semester sessions of the unit.

Theme 1. *Geometry as a practical science used to solve real problems.* The students were made aware of the problems the Egyptians and Chinese faced in determining the areas of irregular shaped polygons. The students were set the task [see Figure 2] of estimating the lengths of the sides of an irregular quadrilateral or pentagon on the lawns near their lecture hall using only pacing. They were then expected to check their estimates against those of their colleagues in their group by converting their paces to standard metric measures. Subsequently the use of a scale drawing and historical methods of working with polygon areas were used to allow an area calculation. One method considered was the construction problem (which was solved by the Chinese and the Greeks) of transforming an irregular polygon into a rectangle or triangle of equal area. From this exercise it was natural to look at more sophisticated techniques used in China and the West in the Middle Ages to carry out land surveys.

Theme 2. *Geometry as a constructive and aesthetic medium where patterns, transformations and geometrical relationships predominate.* Rather than follow the traditional approach of using geometry as an introduction to deductive proof, we focused on the Islamic use of geometrical patterns for decorations such as those shown in El-Said and Parman's *Geometric Concepts in Islamic Art* and in slides taken by V. Mohan-Ram in Pakistan. The students were then encouraged to construct designs (see Figure 3) based on these patterns.

In this way the students were involved in identifying symmetric patterns in their surroundings as well as in designing their own patterns. They were led to reflect on the symmetrical properties and isometric transformations of simple regular polygons. These activities served as a forerunner for their study of group structures to be taken up in the second half of the unit in Semester 2.

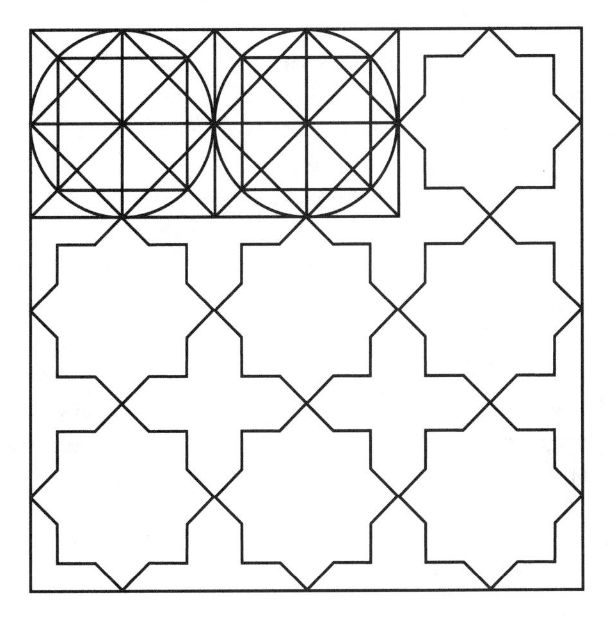

FIGURE 3
Design of Tiling Pattern seen in the Iran Bastan Museum [after El-Said, p. 13]

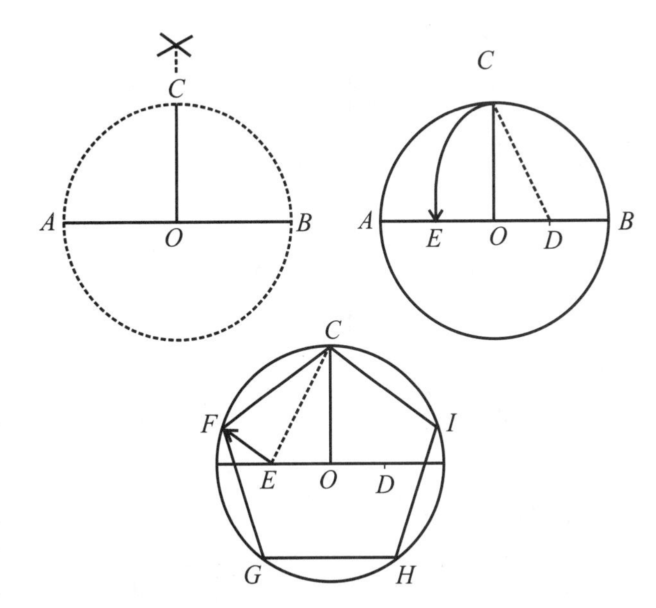

FIGURE 4
Construction of the regular pentagon

Theme 3. *The ritualistic requirements for accurate constructions.* The construction of various shapes used in religious and occult practices such as the pentagon [see Figure 4] were performed by the students under the guidance of the lecturer. The determination of the dimensions of the trapezium [see Figure 5] used in some Hindu altars was also demonstrated by the lecturer.

For most rituals to have significance, it was believed that the dimensions of the ritualistic structures needed to be exact. Inexactitude implied imperfection and the resulting impurity of the ritual. We suggested that this need for perfection was probably the initial motivation for the Hindu and Chinese mathematicians to develop algorithms for the accurate construction of the basic polygons. However, some mathematicians went beyond just satisfying the ritualistic requirements of their religions and developed algorithms for solving construction problems of no immediate practical value.

For example, Hindu mathematicians posed, and solved, the problem of how to construct a square equal in area to that of the sum of two other squares. The students were asked to conjecture how this could be done if the two squares had areas of 2 cm^2 and 9 cm^2. This problem is very similar to that described by Joseph (1990, p. 231) in *The Crest of the Peacock*. This problem was also used as one of the examples to contrast the practical algorith-

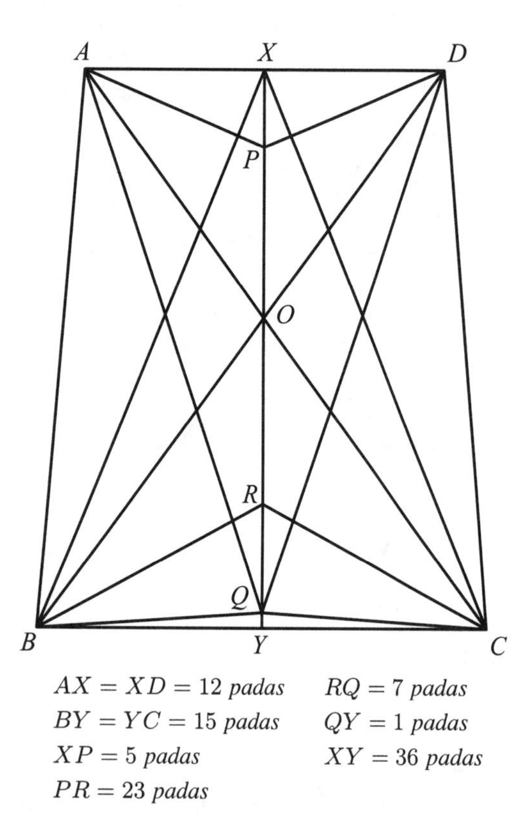

FIGURE 5
The layout of a *samassana sacrificial* altar [after Joseph (1990), p. 230]

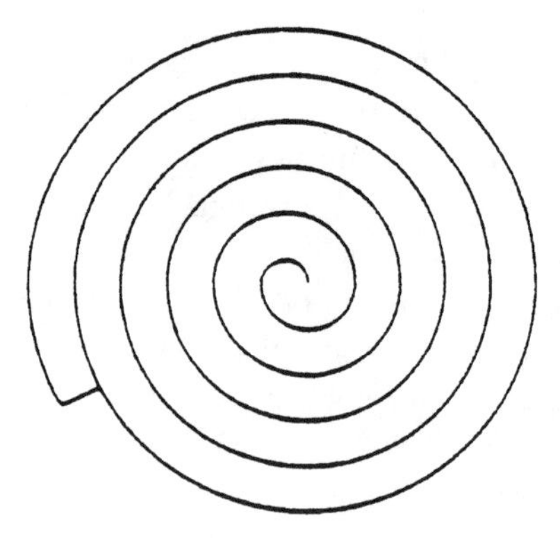

FIGURE 6
The Snake Curve [after Gerdes (1985) p. 262]

mic problem solving approach of the Chinese and Hindu mathematicians to the theoretical deductive approach of the Greek mathematicians.

Theme 4. *Measurement as an introduction to numbers which are not rational.* The construction of squares whose sides were not rational numbers was a challenging task for most of the students. Rather than solving the problem posed by Socrates in Plato's *Meno* of constructing a square of area twice that of 4 square units, the problem was modified to finding the area of a square of area half of 4 cm^2 [see Figure 1 above]. For most of the students it was now a much easier task to construct a square of area 8 cm^2. This naturally led to the attempt to determine the square root of numbers like 2 and 8. The students attempted it in three ways: (i) by measuring the side of a square of area 2 cm^2; (ii) by sketching; using the geometrical approach of Apastamba in his *Sulbasutra* [see Joseph (1990), pp. 234–236]; and (iii) by using trial-and-error with a calculator. These explorations led us to the concept of an irrational number. However this type of number was not fully explored at this time, but left to the second semester when it was to be reexamined as part of the study of 'Numbers.'

The study of areas of polygons led us to consider the areas of regular polygons and the areas of circles. The conjecture of Gerdes (1985) was introduced regarding the *snake curve* [see Figure 6]. Constructing the *snake curve* using a string was the task the students performed to arrive at the Egyptian approximation for π [$= 4(8/9)^2$].

The students next attempted to determine the value of π by calculating the ratio of the circumference to the diameter of selected circular discs of diameter 3.5 cm, 7.0 cm, 10.5 cm and 14.0 cm. They then went on to look at Archimedes' method of 'exhausting' the circle, by considering the perimeters of 6- and 12-sided regular polygons inscribed in a circle as an approximate measure of the circumference of the circle. The Chinese value of π ($= 355/113$) as determined by Tsu [Joseph (1990), p. 195] was also mentioned. Once again we were brought into contact with a number which seemed to be not rational. The question was posed: Is it a number like the square root of 2 or is it a new sort of number? No attempt was made to answer the question at this point but the promise was made to come back to it in the following semester.

Theme 5. *Logical justifications in geometry.* The lecturers, however, did not evade the need for general proofs when we felt they were necessary. For example the Chinese proof [see Figure 7] based on decomposing the square of side $(a + b)$ was modified [see Figure 8] to show that Pythagoras' theorem is true [see Joseph (1990), pp. 180–181].

Assessment of the unit

Students were assessed on three sets of activities for this unit: (i) an oral presentation based on a professional journal article which related to some aspect of the unit; (ii)

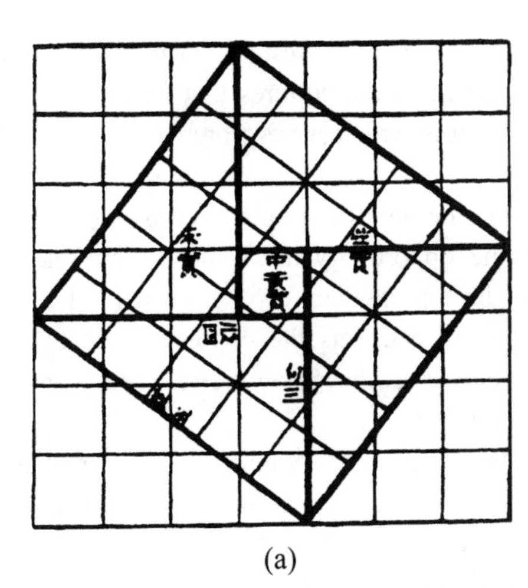

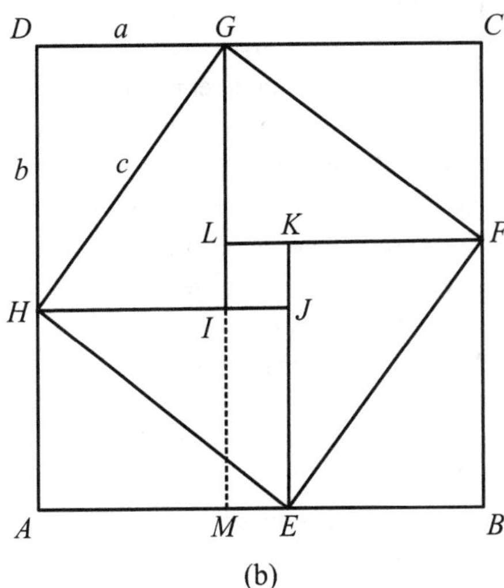

FIGURE 7
(a) The *kou ku* theorem and (b) the modern 'translation' (after Joseph (1990) pp. 180–181)

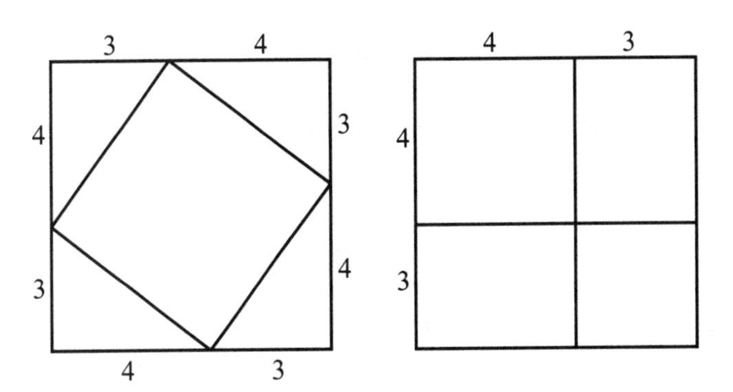

FIGURE 8
The modified version of the *kou ku* theorem discussed with the students

'Journal' entries which consisted of their reflections and reactions to the unit and a 'working log' showing their attempts to complete the tasks set by the lecturers; (iii) exercises based on the learning outcomes for the topics 'Working Mathematically,' 'Measurement' and 'Space,' from the Australian Education Council's (1994) document *Mathematics—a Curriculum Profile for Australian Schools*. The requirement that the oral presentation be related to some aspect of the unit had to be relaxed as students had great difficulty in finding articles dealing with the history of geometry in the journals available in the University library. For the exercises based on the learning outcomes, the students were able to select how many examples they wished to attempt from each category. To achieve a Pass grade students needed to do no more than 1 or 2 exercises from each of the three categories, whereas students who wished to aim for a grade of High Distinction were expected to do outstanding work on a minimum of 4 exercises from each category.

Student Reactions

In their journals the students were required to record (i) their attempts at the tasks set as class activities and extensions of these class activities; (ii) their perceptions and reactions to the lectures and tasks, and (iii) their reflections on the nature of mathematics and its relevance to them.

At the end of the semester we asked the students to give us their opinions on the unit using a Likert style questionnaire. In particular we were interested in (i) their current beliefs about the nature of mathematics as compared with those at the start of this unit; and (ii) their views about the significance of mathematics in solving quantitative problems in ancient societies.

At the start of the unit, and during the development of the unit (as students became more confident in expressing their opinions), many students (about a third) expressed in their journals their concern at the apparent lack of relevance of the content and activities to their future role as primary teachers. Typical of this group is the following journal entry made in the first four weeks of the course:

> Although I find this information/History interesting it usually tends to confuse me. I also have difficulty relating it to current mathematics. How is it relevant to primary school maths?

Others commented on the difference between this mathematics and the mathematics they had done in school. As one student put it,

> I have never been taught maths in this way before, rather I have always been given a set of problems and been told to solve them, which has

> made maths a boring subject. Yet when faced with an answer, question, and the history behind a particular thing it makes further questions easier to handle and I found it stays ingrained in the memory better.

The remaining students, about 33%, did not make any direct comments about the nature of the mathematics explored in this course. They were either non-committal or they made vague brief statements such as, 'it was interesting' or 'it is confusing.'

At the end of the semester 35% agreed that 'the unit seemed unrelated to the maths I will need to teach in primary schools,' 23% were undecided whether to agree or disagree, and 42% disagreed with the statement. A majority (57%) felt that the unit had helped them to change their attitude towards mathematics. However, a significant minority (25%) were undecided and the remaining respondents (18%) disagreed.

Conclusion. Only one student (and a very able one at that) was capable of expressing a view congruent with that of the lecturers; she wrote:

> I perceive this unit as giving students the opportunity to enhance their existing mathematical skills, which in time will enable them to construct a more reflective attitude towards teaching a subject that has been traditionally seen as requiring genius mentality.
>
> Three common myths associated with Mathematics have been based on the Genius Mentality, that mathematics has nothing or very little to do with reality (life) and that you have to do everything mentally and fast and if you cannot then you are obviously not mathematical. This unit has worked towards discrediting these myths.... It also dispelled the myth that mathematics is a purely Western Phenomena, it has given credit to the cultures that have originated the concepts which were prompted by their life styles.
>
> Another contributing aspect of this unit has been the opportunity the students have experienced in setting their own problems... [it has] given many students the opportunity to experience success as they have dictated the problems to match them....
>
> As an introductory unit for a universal subject, it has been my pleasure to have witnessed the "beauty" of Mathematics in Action.

The lecturers concluded that, for the majority of students, they still had some convincing to do. We recognized that, at the start of the second semester, we would again have to emphasize the need for teachers to have a broader and deeper view of the subject than their pupils. We also agreed to explicitly link the topic of 'Numbers and Numeration' to primary school arithmetic in the second half of the unit. Likewise, when the unit was repeated in the next year, we would attempt to clearly relate the geometrical ideas to the 'Space' topics in the primary school curriculum.

References

Australian Education Council, 1994: *Mathematics—a Curriculum Profile for Australian Schools,* Melbourne: Curriculum Corporation.

El-Said, Issam and Ayse Parman, 1976: *Geometric Concepts in Islamic Art,* Palo Alto, CA: Dale Seymour Publications.

Gerdes, Paulus, 1985: "Three alternate methods of obtaining the ancient Egyptian formula for the area of a circle," *Historia Mathematica* 12, 261–268.

Joseph,George C., 1990: *The Crest of the Peacock: Non-European Roots of Mathematics,* London: Penguin Books.

Jowett, B., 1953: *The Dialogues of Plato: The Meno,* Oxford: The Clarendon Press.

The Analysis of Regula Falsi as an Instance for Professional Development of Elementary School Teachers

Greisy Winicki
Oranim—School of Education of the Kibbutz Movement
ISRAEL

According to our school curricula, mathematics teachers are not supposed to teach history of mathematics for its own sake. So, why are more and more universities giving courses on history of mathematics to future teachers? "Without it, educators can't teach students to love or even like, to appreciate or even to understand, mathematics" (Tymoczko, 1993, p. 14). I believe that such courses can have a huge impact not only on the university students' formation as future mathematics teachers, but also on the teachers' professional development. The purposes of this paper are: a) to describe a concrete approach of introducing history of mathematics to teacher training programs, and b) to address some benefits from this introduction to the mathematics teacher's professional development.

The activity described below is one of the workshops of a course called *Some Chapters in the History of Mathematics*. This course is part of a wider program for elementary school teachers which has as its principal aim to qualify *professional* mathematics teachers for the elementary school. In Israel, elementary school teachers are, in general, not specialized in mathematics education; those who want to teach mathematics are encouraged to participate in professional development programs. Several schools of education around Israel adopted the program described here; one of the courses taught is the course mentioned above. There is a general syllabus for the course but all lecturers design their own activities. Eighteen teachers participated in the course. All of them were elementary school teachers and they had at least five years of experience in teaching mathematics. At the end of the course, the participants were asked to devise a didactic unit in which they integrated the history of some mathematical topic to its teaching.

The topic of the workshop described in this paper was equations, specifically, the methods known as *Regula Falsi* or *Rule of False Position* (Ofir and Arcavi, 1992; Eaton, 1967; Eves, 1958). Although elementary school teachers are not supposed to teach linear equations nor linear functions to their students, they are supposed to understand the mathematical principles behind them since they do teach ratio and proportion ideas, and they also teach how to deal with proportional situations.

The participants were organized in pairs. Each member of the pair was given a different card (see Appendices A and B) and had twenty minutes to work individually and finish the tasks on it. Each card was divided into three parts:

a. Reading Material: A short historical background, the problem formulation and a verbal description of a method to solve the problem.

b. Problems to solve: An historical word problem to be translated into an equation solved by using the method described in part (a), and some other linear equations to be solved using the same method.
c. Reflection Questions: A set of five questions about the method.

While working on the card, the participants were asked to write down all their conjectures, answers, questions, and proofs, as well as examples and counterexamples. After twenty minutes of individual work, they were asked to explain to their partner the method they had studied. They worked in pairs for almost twenty minutes.

Following the card tasks, a very 'hot' discussion took place. In the following lines, the description and analysis of that inspiring discussion appear as follows. The main questions that directed the analysis were:

1. What factors do the teachers find relevant to the ancient methods used to solve equations?
2. Are the teachers aware of the improvement that symbolic notation brings to the development of mathematics in general, and to solving equations in particular?

The participants were asked to discuss the following questions:

- Does the method work for all the above equations?
- Does the final answer depend on the guesses?
- Will the method work for any linear equation?
- Will the method work for any equation?
- Why do you think such a method was invented?

The first four questions were answered correctly by the participants. However, almost all of them were unable to justify why the *Rule of False Position* worked nor why the *Rule of Double False Position* did. This was an appropriate occasion to discuss mathematical concepts such as *proportionality, linear functions* and *linear equations,* and the *rule of three.* In this sense, the historical material allowed a review of mathematical content without causing embarrassment to the teachers. I believe that during this review, the participants deepened their own conceptual understanding rather than the procedural skills involved in proportional reasoning.

While the first four questions led to consensus, the last one was found to be a difficult question to answer since the participants felt they were not used to these kinds of questions. Some of the answers received were:

- "because it is very intuitive"
- "because they didn't like fractions"
- "because they didn't like to divide"
- "because they didn't have 0"

Nobody mentioned anything related to symbolic notation or to algebra. This was an appropriate occasion to introduce some historical content like Egyptian rhetorical algebra, Greek geometric algebra, Diophantus' contribution to the introduction of symbolism, the Islamic extension of Babylonian and Greek algebra and their own geometric proofs, Viete's use of abbreviations or words to represent quantities, and his adoption of certain letters to represent knowns and others to represent unknowns, and so on.

The participants were also asked about their general opinion about the methods. Some of them were in favor of it as can be seen in their own words:

> Even if we don't teach such methods to our students, it is an historical fact that at least someone thought about it. So, we are in front of a mathematical problem, that can be solved with quite elementary tools.
>
> The Regula Falsi method seems very intuitive to me, and I see that it also was intuitive to other people. I wonder what's the justification of its correctness.
>
> I liked the guessing stuff, but I'm curious to see how children react to it.

But others did not find those methods and the exposure to them so valuable:

> I don't think it will be suitable to bring those methods to the class. They are not very clear and they rely on guessing. I don't think there is a place for guessing in mathematics.
>
> How can you be sure that your guess is a 'good' one? Maybe you are lucky once, but you have to develop a general method and not to rely on luck.
>
> Now, we have better methods to solve linear equations. So, what would be the utility of teaching our students such methods?

This experience enabled the participants to expose explicitly—but mainly implicitly—their conceptions about what mathematics is about, how it develops, how it is learned and how it can be effectively taught. The discussion that took place at the end of the workshop gave all the opportunity to analyze important didactic issues. In that sense, this workshop shows that historical problems can lead teachers to discuss not only mathematics content but also didactic issues. As an example, the workshop described here focused on the following didactic issues:

1. learning by reading;
2. learning from examples;
3. explaining ideas through examples;

4. using different representations of the same concept;
5. proving and the role of proof;
6. using historical material (not first sources but historical problems) as a trigger for meaningful discussion;
7. using historical material directly connected to the content taught in order to provide a humanistic perspective to the content.

For instance, the participants pointed out that while trying to justify the *Rule of False Position,* some of them used the traditional proportion algorithm (cross-multiply and divide); others tried to look for rules with the form $y = mx$ in which m is the constant factor that relates the two quantities x and y; and others used a verbal approach. When they were asked to justify the *Rule of Double False Position,* teachers used several representations of the concept of function: table, ordered pairs, graph, and the equation. They found that the rationale of this method could be well understood if one considered that through two different given points, a unique straight line can be drawn. They asked for a proof to the correctness of the method. A teacher very familiar with the algebraic language gave the following explanation:

> If we have to solve the equation $ax + b = 0$, we can consider the function $y = ax + b$ and try to look for its root. We know it is $-b/a$ and we will show that this is what you get if you follow the instructions [the Double False Position Method]. If p is the first guess, when you substitute in the function, you get $ap+b$. So, the point $(p, ap+b)$ belongs to the graph of the linear function. Let us say $(q, aq + b)$ is the other point. According to the instructions we should calculate
>
> $$\frac{(ap+b)q - (aq+b)p}{(ap+b) - (aq+b)}.$$
>
> After the corresponding simplifications, you get that the root is indeed $-b/a \ldots.$

After this explanation, the participants discussed if, and in what way, this explanation constituted a proof and what might be the prerequisites to understand it.

Since the utilitarian face of mathematics is widely acknowledged, during this workshop the participants were exposed to another face of the subject, the one usually called a humanistic face, a face that reminds us that mathematics is an integral part of our culture. This aspect of mathematics must be communicated to the students and it must have an influence in the ways teachers teach mathematics. The role that mathematics plays in our culture should be exposed, and teachers have a major responsibility in that task.

The participants were also encouraged to think, to talk, and to discuss about mathematics, essential experiences to the development of their metamathematical knowledge (Vollrath,1992). They considered this workshop to be a meaningful type of experience involving topics in the history of mathematics. It also reinforced the idea that these kinds of activities may provide an opportunity to expose and to discuss the teachers' own beliefs about mathematics, mathematics teaching, and mathematics learning.

References

Easton, J.B.: 1967, "The Rule of Double False Position," *The Mathematics Teacher* 60 (1) pp. 56–58.

Eves, H.: 1958, "On the Practicability of the Rule of False Position," reprinted in Swetz, F. J. (ed.) *From Five Fingers to Infinity: A Journey through the History of Mathematics,* Chicago: Open Court, pp. 149|150.

Ofir, R. and Arcavi, A.: 1992, "Word problems and equations: an historical activity for the algebra classroom," *The Mathematical Gazette* 76 (March) 69–84.

Tymoczko,T.: 1993, "Humanistic and Utilitarian Aspects of Mathematics," in White, A.W. (ed.), *Essays in Humanistic Mathematics,* Washington, DC: The Mathematical Association of America, pp. 11–14.

Vollrath, H.J.: 1992, "About the appreciation of theorems by students and teachers," *Selected Lectures from the 7th International Conference on Mathematics Education,* Quebec: University of Laval, pp. 353–363.

Appendix A

Card I: Regula Falsi Method
Subject: Solution of Equations

The following problem originated in ancient Egypt (Rhind Papyrus, 2000 BC):

> The sum of a certain number and a seventh of the same number equals 16. Compute the number.

The solution is explained verbally on the papyrus:

> If the number was 7, the answer would be 8 since a seventh of 7 is 1 and seven plus one equals eight. If we multiply 8 by 2 we get 16, therefore the solution is 7 (the guess) multiplied by 2 (the correction factor).

This is the *Regula Falsi* Method.

Tasks:

1. a. Write down the Egyptian's equation
   b. Check that the given answer is correct.
2. The above method for solving equations was also used in the Middle Ages and appeared in the book *Liber Abaci* by Fibonacci (in the XII century). The following problem comes from this book.
   A grocer bought a certain amount of apples and paid one dinar for every 7 apples. The following day he sold all the apples at 1 dinar each 5 apples. His net profit was 12 dinar. How much money did he invest in the apples to begin with?
   a. Algebraically we should solve the equation ____________ using x as the amount invested.
   b. Solve the equation you just wrote by using the method described above.
      guess:
      result:
      solution according to the method:
      verification:
3. Solve the following equations using the same method:
   a. $x + \frac{x}{3} = 24$
      guess:
      result:
      solution according to the method:
      verification:
   b. $2x + \frac{5x}{3} = 33$
      guess:
      result:
      solution according to the method:
      verification:
   c. $x + 4x = 12$
      guess:
      result:
      solution according to the method:
      verification:
4. a. Does the method work for all the above equations?
   b. Does your final answer depend on your guess?
   c. Will the method work for any linear equation?
   d. Will the method work for any equation?
   e. Why do you think such a method was invented?

Appendix B

Card II: Double False Position Method
Subject: Solution of Equations

The following method for solving equations was used during the Middle Ages. It's called the *Double False Position Method.* In order to solve an equation such as $5x + 10 = 22$, we first have to write an equivalent equation of the form $ax + b = 0$; in this case we get $5x - 12 = 0$.

- If we make a guess and say $x = 1$, we will get -7, instead of 0 (the value -7 comes from the substitution of 1 in the pattern $5x - 12$).
- If we substitute 5 for x we will get 13 instead of 0.

  This is when we use the *Double False Position Method,* which says that the solution to the equation $5x - 12 = 0$ is

$$x = \frac{1 \cdot 13 - 5 \cdot (-7)}{13 - (-7)}$$

Tasks:

1. Check that the given answer is correct.
2. There are ancient collections of math problems. The following problem was taken from a collection written by Chuquet in 1484.
   A merchant went to 3 markets. At the first market he doubled his money and spent 30 francs. At the second market he tripled his money and spent 54 francs. At the third he quadrupled his money and spent 72 francs. When he was finished he had 48 francs. How much money did he start off with?
   a. Algebraically we should solve the equation ____________ using x as the amount he had at the beginning.
   b. Solve the equation you just wrote by using the method described above.
      guess #1:
      result:
      guess #2:
      result:
      solution according to the method:
      verification:
3. Solve the following equations using the same method:
   a. $x + \frac{x}{3} = 24$
      guess #1:
      result:
      guess #2:
      result:
      solution according to the method:
      verification:
   b. $x + 4x = 12$
      guess #1:
      result:
      guess #2:
      result:
      solution according to the method:
      verification:
4. a. Does the method work for all the above equations?
   b. Does your final answer depend on your guesses?
   c. Will the method work for any linear equation?
   d. Will the method work for any equation?
   e. Why do you think such a method was invented?

Mathematics and Its History: An Educational Partnership

Maxim Bruckheimer and Abraham Arcavi
Weizmann Institute of Science
Rehovot, Israel

Introduction

About 25 years ago we wrote (in an introduction to Popp, 1975): "We make far too little use of the history of mathematics in our everyday teaching at all levels. But before an effective use of it can be made, there must be a wider knowledge of the elementary 'facts'. Then we can proceed to the next step of using and integrating history into the general process of teaching and learning mathematics."

It seemed to us then that history of mathematics did not have the place it deserves in mathematics education. Therefore we thought it appropriate to start with a small text to provide the basics of the history of some of the topics in the school curriculum. That was a first step in a long-term and ongoing program which we describe in the following.

As a non-funded companion to larger funded curriculum development projects in mathematics, our work has suffered from some objective constraints which have made it a low key rather than a full-fledged program. Nevertheless, it can be called a program because:

- it is based on a rationale for the use of history of mathematics in mathematics education;
- it has internal coherence;
- it has produced a considerable amount of material;
- its content is intimately related to a large ongoing curriculum development project;
- it has a very particular population as its target, the middle school years (grades 7 to 9) studying the official syllabus in Israel;
- it addresses both teachers' and students' needs and interests;
- it includes implementation and dissemination of the products in teacher courses and classrooms; and finally,
- it continues.

The following description is an overview of the whole program, and we refer to our already published reports for details on particular issues. The description is divided into two main parts: history for teachers and history for students.

History for teachers

Our starting point was based on an intuitive assumption, which is in agreement with claims in the literature, that teachers can profit from studying the history of mathematics (see, for example, reports by the MAA in 1935, or by the British Ministry of Education in 1958, and many individual authors such as Barwell, 1913; Jones, 1969; Shevchenko, 1975; Grattan-Guinness, 1978; Rogers, 1980; Struik, 1980 and Arcavi *et al.,* 1982). Later on, we confirmed empirically some initial claims and intuitions and also refined or redefined some of the goals, as we describe below. The following are the main characteristics of our work with teachers, as we established them *a priori.*

Active participation. The book by Popp (1975) was itself primarily devoted to teachers and designed to be read, in order to create a "wider knowledge of the elementary facts". However, our concept of historical work with teachers (as well as with students) was more in the direction of "workshops", in which teachers read and work. Thus the reading is not an end in itself; rather it is the raw material for "doing" mathematics and for oral and written discussions. Our aim was, and is, that learning should be achieved by "doing" and "communicating".

"Conceptual" history. We decided to concentrate, as far as possible, on the history of the development of a mathematical topic. In other words, we attempted to trace (at least in outline) the evolution of a concept or an idea, the different ways mathematicians in the past approached it, the difficulties, the gradual process of symbolization and formalization, and so on. Facts such as dates, biographical information, some historico-mathematical context, and anecdotes were included, as important in themselves to develop historical literacy, and as necessary background to the "conceptual" evolution, but not as a main goal. These remarks are exemplified later when we describe the content of the Negative Numbers sequence of worksheets.

Relevance. After deciding on our target teacher population (grade 7-9 teachers who teach according to the national syllabus), we decided that studying history would be a relevant and motivating experience if we concentrate on mathematical topics central to the syllabus they teach. Moreover, what they study should be potentially applicable to their work in the classroom. Thus, we were able to pursue a strictly mathematical objective—to deepen teachers' understanding of subtleties of a topic and even to address possible misunderstandings. In our view, relevance is also related to didactical objectives—to discuss various approaches to the topic and increase teacher awareness of possible student difficulties, which in some cases may parallel those found in the history of the topic.

Although we made relevance a necessary condition, it was not sufficient. Within the relevant topics, we chose those for which we found readable sources (see below), and whose history, on the one hand lends itself well to the mathematical and pedagogical objectives described above, and on the other, is not too complicated.

Primary sources. We decided to build the workshops around primary sources and historical documents. The symbols, the language and the approach of primary sources not only provide a genuine flavor of the past, but they also constitute an opportunity for some non-mediated encounters with the very stuff of history. Moreover, in many cases, secondary sources have their problems (e.g., May, 1975, Bruckheimer & Arcavi, 1995a).

However, primary sources are often difficult to read. Thus we were very careful in our selection, taking length and content complexity into account. In addition, we invested considerable effort in the design of worksheets to support the reading of these sources, as described in the following.

There may be significant didactical benefit to be gained from studying primary sources, in that they can be used to enhance our sensitivity to ways of thinking and communicating mathematics other than those to which we are used. Since the mathematical topic chosen is known to the teachers (see relevance above), and they know they are facing a correct and legitimate, but different approach, they have to set aside for a while their own views in order to understand a meaningful text. We believe that if this kind of activity is exercised often enough, it may sensitize teachers towards the search for coherence, structure, and even "correct" conceptualizations present in much of what students say.

As teachers, we very often and very quickly tend to dismiss student arguments when they do not exactly follow the lines we expect. Learning to listen to what our students say needs the kinds of de-centering we can exercise through reading and trying to understand historical sources.

The worksheets. We created three sequences of worksheets—negative numbers (described in Arcavi *et al.,* 1982), irrational numbers (Arcavi *et al.,* 1987), and linear and quadratic equations (Arcavi, 1985).

With a few exceptions, each worksheet is based on one or more short extracts from primary sources, preceded by an introduction to set the scene and to provide some historical information. The extract is followed by leading questions to scaffold the reading: there are questions to help overcome the difficulty of unfamiliar (or even obscure) language and notation, followed by questions to apply the mathematics to the examples in the text or to other examples, and finally questions about the mathematics involved as compared to what we "know" about it today (see, for example, Arcavi and Bruckheimer, 1991).

In order to provide an ordered record of the activity, an extensive summative discussion of the ideas in the worksheets, and additional historical information, we prepared detailed answer sheets.

An appendix follows with an example of a worksheet from the sequence on the history of negative numbers: "François Viète." The contents of the three sequences of worksheets (each with its corresponding answer sheets) is as follows.

Negative Numbers

1 Introduction
2 François Viète (see appendix)
3 René Descartes
4 Contradictions that arose in the use of negative numbers
5 Nicholas Saunderson
6 Leonhard Euler
7 William Frend
8 George Peacock (brought in full in Arcavi *et al.,* 1982)
9 Formal entrance of the negative numbers to mathematics
10 Summary

Irrational Numbers

1 The Pythagoreans
2 Euclid and the Elements
3 Irrationals in the 16th and 17th centuries
4 Rafael Bombelli (discussed in detail in Arcavi & Bruckheimer, 1991)
5 Nicholas Saunderson
6 Dedekind and the definition of irrationals

Linear and Quadratic Equations

1 The Rhind Papyrus
2 The Rule of False Position
3 Babylonian Mathematics
4 Euclid and the Elements
5 Al-Khwarizmi.

The following notes on the Negative Numbers sequence illustrate the conceptual story in these sequences. The concept of negative numbers arose from their need in application contexts (e.g., as a model for debts, etc., as dealt with in the "Introduction") or by the extension of previous ideas (e.g., generalization in equation solving as dealt with also in the worksheet "René Descartes"). In the process, there were attempts i) to cope with apparent contradictions between negative numbers and the concept of proportion (as dealt with in the worksheet "Contradictions in the use of negative numbers"); ii) to extend and justify laws of operations (as in the worksheets "Leonhard Euler" and "Nicholas Saunderson"); iii) to reject negatives altogether and as a consequence the attempts to define them properly (as in the worksheets "Frend" and "Peacock"). The formal definition arises as a need to mathematize and to "tighten up loose ends" (as dealt with in the worksheet "The Formal entry of Negative Numbers into Mathematics").

Implementation. The sequences of worksheets were first used with teachers attending in-service workshops. After a brief introduction (in which we asked them to complete a questionnaire and sometimes conducted a dialogue with the participants about their previous experiences with history), we distributed the first worksheet, and encouraged teachers to work in small groups. While they worked, we wandered among the groups, with occasional interventions to add extra questions or a helpful hint if needed. When most of the teachers had finished, we convened a general discussion, after which we distributed the answer sheet, which was read before moving on to the next worksheet. At the end of the sequence, we conducted a general summary, and asked them to complete another questionnaire.

Following these first experiences, we extended the use of the materials to pre-service courses in teacher colleges, and as a course for university mathematics undergraduates. Given the form of the materials (questions and problems to work, and extensive printed answer sheets), we realized that they were also suitable for long distance learning. Thus we conducted correspondence courses, in which we commented in writing on the answers and solutions mailed to us by the participants, as well as sending them our prepared answer sheets. Recently, the negative numbers sequence is being adapted for another long distance learning setting: the internet. One of the advantages of this environment is the hypertext, which enables participants to access historical or mathematical information according to their needs and timing. A first trial is currently under way.

Some results. The response of the teachers to these experiences has been favorable. All the workshops and courses were well attended (30 to 50 participants), and the materials are used in teacher colleges in Israel. We have also had requests from abroad with occasional informal feedback.

As mentioned above, we collected some data in the form of questionnaires about teachers' opinions and knowledge before and after the courses. A summary of these data is brought in, for example, Arcavi *et al.* (1982) and Arcavi *et al.* (1987). Here we bring some findings that we found particularly revealing.

Two of the questions posed to participants before starting the course on negative numbers were:

I. When were the negative numbers formally defined?
   a) Before the Common Era (Babylonians, Greeks, etc.)
   b) Early Middle Ages (Hindus, Arabs)
   c) Between 1300–1600 (Europeans)
   d) Between 1600–1800 (Europeans)
   e) Between 1800–1900 (Europeans).

II. When did mathematicians start to use negative numbers freely?
   a) Before the Common Era (Babylonians, Greeks, etc.)
   b) Early Middle Ages (Hindus, Arabs)
   c) Between 1300–1600 (Europeans)
   d) Between 1600–1800 (Europeans)
   e) Between 1800–1900 (Europeans).

In general, a first reaction to these questions was "I don't know", or "I am not sure". We encouraged participants to make guesses. Many assigned the emergence of negative numbers to periods much earlier than was the case. But most interestingly, a large number of participants assigned an earlier date to the formal mathematical definition than to the 'free use'. We interpreted these findings as an indication that teachers may view mathematics and the evolution of mathematical ideas mostly through the lens of their teaching practices and curriculum organization. The following might seem to be the common underlying assumptions of their views.

a) "Elementary school topics appeared in early history—secondary and university topics appeared much more recently." Negative numbers are usually taught in elementary school (or early middle school), and even though students may have some calculation difficulties, this is indeed an elementary topic—thus it may have appeared quite early in history.

b) "First you define concepts, then you exercise them." In many classroom practices, concepts are not generally exercised before they have been presented, defined or explained. Assigning this 'chronology' to the evolution of the concept would suggest that a formal definition of negative numbers preceded their free use.

Indeed, alternative explanations can be offered, such as: some teachers may not be fully aware of what a formal definition is, or the order in which the questions were posed biased the answers, etc. Nevertheless, we suggest that our interpretation was valid for at least some of the participants, and there was confirmation of this in the answers to the questionnaire after the workshop. Here are some quotes from the responses to the question "What did you learn in the workshop from the point of view of (a) history, (b) didactics and (c) mathematics?"

> A fascinating illustration of the fact that in many cases the use of concepts far preceded the possibility of human thought to define those concepts in a correct way.
>
> It became clear to me that one should distinguish between didactics of mathematics and pure mathematics.
>
> The informal explanation of the formulae are very intuitive and speak to your common sense.
>
> The definition of negative numbers.
>
> I knew about the development from the naturals to the whole numbers, but I never imagined that there were arguments and battles even when the necessity for extension had been recognized.

Such findings helped us to distill a goal of history of mathematics as a way to provide a more appropriate view of mathematics and mathematical activity.

Another goal we re-established as a consequence of the workshop experiences, relates to teacher sensitivity to alternative ways of doing mathematics. In the answers to the questionnaire we also found the following: "I received support for the view that no student answer should be dismissed, but one should relate to them all, think about them and discover their rationale". It seems that, by reading primary sources in which notation and justification arguments are very different from ours, and by being compelled to understand them, some teachers developed more sensitivity and openness towards idiosyncratic ways of doing and expressing mathematics.

Finally, verbal explanations of mathematical properties seem to have didactical appeal and can enrich the teachers' repertoire of explanations to be used when needed.

Less comprehensive developments. In addition to the sequences of worksheets, history is often a component in many in-service activities, if not the major component certainly as background. In this role, we used history in two ways:

1) "History" as information about faces and lives behind the names of theorems (or concepts) and the "antiquity" of the issues, or
2) "History" as a source of inspiration for mathematical and didactical developments.

An instance of 1) is the following. Towards the end of a two-week teacher workshop on Euclidean geometry that integrates computer explorations, we devoted two sessions to the Euler line and the Feuerbach (nine-point) circle. In the introduction and conclusion we discussed the history of the two topics and included anecdotal material about Euler and the Feuerbach family as briefly described below. The center of the nine-point circle was not on the Euler line if only because the nine-point circle had not yet been revealed. The Feuerbach circle had only six points on it; the other three had already been revealed a couple of years earlier, but Feuerbach apparently did not know about it. Nevertheless, we argued in favor of calling the nine-point circle the Feuerbach circle (as it is usually known on the continent of Europe) because of his beautiful theorem—the Feuerbach circle of a triangle is tangent to the inscribed and the three escribed circles of the triangle. A little about the gifted and successful members of the Feuerbach family adds to the pathos of Feuerbach's short and sad life, including the story of his last appearance as a classroom teacher with a drawn sword, threatening to behead any student who could not solve the questions he had written on the board (see, for example, Guggenbuhl, 1955). The elegant mathematics of this topic is thus made even more attractive and lively.

An example of 2) is our long-standing romance with fractions, especially unit fractions, and the combination of fractions defined by $(a/b)\oplus(c/d) = (a+c)/(b+d)$. This is certainly not a well-defined operation on the rational numbers (i.e., equivalence classes of fractions), because different representatives (fractions) of the same rational number yield different results (for example, $(1/2) \oplus (4/7) = 5/9$ and $(1/2)\oplus(8/14) = 9/16$.) However, it is well defined on individual fractions (where $4/7 \neq 8/14$) and models "real" situations. For example, the probability of drawing a white cube from my right-hand pocket which contains 2 identical cubes, one white and one red, is $1/2$. The probability of drawing a white cube from my left-hand pocket which contains 14 identical cubes, 8 white and 6 red, is $8/14$. And when we put all the cubes in a hat, the probability of drawing a white cube becomes $(1/2) \oplus (8/14) = 9/16$.

The operation $\oplus$ is also a useful technique for finding a number which is between the two fractions combined by $\oplus$. Chuquet in his *Triparty*[1] used this property of $\oplus$ to approximate roots of equations. This topic has led us across history from ancient Egypt to very modern times, and we have ranged widely across mathematics.

It all started with some work with teachers on Egyptian unit fractions and the $2/n$ table on the recto of the Rhind Papyrus. This led, via Gillings (1972) and Bruckheimer and Salomon (1977), to the infinite number of unit fraction expressions for any given fraction, to the Fibonacci-Sylvester[2] algorithm for obtaining one such expression. This algorithm is direct and "obvious", but the Farey sequence algorithm was much more surprising. The Farey sequence[3] of order n consists of all the reduced fractions between 0 and 1, whose denominators do not exceed n, arranged in order. These sequences have two fundamental properties: one is that two adjacent fractions differ by a unit fraction, which is the property we need in the above algorithm. The second is that, if a/b, c/d, e/f are consecutive fractions in a Farey sequence, then $c/d = (a+e)/(b+f)$, that is, the combination of fractions defined above. In trying to supply some historical background on Farey and these sequences, we not only uncovered, once more, a tangled tale of attribution (Bruckheimer and Arcavi, 1995a), but also a remarkable paper by Georg Pick (1859–1943) containing Pick's (area) theorem and some geometrical proofs of the Farey sequence properties. We added some of the missing proofs and thus developed an interesting application of simple Euclidean geometry to algebra (Bruckheimer and Arcavi, 1995b).

This was a "meta-use" of history, namely, the historical search we undertook as background to the preparation of materials, inspired us to a much more interesting presentation of the material itself and the mathematical interconnections thereof.

We came across the combination of fractions also in connection with some historical approximations to π (Bruckheimer and Markovits, 1986) and Simpson's Paradox in statistics (e.g., Lord, 1990).

None of this has, as yet, been written up as a sequence of worksheets in the form of those described earlier, and little in the form of material for the classroom as described in the next section, but it clearly can serve very well the objectives we have for both—and one more objective, at least—the interconnectedness of mathematics when viewed in an appropriate way, even at the level of grades 7–9. It is also remarkable, and has surprised many teachers, that something like the combination of fractions, which many regard solely as a prevalent student error, to which an irrelevant (and unhelpful) response is commonly something like "that's not the way to add" with an example to show that it contradicts what we expect of addition, should prove to be so rich an historical and mathematical topic.

History in the classroom

We wrestled with the form of historical materials for the classroom for a long time and our work in this area is more recent. The historical aside in the regular mathematics texts may have a non-negligible amusement value and perhaps provide some motivation. Both motivation and amusement in the mathematics classroom have much to be said for them, but we felt that history has more to offer.

At the other extreme, an integrated historical approach to the whole curriculum, even though it may appeal to many of us, may not be practical, appropriate, or efficient at the present time, if at all. Therefore, we decided to develop relatively short (2–3 lesson periods at most) historical "happenings" with strong ties to the curriculum. One such is described in its developmental form in Ofir and Arcavi, 1992; the "final" form is not significantly different.

These historical happenings do not eliminate either the use of primary sources (where possible) or the active participation of the "learner", components which were so important to us in our work with teachers, but the doses of both components are in accordance with the needs and capabilities of grade 7–9 students. The primary sources are very short, often no more than a question or a 3–4 line quotation. The whole is meant to be teacher "driven" but not directly taught. Rather the teacher is there to set the scene and to discuss the ideas and problems with the class, but with much interspersed student activity. We provide the teacher with a detailed text of the activity and its background and all the illustrative material in the form of overhead projector transparencies. The latter are either what we call "wallpaper"—such as reproductions from the original texts, pictures of the mathematicians, etc.—or an integral part of the activity—quotations, problems and their solutions, and issues for discussion. Each happening is narrowly focused on a very small topic in, or relevant to, the curriculum. The package the teacher receives is "self-contained" in the sense that it covers not only the activity itself, but also historical and didactical background.

The overall general objectives of the classroom materials are to enrich the mathematics in the curriculum, to present mathematics as a developing dynamic activity, to enhance the understanding of curriculum topics by comparing and contrasting them with the same or similar topics in previous forms, and to see the curriculum itself as a dynamic entity. Most students have the impression that the mathematics they are learning was and will always be so, and this view may only be modified a generation later when faced with their children's homework.

So far we have developed five such happenings:

- Ancient numerals and number systems
- Arithmetic in ancient Egypt
- π and the circumference of the circle
- Word problems and equations
- Casting out nines

The first two present and discuss very different number systems from ours, the way operations were performed in those systems, and touch upon the beginning of the fractions story described in the previous section. One of the specific objectives here is to consider (in greater depth than the curriculum itself, which also touches on other numeration systems) our decimal system in the light of others such as those of the Babylonians, Egyptians, Romans and the Maya. The characteristics of our decimal system with its advantages and possible disadvantages become more explicit when they are contrasted with other systems. So far, we have only anecdotal evidence on the implementation of these first two happenings, but it is worth quoting a student. She was amazed by the Egyptian multiplication system (which can be applied to our decimal system as well)—doubling, selecting the appropriate "doubles" and adding—to the point that she engaged the teacher in a discussion of why they were not taught this method before. In this case history was much more than motivation or amusement; it was the background against which she had a serious opportunity to talk about mathematics and to challenge the characteristics and convenience of a certain notation system and its profound implications for performing arithmetical operations.

The number π gets scant treatment in our curriculum and is sometimes misunderstood (see, for example, Arcavi *et al.*, 1987). So "π and the circumference of the circle", above all else closes a gap. We consider Babylonian, Greek and Chinese approximations. Pythagoras's Theorem is applied often and usefully and irrationality becomes a subject for informal discussion. The happening has a particular local flavor in its final section, in which we discuss Hebrew sources. For example, in the Bible π is apparently taken to be 3, but there are other theories. In addition, we use an extract from the twelfth century writings of Maimonides, from which it seems clear that he "knew" that π is irrational.

"Word problems and equations" is described in detail in Ofir and Arcavi (1992). It is essentially a discussion of problems in pre-algebra which are equivalent to solving $ax = b$ in algebra. Again the comparisons and contrasts between the methods used in the past and those the students have recently learned are designed to clarify and give more meaning to the latter. In particular, the objective is to regard the study of algebra in a different light, by showing what immense power is packed "into that little x".

"Casting out nines" may, at first sight, seem a surprising choice. After all there are plenty of mathematics topics which have dropped out of the curriculum, and this would seem to be just another one. But when we started working on it, we were surprised at its richness and relevance to the curriculum. We begin the activity with the remark that it is "the story of how you [the student] would probably have been doing arithmetic if you had been born some sixty years ago [or earlier]." So, it is "history" but not only in the sense of an outdated and no longer in use method. The activities are organized around bygone textbooks which are, in this case, the primary sources, and show clearly that textbooks have changed and are changing significantly.

There is a lot of "good" mathematics in this topic, which is also elementary enough to be kept well within the limits of grade 7–9. To give some idea of the content, we describe it here briefly and somewhat abstractly (more details can be found in Bruckheimer *et al.,* 1995).

First, we work on questions around the theme $a \pmod 9 \cdot b \pmod 9 = ab \pmod 9$ and its use as a check for the calculation of ab. The next stage is to show that we do not need to divide by 9 to find $a \pmod 9$—it can be found by adding the digits of a (mod 9 if necessary)—hence the name "casting out nines". Throughout we quote texts from earlier this century and earlier centuries; the quotations are usually statements (often there is no more in the original) which the students are asked to test, discuss, justify or reject.

Once the method is understood and justified, we consider the question of its reliability. Casting out nines provides a necessary check for correctness, but it is not sufficient. We use the obvious ways to demonstrate this by presenting a calculation that is incorrect but casting out nines checks out correctly. This leads to the question of what sort of error would remain undetected—for example, a copying error in which two digits are interchanged, or incorrect positioning of a row in the working of a long multiplication. In this section of the activity, we come to the conclusion that casting out nines has the advantage of simplicity, but it is not entirely reliable. We then raise a new question. Perhaps using the same idea, but a number other than 9, will give us a more reliable check, without undue loss of simplicity. To open the discussion we found an appropriate primary source in Hatton's *Intire System of Arithmetic* (1731, p. 54).

Checks by 2 and 10 are indeed simple, but how reliable are they? There is plenty of classroom material here. The final section considers, in particular, checks using 7 (which is not simple, but more reliable in a certain sense) and 11 (which is simple and at least as reliable as 9, if not more so), and yet 11 was never popular. The appendix to the activity brings a remarkable primary source—a letter from Rev. C. L. Dodgson to *Nature*—about how one can find the quotient on division by 9 or 11, not just the remainder, without division. An interesting historical snippet.

Epilogue

We live in an era in which our approach to information and knowledge are undergoing significant and rapid transformation. The amount of knowledge we need to cope with is several orders of magnitude greater than it was in the not so distant past. As a consequence, education in general, and especially mathematics education, is less committed to specific content and more oriented towards strategies, tools, heuristics and dispositions to cope with developing subject matter.

This tendency is explicitly spelled out in the *Professional Standards for Teaching Mathematics* (NCTM, 1991), according to which some of the major shifts in mathematics education consist of moving away from mechanistic answer finding, toward conjecturing, inventing and problem solving; away from merely memorizing procedures, toward mathematical reasoning; away from treating mathematics as a body of isolated concepts and procedures, toward connecting mathematics and its ideas; away from the teacher as the sole authority for right answers, toward mathematical evidence as verification. The *Standards* frame these shifts in terms of questions that should be not only legitimately asked in the classroom, but should also become common practice. For example, "Do you agree?", "Does anyone have the same answer but different ways to explain it?", "What is alike and what is different about your method of solution and hers?", "Do you understand what they are saying?", "Does that make sense?", "Does

to pretend to prove Multiplication by caſting out the Nines, is a Miſtake, as I have elſewhere demonſtrated; for why divide by 9 more than 2, or any other Digit, which would prove the Work as well? But the eaſieſt way is to divide the Factors by 10, and the Product of the Remainers by 10, which will leave a Remainer equal to that of the Product divided by 10.

FIGURE 1

that always work?", "Can you think of a counterexample?", "What assumptions are you making?", "How does this relate to...?", and so on.

The historical materials we have developed share this spirit. These types of question abound in both our materials for students and teachers. Therefore, we firmly believe that history is now (possibly even more that it used to be) a rich source for the ongoing development of educational materials which pursue the above objectives.

References

Arcavi, A., 1985, *History of Mathematics as a Component of Mathematics Teachers Background,* Unpublished PhD Thesis, Weizmann Institute of Science, Israel.

Arcavi, A., M. Bruckheimer & R. Ben-Zvi: 1982, "Maybe a Mathematics Teacher can Profit from the Study of the History of Mathematics," *For the Learning of Mathematics,* 3 (1), pp. 30–37.

——: 1987, "History of Mathematics for Teachers: the Case of Irrational Numbers," *For the Learning of Mathematics,* 7 (2), pp. 18–23.

Arcavi, A. & M. Bruckheimer: 1991, "Reading Bombelli's x-purgated Algebra," *College Mathematics Journal,* 22 (3), pp. 212–219.

Barwell, M.: 1913, "The Advisability of Including Some Instruction in the School Course on the History of Mathematics," *The Mathematical Gazette,* 7, pp. 72–79.

British Ministry of Education, 1958, *Pamphlet 36,* H.S.M.O., pp. 134–154.

Bruckheimer, M. & Y. Salomon: 1977, "Some Comments on R. J. Gillings's Analysis of the 2/n table in the Rhind Papyrus," *Historia Mathematica,* 4, pp. 445–452.

Bruckheimer, M. & Z. Markovits: 1986, "Several Incarnations of McKay," *Mathematics Teacher,* 79 (4), 294–297.

Bruckheimer, M. & A. Arcavi: 1995a, "Farey Series and Pick's area theorem," *The Mathematical Intelligencer,* 17 (4), pp. 64–67

——: 1995b, "A visual approach to some elementary number theory," *Mathematical Gazette,* 79 (486), pp. 471–478.

Bruckheimer, M., R. Ofir & A. Arcavi: 1995, "The case for and against casting out nines," *For the Learning of Mathematics,* 15 (2), pp. 24–35.

Gillings, R. J.: 1972, *Mathematics in the Time of the Pharaohs,* MIT Press: Cambridge, MA.

Grattan-Guinness, I.: 1978, "On the Relevance of the History of Mathematics to Mathematics Education," *International Journal of Mathematics Education in Science and Technology,* 8, pp. 275–285.

Guggenbuhl, L.: 1955, "Karl Wilhelm Feuerbach, Mathematician," *The Scientific Monthly,* 81, pp. 71–76.

Hardy, G. H. & E. M. Wright: 1945, *An Introduction to the Theory of Numbers,* Second Edition, Clarendon Press, London.

Jones, P., 1969, "The History of Mathematics as a Teaching Tool," In *Historical Topics for the Mathematics Classroom,* Thirty First Yearbook, NCTM: Reston, VA, pp. 1–17.

Lord, N., 1990, "From Vectors to Reversal Paradoxes," *The Mathematical Gazette,* 74(467), pp. 55–58.

Mathematical Association of America (MAA), 1935, "Report on the Training of Teachers of Mathematics," *American Mathematical Monthly,* 42, pp. 57–61.

May, K. O.: 1975, "Historiographic Vices. I. Logical Attribution," *Historia Mathematica,* 2, pp. 185–187.

Ofir, R. and A. Arcavi: 1992, "Word Problems and Equations—An Historical Activity for the Algebra Classroom," *Mathematical Gazette,* 76 (475), pp. 69–84.

Popp, W.: 1975, *History of Mathematics. Topics for Schools,* translated from the German by Maxim Bruckheimer, Transworld Publishers: London, England.

Rogers, L.: 1980, *Newsletter of the International Study Group on the Relationship between History and Pedagogy of Mathematics.*

Shevchenko, I. N.: 1975, "Elements of the Historical Approach in Teaching Mathematics," *Soviet Studies in Mathematical Education,* 12, pp. 91–139.

Struik, D. J.: 1980, "Why Study the History of Mathematics?," *The Journal of Undergraduate Mathematics and its Applications,* 1, pp. 3–28.

Endnotes

[1] The date for the original manuscript *Le Triparty en la Science de Nombres* is 1484. It remained unpublished until 1880, when its first section appeared in vol. XIII of the *Bullettino di Bibliografia e di Storia delle Scienze Matematiche e Fisiche Publicato da B. Boncompagni.* A recent English version was published in Flegg, Hay and Moss (1985) to commemorate 500 years of the original manuscript.

[2] Again it is interesting to speculate on the strange and often tangled tale of attribution and the historical application of name plaques. Apparently Fibonacci (c. 1170–c. 1250) described the algorithm and so did J. J. Sylvester (1814–1897) independently, some 700 years later. The algorithm is 'pure common sense'. To find the unit fraction expression for a/b, subtract the nearest unit fraction not greater than a/b and continue thus with the remainder until the remainder is a unit fraction. Sylvester proved that the method was in fact an algorithm.

[3] Apparently it was A. L. Cauchy (1789–1857) who caused the sequences to be labelled with the name of J. Farey, an English mineralogist. Cauchy saw a translation of a letter to the *Philosophical Magazine* (1816), in which Farey noticed one property of the sequence (and in that he was not the first) without proving it. Cauchy supplied the proof quoting the conjecture from Farey's letter and since then the name tag has remained.

Worksheet: François Viète

François Viète (or Franciscus Vieta in the Latin spelling), a Frenchman was born in 1540 and died in 1603. Viète studied law and most of his life worked in public service, devoting his spare time to mathematics, in which he made his fame. He wrote books on algebra and geometry. His important work, *In Artem Analyticem Isagoge* (*Introduction to the Analytic Art*), appeared in 1591.

One of the innovations attributed to Viète is the use of letters, not only for variables, but also for coefficients. He thus made it possible to deal with general algebraic forms. Even though the notation in his work is more complicated than that in use today, his contribution to the advance of algebra is highly rated and he has been dubbed the "father of modern algebra".

The following are excerpts from the above book. The text is taken from the English translation brought by J. Klein, *Greek Mathematical Thought and the Origin of Algebra* (MIT Press, 1968), where also more details on Viète and his contribution to mathematics can be found.

VIETA'S *ANALYTIC ART*

Precept II

To subtract a magnitude from a magnitude

Let there be two magnitudes A and B, and let the former be greater than the latter. It is required to subtract the less from the greater. ... subtraction may be fittingly effected by means of the sign of the disjoining or removal[24] of the less from the greater; and disjoined, they will be A "minus" B,...

Nor will it be done differently if the magnitude which is subtracted is itself conjoined with some magnitude, since the whole and the parts are not to be judged by separate laws; thus, if "B 'plus' D" is to be subtracted from A, the remainder will be "A 'minus' B, 'minus' D," the magnitudes B and D having been subtracted one by one.

But if D is already subtracted from B and "B 'minus' D" is to be subtracted from A, the result will be "A 'minus' B 'plus' D," because in the subtraction of the whole magnitude B that which is subtracted exceeds by the magnitude D what was to have been subtracted. Therefore, it must be made up by the addition of that magnitude D.

The analysts, however, are accustomed to indicate the performance of the removal by means of the symbol —....

But when it is not said which magnitude is greater or less, and yet the subtraction must be made, the sign of the difference is: $=$, i.e., when the less is undetermined; as, if "A square" and "B plane" are the proposed magnitudes, the difference will be: "A square$=B$ plane," or "B plane$=A$ square."

VIETA'S *ANALYTIC ART*

Precept III

To multiply a magnitude by a magnitude

Let there be two magnitudes A and B. It is required to multiply the one by the other. ... their product will rightly be designated by the word "*in* or "*sub*," as, for example, "A in B," by which it will be signified that the one has been multiplied by the other...

If, however, the magnitudes to be multiplied, or one of them, be of two or more names, nothing different happens in the operation.[27] Since the whole is equal to its parts, therefore also the products under the segments of some magnitude are equal to the product under the whole. And when the positive name[28] (nomen adfirmatum) of a magnitude is multiplied by a name also positive of another magnitude, the product will be positive, and when it is multiplied by a negative name (nomen negatum), the product will be negative.

From which precept it also follows that by the multiplication of negative names by each other a positive product is produced, as when "$A=B$" is multiplied by "$D=G$"...

... since the product of the positive A and the negative G is negative, which means that too much is removed or taken away, inasmuch as A is, inaccurately, brought forward (producta) as a magnitude to be multiplied...

... and since, similarly, the product of the negative B and the positive D is negative, which again means that too much is removed, inasmuch as D is, inaccurately, brought forward as a magnitude to be multiplied... ... Therefore, by way of compensation, when the negative B is multiplied by the negative G, the product is positive.

Questions

1. Translate all the formulae of Precept II and Precept III into modern notation.
2. Viète explains some of the formulae. Compare his explanations with those which you (or the textbook you use) give.
3. According to this passage, how does Viète regard negative numbers?
4. The main mathematical preoccupation of the Greeks was geometric. How would they have demonstrated (or justified) the rules in Precept III?

Answer sheet: François Viète

1. Precept II

Viète	In modern notation
"Let there be two magnitudes A and B, and let the former be greater than the latter"	A, B $A > B$
"...and disjoined, they will be A "minus" B ..."	$A - B$
"...if "B 'plus' D" is to be subtracted from A, the remainder will be "A 'minus' B, 'minus' D"..."	$A - (B + D) = A - B - D$
"But if D is already subtracted from B and "B 'minus' D" is to be subtracted from A, the result will be "A 'minus' B 'plus' D"..."	$A - (B - D) = A - B + D$
"A square = B plane" or "B plane = A square"	$\mid A^2 - B \mid$

Precept III

Viète	In modern notation
"A in B"	AB, $A \cdot B$ or $A \times B$
"If, however, the magnitudes to be multiplied, or one of them, be of two or more names, ... when the positive name of a magnitude is multiplied by a name also positive of another magnitude, the product will be positive, and when it is multiplied by a negative name the product will be negative"	$A(B + C)$ or $(A + B)(C + D)$ $\underline{A}(B\underline{+C}) = AB\underline{+AC}$ $\underline{A}(B\underline{-C}) = AB\underline{-AC}$

(Note: from the text in Precept II, it is clear that, in this passage, Viète does not mean $-C$ standing on its own, but rather $B - C$, (or $B = C$), in which he will call, as later, "– C" the "negative C".)

Viète	In modern notation
"...as when "A = B" is multiplied by "D = G", since the product of the positive A and the negative G is negative... the product of the negative B and the positive D is negative..."	$(A - B)(D - G)$ $\underline{A}(D\underline{-G}) = AD\underline{-AG}$ $(A\underline{-B)D} = AD\underline{-BD}$

(Note: By "inasmuch as A is, inaccurately, brought forward (producta) as a magnitude to be multiplied...", Viète means that we are required to multiply, not by A, but by $A - B$ which is less than A. Hence in the product $A(D - G)$ "too much is removed".)

Viète	In modern notation
"Therefore, by way of compensation, when the negative B is multiplied by the negative G, the product is positive"	$(A\underline{-B})(D\underline{-G}) = AD - BD - AG\underline{+BG}$

Note: Viète's justification, although based on arguments of magnitude, seems to be more qualitative than quantitative: we twice subtracted too much, so we have to add some on, but there is no explicit justification that the "compensation" is exact.)

2. The explanation will depend on the textbook and the level of the students. Thus $A - (B + D) = A - B - D$ can be justified, for example, by movements on the number line, or by regarding the '$-$' in front of the brackets as equivalent to multiplication by -1, and then using the distributive law. The significance of this question is in the discussion of different approaches.

3. This point has been touched upon in the notes written in answer to Question 1, but we will detail it again here. If one reads the quote from Viète superficially, one may get the impression that he deals with negative numbers and that he even "deduces" the law of signs in Precept III. But Viète did not admit negative numbers in any sense, as can be seen from

i) In order to perform the subtraction "A 'minus' B", the first magnitude must be greater than the second.

ii) If one does not know which of the magnitudes is the greater, then Viète uses the special symbol $=$, to indicate that the smaller of the two is to be subtracted, i.e., to avoid obtaining a negative result.

iii) Stating the law of signs does not imply recognition of negative numbers.

The Greeks were aware of the extended distributive law (see also the answer to Question 4) and that the product of two subtracted numbers (as opposed to negative numbers) gives a result which has to be added and not subtracted. Thus the use of the law of signs does not imply acceptance of negative numbers. In all our reading of historical source material, we have to be careful not to attach unjustified modern meaning to what we read. The strength of our modern symbolism is its generality, but in deducing that Viète's Precept III can be written in the form $(A-B)(D-G) = AD - BD - AG + BG$, we have to be careful to remember the limitations he imposes, i.e., $A > B$, $D > G$.

4. In answer to this question we do not expect a historically faithful demonstration (see later), but rather something like the following, which is also found in modern school texts.

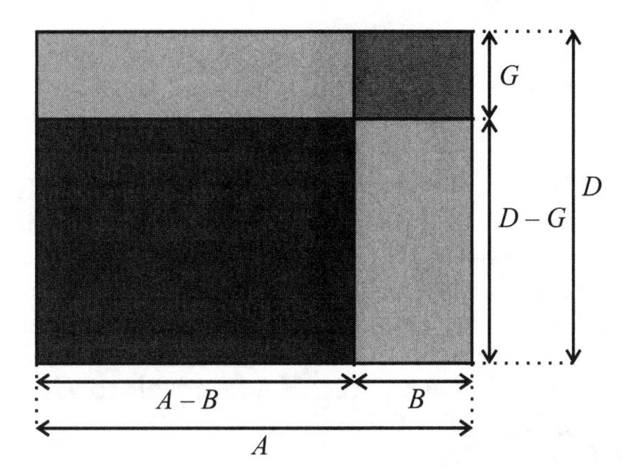

The area of the rectangle whose sides are of length $A-B$ and $D-G$ is $(A-B)(D-G)$. But from the figure, we can obtain this area as the area of the large rectangle $(A \times D)$, less the area of the two bounding rectangular strips ($A \times G$ and $B \times D$), except that we have removed the small rectangle $(B \times G)$ twice. So we have to add it back on once.

Since Viète was influenced by Diophantus,* it seems reasonable to suggest that he described in words and "rudimentary" symbolism, what the Greeks drew, so helping in the process of giving algebra an existence separate from geometry.

To get the flavor of an original proof of what we today regard as an algebraic result, we show Proposition 7 from Book II of Euclid's *Elements,* as it appears in the 7th edition of the English version of *Chales' Euclid* (London, 1726, p. 106).

The result is algebraically equivalent to $a^2 + b^2 = 2ab + (a-b)^2$, where a is the whole segment, and b is the first part.

* Diophantus of Alexandria lived in the third century CE. According to Kline, M., *Mathematical Thought from Ancient to Modern Times,* Oxford University Press, 1972, "The highest point of Alexandrian Greek algebra is reached with Diophantus." His major extant work is the apparently incomplete *Arithmetica.*

PROPOSITION VII.

A PROBLEM.

If a Line be divided, the Square of the whole Line with that of one of its Parts, is equal to two Rectangles contain'd under the whole Line, and that first Part, together with the Square of the other Part.

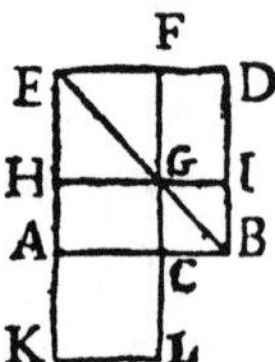

LET the Line AB be divided any where in C; the Square AD of the Line AB, with the Square AL, will be equal to two Rectangles contain'd under AB and AC, with the Square of CB. Make the Square of AB, and having drawn the Diagonal EB, and the Lines CF and HGI; prolong EA so far, as that AK may be equal to AC; so AL will be the Square of AC, and HK will be equal to AB; for HA is equal to GC, and GC, is equal to CB, because CI is the Square of CB *by the Coroll. of the* 4.)

Demonstration.

'Tis evident, that the Squares AD and AL are equal to the Rectangles HL and HD, and the Square CI. Now the Rectangle HL is contain'd under HK equal to AB, and KL equal to AC. In like manner the Rectangle HD is contain'd under HI equal to AB, and HE equal to AC. Therefore the Squares of AB and AC are equal to two Rectangles contain'd under AB and AC, and the Square of CB.

In Numbers.

Suppose the Line AB to consist of 9 Parts, AC of 4, and CB of 5. The Square of AB 9 is 81, and that of AC 4 is 16; which 81 and 16 added together make 97. Now one Rectangle under AB and AC, or four times 9, make 36, which taken twice, is 72; and the Square of CB 5 is 25; which 72 and 25 added together make also 97.

Part V

The History of Mathematics

Mesopotamian Mathematics: Some Historical Background

Eleanor Robson
University of Oxford

Introduction

> When I was young I learned at school the scribal art on the tablets of Sumer and Akkad. Among the high-born no-one could write like me. Where people go for instruction in the scribal art there I mastered completely subtraction, addition, calculation and accounting.[1]

Most mathematicians know at least a little about 'Babylonian' mathematics: about the sexagesimal place value system, written in a strange wedge-shaped script called cuneiform; about the very accurate approximation to $\sqrt{2}$;[2] and about the famous list of Pythagorean triples, Plimpton 322.[3] This kind of information is in most math history books. So the aim of this article is not to tell you about things which you can easily read about elsewhere, but to provide a context for that mathematics—a brief overview of nearly five millennia of mathematical development and the environmental and societal forces which shaped those changes.[4]

So where are we talking about, and when? The Greek word 'Mesopotamia' means 'between the rivers' and has referred to the land around the Tigris and Euphrates in modern day Iraq since its conquest by Alexander the Great in 330 BCE. But its history goes back a good deal further than that. Mesopotamia was settled from the surrounding hills and mountains during the course of the fifth millennium BCE. It was here that the first sophisticated, urban societies grew up, and here that writing was invented, at the end of the fourth millennium, perhaps in the southern city of Uruk. Indeed, writing arose directly from the need to record mathematics and accounting: this is the subject of the first part of the article. As the third millennium wore on, counting and measuring systems were gradually revised in response to the demands of large-scale state bureaucracies. As the second section shows, this led in the end to the sexagesimal, or base 60, place value system (from which the modern system of counting hours, minutes and seconds is ultimately derived).

By the beginning of the second millennium, mathematics had gone beyond the simply utilitarian. This period produced what most of the text-books call 'Babylonian' mathematics, although, ironically, it is highly unlikely that any of the math comes from Babylon itself: the early second millennium city is now deep under the water table and impossible to excavate. The third part of this article examines the documents written in the scribal schools to look for evidence of how math was taught at this time, and why it might have moved so far from its origins. But after the mid-second millennium BCE we have almost no knowledge of mathematical activity in Mesopotamia, until the era of the Greek conquest in the late fourth century BCE—when math from the city of Babylon is known. The fourth and final part looks at why there is this enormous gap in the record: was there really very little math going on, or can we find some other explanations for our lack of evidence?

Counting with clay: from tokens to tablets

But now let us start at the beginning. The Tigris-Euphrates valley was first inhabited during the mid-fifth millennium BCE. Peoples who had already been farming the surrounding hills of the so-called 'Fertile Crescent' for two or three millennia began to settle, first in small villages, and then in increasingly large and sophisticated urban centres. The largest and most complex of these cities were Uruk on the Euphrates, and Susa on the Shaur river. Exactly why this urban revolution took place need not concern us here; more important to the history of mathematics are the consequences of that enormous shift in societal organisation.

Although the soil was fertile and the rivers full, there were two major environmental disadvantages to living in the southern Mesopotamian plain. First, the annual rainfall was not high enough to support crops without artificial irrigation systems, which were in turn vulnerable to destruction when the rivers flooded violently during each spring harvest. Second, the area yielded a very limited range of natural resources: no metals, minerals, stones or hard timber; just water, mud, reeds and date-palms. Other raw materials had to be imported, by trade or conquest, utilised sparingly, and recycled. So mud and reeds were the materials of everyday life: houses and indeed whole cities were made of mud brick and reeds; the irrigation canals and their banks were made of mud reinforced with reeds; and there were even some experiments in producing agricultural tools such as sickles from fired clay.

It is not surprising then that mud and reeds determined the technologies available for other everyday activities of urban society, such as managing and monitoring labour and commodities. The earliest known method of controlling the flow of goods seems to have been in operation from the time of the earliest Mesopotamian settlement, predating the development of writing by millennia [Nissen, Damerow and Englund, 1993: 11]. It used small clay 'tokens' or 'counters', made into various geometric or regular shapes. Each 'counter' had both quantitative and qualitative symbolism: it represented a specific number of a certain item. In other words it was not just a case of simple one-to-one correspondence: standard groups or quantities could also be represented by a single token. It is often impossible to identify exactly which commodity a particular token might have depicted; indeed, when such objects are found on their own or in ambiguous contexts, it is rarely certain whether they were used for accounting at all. The clearest evidence comes when these tokens are found in round clay envelopes, or 'bullae', whose surfaces are covered in impressed patterns. These marks were made, with an official's personal cylinder seal, to prevent tampering. The envelope could not be opened and tokens removed without damaging the pattern of the seal. In such a society, in which literacy was restricted to the professional few, these cylinder-seals were a crucial way of marking individual responsibility or ownership and, like the tokens, are ideally suited to the medium of clay.

Of course, sealing the token-filled envelopes meant that it was impossible to check on their contents, even legitimately, without opening the envelope in the presence of the sealing official. This problem was overcome by impressing the tokens into the clay of the envelope before they were put inside. It then took little imagination to see that one could do without the envelopes altogether. A deep impression of the tokens on a piece of clay, which could also be sealed by an official, was record enough.

At this stage, c. 3200 BCE, we are still dealing with tokens or their impressions which represent both a number and an object in one. A further development saw the separation of the counting system and the objects being counted. Presumably this came about as the range of goods under central control widened, and it became unfeasible to create whole new sets of number signs each time a new commodity was introduced into the accounting system. While we see the continuation of *impressions* for numbers, the objects themselves were now represented on clay either by a drawing of the object itself or of the token it represented, *incised* with a sharp reed. Writing had begun.[5]

Now mathematical operations such as arithmetic could be recorded. The commodities being counted cannot usually be identified, as the incised signs which represent them have not yet been deciphered. But the numerals themselves, recorded with impressed signs, can be identified with ease. For instance, one tablet displays a total of eighteen D-shaped marks on the front, and three round ones, in four separate enclosures. On the back are eight Ds and four circles, in one enclosure.[6] We can conclude that the circular signs must each be equivalent to ten Ds. In fact, we know from other examples that these two signs do indeed represent 1 and 10 units respectively, and were used for counting discrete objects such as people or sheep.

Using methods like this, a team in Berlin have identified a dozen or more different systems used on the ancient tablets from Uruk [Nissen, Damerow and Englund, 1993:

28–29]. There were four sets of units for counting different sorts of discrete objects, another set for area measures, and another for counting days, months and years. There were also four capacity measure systems for particular types of grain (apparently barley, malt, emmer and groats) and two for various kinds of dairy fat. A further system is not yet completely understood; it may have recorded weights. Each counting or measuring system was context-dependent: different number bases were used in different situations, although the identical number signs could be used in different relations within those contexts. One of the discrete-object systems was later developed into the sexagesimal place value system, while some of the other bases were retained in the relationships between various metrological units. It is an enormously complex system, which has taken many years and a lot of computer power to decipher; the project is still unfinished.

It is unclear what language the written signs represent (if indeed they are language-specific), but the best guess is Sumerian, which was certainly the language of the succeeding stages of writing. But that's another story; it's enough for our purposes to see that the need to record number and mathematical operations efficiently drove the evolution of recording systems until one day, just before 3000 BCE, someone put reed to clay and started to write mathematics.

The third millennium: math for bureaucrats

During the course of the third millennium writing began to be used in a much wider range of contexts, though administration and bureaucracy remained the main function of literacy and numeracy. This restriction greatly hampers our understanding of the political history of the time, although we can give a rough sketch of its structure. Mesopotamia was controlled by numerous city states, each with its own ruler and city god, whose territories were concentrated on the canals which supplied their water. Because the incline of the Mesopotamian plain is so slight—it falls only around 5 cm in every kilometre—large-scale irrigation works had to feed off the natural watercourses many miles upstream of the settlements they served. Violent floods during each year's spring harvest meant that their upkeep required an enormous annual expenditure. The management of both materials and labour was essential, and quantity surveying is attested prominently in the surviving tablets.

Scribes had to be trained for their work and, indeed, even from the very earliest phases around 15% of the tablets discovered are standardised practice lists—of titles and professions, geographical names, other sorts of technical terminology. From around 2500 BCE onwards such 'school' tablets—documents written for practice and not for working use—include some mathematical exercises. By this time writing was no longer restricted to nouns and numbers. By using the written signs to represent the *sounds* of the objects they represented and not the objects themselves, scribes were able to record other parts of human speech, and from this we know that the earliest school math was written in a now long-dead language called Sumerian. We currently have a total of about thirty mathematical tablets from three mid-third millennium cities—Shuruppak, Adab and Ebla—but there is no reason to suppose that they represent the full extent of mathematical knowledge at that time. Because it is often difficult to distinguish between competently written model documents and genuine archival texts, many unrecognised school tablets, from all periods, must have been published classified as administrative material.

Some of the tablets from Shuruppak state a single problem and give the numerical answer below it [Powell, 1976: 436 n19]. There is no working shown on the tablets, but these are more than simple practical exercises. They use a practical pretext to explore the division properties of the so-called 'remarkable numbers' such as 7, 11, 13, 17 and 19, which are both irregular (having factors other than 2, 3 and 5) and prime [cf. Høyrup, 1993]. We also have a geometrical diagram on a round tablet from Shuruppak and two contemporary tables of squares from Shuruppak and Adab which display consciously sexagesimal characteristics [Powell, 1976: 431 & fig. 2]. The contents of the tablets from Ebla are more controversial: according to one interpretation, they contain metrological tables which were used in grain distribution calculations [Friberg, 1986].

Mesopotamia was first unified under a dynasty of kings based at the undiscovered city of Akkad, in the late twenty-fourth century BCE. During this time the traditional metrological systems were overhauled and linked together, with new units based on divisions of sixty. Brick sizes and weights were standardised too [Powell, 1987–90: 458]. The new scheme worked so well that it was not substantially revised until the mid-second millennium, some 800 years later; indeed, as we shall see, some Akkadian brick sizes were still being used in the Greek period, in the late fourth century BCE.

There are only eight known tablets containing mathematical problems from the Akkadian period, from Girsu and Nippur. The exercises concern squares and rectangles. They either consist of the statement of a single problem and its numerical answer, or contain two stated problems which are allocated to named students. In these cases the answers are not given, and they appear to have been written by an instructor in preparation for teaching. Indeed, one of these assigned problems has a solved counterpart amongst the problem texts. Certain numerical errors suggest that the sexagesimal place system was in use for calculations, at least in prototype form [Whiting, 1984].

A round tablet from Nippur shows a mathematical diagram which displays a concern with the construction of problems to produce integer solutions. The trapezoid has a transversal line parallel to the base, dividing it into two parts of equal area. The lengths of the sides are chosen in such a way that the length of the transversal line can be expressed in whole numbers [Friberg, 1987–90: 541]. No mathematical tables are known from this period, but model documents of various kinds have been identified, including a practice account from Eshnunna and several land surveys and building plans [Westenholz, 1977: 100 no. 11; Foster, 1982: 239–40]. In working documents too, we see a more sophisticated approach to construction and labour management, based on the new metrological systems. The aim was to predict not only the raw materials but also the manpower needed to complete state-funded agricultural, irrigation and construction projects, an aim which was realised at the close of the millennium under the Third Dynasty of Ur.

The Ur III empire began to expand rapidly towards the east in the second quarter of the 21st century BCE. At its widest extent it stretched to the foothills of the Zagros mountains, encompassing the cities of Urbilum, Ashur, Eshnunna and Susa. To cope with the upkeep of these new territories and the vastly increased taxation revenues they brought in, large-scale administrative and economic reforms were executed over the same period. They produced a highly centralised bureaucratic state, with virtually every aspect of its economic life subordinated to the overriding objective of the maximisation of gains. These administrative innovations included the creation of an enormous bureaucratic apparatus, as well as of a system of scribal schools that provided highly uniform scribal and administrative training for the prospective members of the bureaucracy. Although little is currently known of Ur III scribal education, a high degree of uniformity must have been essential to produce such wholesale standardisation in the bureaucratic system.

As yet only a few school mathematical texts can be dated with any certainty to the Ur III period, but between them they reveal a good deal about contemporary educational practice. There are two serious obstacles to the confident identification of school texts from the Ur III period when, as is often the case, they are neither dated nor excavated from well-defined find-spots. Firstly, there is the usual problem of distinguishing between competently written practice documents and those produced by working scribes. Secondly, palaeographic criteria must be used to assign a period to them. In many cases it is matter of dispute whether a text is from the late third millennium or was written using archaising script in the early second millennium. In particular, it was long thought that the sexagesimal place system, which represents numerals using just tens and units signs, was an innovation of the following Old Babylonian period so that any text using that notation was assumed to date from the early second millennium or later. However, we now know that it was already in use by around 2050 BCE—and that the conceptual framework for it had been under construction for several hundred years. Crucially, though, calculations in sexagesimal notation were made on temporary tablets which were then reused after the calculation had been transferred to an archival document in standard notation [Powell, 1976: 421].[7] We should expect, then, to find neither administrative documents using the sexagesimal system nor sexagesimal school texts which were used to train the scribes (because, in general, they were destroyed after use, and we can hardly distinguish them from later examples).

One conspicuous exception to our expectations is a round model document from Girsu [Friberg, 1987–90: 541]. On one side of the tablet is a (slightly incorrect) model entry from a quantity survey, giving the dimensions of a wall and the number of bricks in it. The measurements of the wall are given in standard metrological units, but have been (mis-)copied on to the reverse in sexagesimal notation. The volume of the wall, and the number of bricks in it, are then worked out using the sexagesimal numeration, and converted back into standard volume and area measure, in which systems they are written on the obverse of the tablet. These conversions were presumably facilitated by the use of metrological tables similar to the many thousands of Old Babylonian exemplars known. In other words, scribal students were already in the Ur III period taught to perform their calculations—in sexagesimal notation—on tablets separate from the model documents to which they pertained, which were written in the ubiquitous mixed system of notation.

The writer of that tablet from Girsu might easily have gone on to calculate the labour required to make the bricks, to carry them to the building site, to mix the mortar, and to construct the wall itself. These standard assumptions about work rates were at the heart of the Ur III regime's bureaucracy. Surveyors' estimates of a work gang's expected outputs were kept alongside records of their actual performances—for tasks as diverse as milling flour to clearing fallow fields. At the end of each administrative year, accounts were drawn up, summarising the expected and true productivity of each team. In cases of shortfall, the foreman was responsible for catching up the following year; but work credits could not be carried over [Englund, 1991]. The constants used in these administrative calculations are found in a few contemporary school practice texts too [Robson, 1999: 31].

Math education in the early second millennium

But such a totalitarian centrally-controlled economy could not last, and within a century the Ur III empire had collapsed under the weight of its own bureaucracy. The dawn of the second millennium BCE—the so-called Old Babylonian period—saw the rebirth of the small city states, much as had existed centuries before. But now many of the economic functions of the central administration were deregulated and contracted out to private enterprise. Numerate scribes were still in demand, though, and we have an unprecedented quantity of tablets giving direct or indirect information on their training. Many thousands of school tablets survive although they are for the most part unprovenanced, having been dug up at the end of the nineteenth century (CE!) before the advent of scientific archaeology. However, mathematical tablets have been properly excavated from a dozen or so sites, from Mari and Terqa by the Euphrates on the Syria-Iraq border to Me-Turnat on the Diyala river and Susa in south-west Iran. A fragment of a multiplication table was even discovered at Hazor in Israel [Horowitz, 1993].

We know of several school houses from the Old Babylonian period, from southern Iraq [Stone, 1987: 56–59; Charpin, 1986: 419–33]. They typically consist of several small rooms off a central courtyard, and would be indistinguishable from the neighbouring dwellings if it were not for some of the fittings and the tablets that were found inside them. The courtyard of one house in Nippur, for instance, had built-in benches along one side and a large fitted basin containing a large jug and several small bowls which are thought to have been used for the preparation and moistening of tablets. There was also a large pile of crumpled up, half-recycled tablets waiting for re-use. The room behind the courtyard had been the tablet store, where over a thousand school tablets had been shelved on benches and perhaps filed in baskets too. But when the house needed repairs, some time in the mid-18th century BCE, the tablets were used as building material and were incorporated into the very fabric of the house itself.

Some of the school tablets were written by the teachers, while others were 'exercise tablets' composed by the apprentice scribes. Sumerian, which had been the official written language of the Ur III state, was gradually ousted by Akkadian—a Semitic language related to Hebrew and Arabic but which used the same cuneiform script as Sumerian. Akkadian began to be used for most everyday writings while Sumerian was reserved for scholarly and religious texts, analogous to the use of Latin in Europe until very recently. This meant that much of the scribal training which had traditionally been oral was recorded in clay for the first time, either in its original Sumerian, or in Akkadian translation, as was the case for the mathematical texts.

Math was part of a curriculum which also included Sumerian grammar and literature, as well as practice in writing the sorts of tablets that working scribes would need. These included letters, legal contracts and various types of business records, as well as more mathematically oriented *model documents* such as accounts, land surveys and house plans. Five further types of school mathematical text have been identified, each of which served a separate pedagogical function [Robson, 1999: 8–15]. Each type has antecedents in the third millennium tablets discussed in the previous section.

First, students wrote out *tables* while memorising metrological and arithmetical relationships. There was a standard set of multiplication tables, as well as aids for division, finding squares and square roots, and for converting between units of measurement. Many scribes made copies for use at work too. *Calculations* were carried out, in formal layouts, on small round tablets—called 'hand tablets'—very like the third millennium examples mentioned above. Hand tablets could serve as the scribes' 'scratch pads' and might also carry diagrams and short notes as well as handwriting practice and extracts from literature. The teacher set mathematical problems from 'textbooks'—usually called *problem* texts in the modern literature—which consisted of a series of (often minimally different) problems and their numerical answers. They might also contain model solutions and diagrams. Students sometimes copied problem texts, but they were for the most part composed and transmitted by the scribal teachers. Teachers also kept *solution lists* containing alternative sets of parameters, all of which would give integer answers for individual problems [Friberg, 1981]. There were also tables of technical constants—conventionally known as *coefficient lists*—many of whose entries are numerically identical to the constants used by the personnel managers of the Ur III state [Kilmer, 1960; Robson, 1999].[8]

Model solutions, in the form of algorithmic instructions, were not only didactically similar to other types of educational text, but were also intrinsic to the very way mathematics was conceptualised. For instance, the problems which have conventionally been classified as 'quadratic equations' have recently turned out to be concerned with a sort of cut-and-paste geometry [Høyrup, 1990; 1995]. As the student followed the instructions of the model solution, it would have been clear that the method was right—because it worked—so that no proof was actually needed.

The bottom line for Old Babylonian education must have been to produce literate and numerate scribes, but those students were also instilled with the aesthetic pleasure of mathematics for its own sake. Although many ostensibly practical scenarios were used as a pretext for setting non-utilitarian problems, and often involved Ur III-style technical constants, they had little concern with accurate mathematical modelling. Let us take the topic of grain-piles as an example. In the first sixteen problems of a problem text from Sippar the measurements of the grain-pile remain the same, while each parameter is calculated in turn.[9] The first few problems are missing, but judging from other texts we would expect them to be on finding the length, then the width, height, etc. The first preserved problem concerns finding the volume of the top half of the pile.

One could imagine how such techniques might be useful to a surveyor making the first estimate of the capacity of a grain-pile after harvest—and indeed we know indirectly of similar late third millennium measuring practices. However, then things start to get complicated. The remaining problems give data such as the sum of the length and top, or the difference between the length and the thickness, or even the statement that the width is equal to half of the length plus 1. It is hardly likely that an agricultural overseer would ever find himself needing to solve this sort of a problem in the course of a working day.

Similarly, although the mathematical grain-pile is a realistic shape—a rectangular pyramid with an elongated apex—even simply calculating its volume involves some rather sophisticated three-dimensional geometry, at the cutting edge of Old Babylonian mathematics as we know it. Further, it appears that at some point the scenario was further refined to enable mathematically more elegant solutions to be used in a tablet from Susa.[10] In both sets of problems the pile is 60 m long and 18–24 m high. It is difficult to imagine how a grain pile this big could ever be constructed, let alone measured with a stick. In short, the accurate mathematical modelling of the real world was not a priority of Old Babylonian mathematics; rather it was concerned with approximations to it that were both good enough and mathematically pleasing.

The evidence for mathematical methods in the Old Babylonian workplace is still sketchy, but one can look for it, for instance, in canal and land surveys. Although these look rather different from their late third millennium precursors—they are laid out in the form of tables, with the length, width and depth of each excavation in a separate column, instead of in lists—the mathematical principles involved are essentially the same. There is one important distinction though; there is no evidence (as yet) for work-rate calculations. This is not surprising; we are not dealing with a centralised 'national' bureaucracy in the early second millennium, but quasi-market economies in which much of the work traditionally managed by the state was often contracted out to private firms bound by legal agreements. One would not expect a consistent picture of quantitative management practices throughout Mesopotamia, even where such activities were documented.

What happened next?
Tracing the path to Hellenistic Babylon

After about 1600 BCE mathematical activity appears to come to an abrupt halt in and around Mesopotamia. Can it simply be that math was no longer written down, or can we find some other explanation for the missing evidence?

For a start, it should be said that there is a sudden lack of tablets of all kinds, not just school mathematics. The middle of the second millennium BCE was a turbulent time, with large population movements and much political and social upheaval. This must have adversely affected the educational situation. But there is the added complication that few sites of this period have been dug, and that further, the tablets which have been excavated have been studied very little. Few scholars have been interested in this period of history, partly because the documents it has left are so difficult to decipher.

But, further, from the twelfth century BCE onwards the Aramaic language began to take over from Akkadian as the everyday vehicle of both written and oral communication. Aramaic was from the same language-family as Akkadian, but had adopted a new technology. It was written in ink on various perishable materials, using an alphabet instead of the old system of syllables on clay. Sumerian, Akkadian and the cuneiform script were retained for a much more restricted set of uses, and it may be that math was not usually one of them. It appears too that cuneiform was starting to be written in another new medium, wax-covered ivory or wooden writing-boards, which could be melted down and smoothed off as necessary. Although contemporary illustrations and references on clay tablets indicate that these boards were in widespread use, very few have been recovered—all in watery contexts which aided their preservation—but the wax had long since disappeared from their surfaces. So even if mathematics were still written in cuneiform, it might well have been on objects which have not survived.

These factors of history, preservation and fashions in modern scholarship have combined to mean that the period between around 1600 and 1000 BCE in south Mesopotamia is still a veritable dark age for us. The light is beginning to dawn, though, and there is no reason why school texts, including mathematics, should not start to be identified, supposing that they are there to be spotted. But, fortunately for us, the art of writing on clay did not entirely die out, and there are a few clues available already. Mathematical and metrological tables continued to be copied and learnt by apprentice scribes; they have been found as far afield as Ashur on the Tigris, Haft Tepe in southwest Iran, and Ugarit and Byblos on the Mediterranean coast. One also finds evidence of non-literate mathematical concepts, which have a distinctly traditional flavour. Not only do brick sizes remain more or less constant—which strongly suggests that some aspects of third millennium metrology were still in use—but there are also some beautiful and sophisticated examples of geometrical decoration. There are, for instance, stunning patterned 'carpets' carved in stone from eighth and seventh century Neo-Assyrian palaces—an empire more renowned for its brutal deportations and obsession with astrology than for its contributions to cultural heritage.

But perhaps more excitingly, a mathematical problem is known in no less than three different copies, from Nineveh and Nippur.[11] Multiple exemplars are rare in the mathematically-rich Old Babylonian period, but for the barren aftermath it may be an indication of the reduced repertoire of problems in circulation at that time. Its style shows that mathematical traditions of the early second millennium had not died out, while apparently new scenarios for setting problems had developed. It is a teacher's problem text, for a student to solve, and it is couched in exactly the sort of language known from the Old Babylonian period. But interestingly it uses a new pretext. The problem ostensibly concerns distances between the stars, though in fact it is about dealing with division by 'remarkable' numbers—a topic which, as we have seen, goes back as far as the mid-third millennium.

Finally we arrive in Babylon itself—a little later than the Persians and Greeks did. By the fourth and third centuries BCE indigenous Mesopotamian civilisation was dying. Some of the large merchant families of Uruk and Babylon still used tablets to record their transactions, but the temple libraries were the principal keepers of traditional cuneiform culture. Their collections included huge series of omens, historical chronicles, and mythological and religious literature as well as records of astronomical observations. It has often been said that mathematics by now consisted entirely of mathematical methods for astronomy, but that is not strictly true. As well as the mathematical tables—now much lengthier and sophisticated than in earlier times—we know of at least half a dozen tablets containing non-astronomical mathematical problems for solution. Although the terminology and conceptualisation has changed since Old Babylonian times—which, after all, is only to be expected—the topics and phraseology clearly belong to the same stream of tradition. Most excitingly, a small fragment of a table of technical constants has recently been discovered, which contains a list of brick sizes and densities. Although the mathematics involved is rather more complicated than that in similar earlier texts, the brick sizes themselves are exactly identical to those invented in the reforms of Akkad around two thousand years before.

Conclusions

I hope I have been able to give you a little taste of the rich variety of Mesopotamian math that has come down to us. Its period of development is vast. There is twice the timespan between the first identifiable accounting tokens and the latest known cuneiform mathematical tablet as there is between that tablet and this book. Most crucially, though, I hope that you will agree with me that mathematics is fundamentally a product of society. Its history is made immeasurably richer by the study of the cultures which have produced it, wherever and whenever they might be.

Bibliography

Bruins, E. M., and Rutten, M.: 1961, *Textes mathématiques de Suse,* (Mémoires de la Délegation en Perse 34), Paris.

Castellino, G. R.: 1972, *Two Shulgi hymns (BC)* (Studi Semitici 42), Rome.

Charpin, D.: 1986, *Le clergé d'Ur au siècle d'Hammurabi,* Paris.

Englund, R.: 1991, "Hard work—where will it get you? Labor management in Ur III Mesopotamia," *Journal of Near Eastern Studies* 50, 255–280.

Foster, B. R.: 1982, "Education of a bureaucrat in Sargonic Sumer," *Archiv Orientální* 50, 238–241.

Fowler, D. H. and Robson, E.: 1998, "Square root approximations in Old Babylonian mathematics: YBC 7289 in context," *Historia Mathematica* 25, 366–378.

Date / Age	POLITICAL PERIODISATION	MATHEMATICAL DEVELOPMENTS	SOCIETY AND TECHNOLOGY	THE REST OF THE WORLD
IRON AGE (*Late Babylonian Period*)	*Parthian (Arsacid) Period 126 BCE–227 CE*	Latest known cuneiform tablets are astronomical records		
0 CE/BCE	*Seleucid (Hellenistic, or Greek) period, 330–127*	Math and astronomy maintained and developed by temple personnel	*Traditional Mesopotamian culture dying under the influence of foreign rulers*	Invention of paper, in China Great Wall of China, 214 The *Elements*
	Persian (Achaemenid) empire, 538–331		*Cotton* *Coinage* *Brass*	Birth of Buddha, c. 570
	Neo-Babylonian empire, 625–539 *Neo-Assyrian empire, 883–612*	Mathematical tradition apparently continues, although the evidence is currently very slight	*Cuneiform Akkadian being replaced by alphabetic Aramaic*	Foundation of Rome, c. 750 Indian mathematics
1000 BCE	*Middle Babylonian period, c. 1150–626*	A few mathematical tables known, from sites on the periphery of Mesopotamia	*Smelted iron, Camels* *Glazed pottery, Glass*	
LATE BRONZE	*Kassite period, c. 1600–1150*			Tutankhamen
MIDDLE BRONZE	*Old Babylonian period, c. 2000–1600*	Best-documented period of math in scribal schools	*Horse and wheel technology improved* *Much Sumerian and Akkadian literature*	Stonehenge completed Rhind mathematical papyrus Earliest recorded eclipse: China, 1876
2000 BCE — EARLY BRONZE AGE	*Ur III empire, c. 2100–2000*	Development of the sexagesimal place value system	*First large empires* *Ziqqurats* *Horses*	
	Kingdon of Akkad, c. 2350–2150	Reform of the metrological systems	*Akkadian written in cuneiform characters* *City states*	Indus Valley civilisation Collapse of the Egyptian Old Kingdom
	Early Dynastic period, c. 3000–2350	Earliest known math tables	*Palaces* *Development of writing into cuneiform (Sumerian)*	
3000 BCE			*High Sumerian culture*	Upper and Lower Egypt united Great Pyramid Stonehenge begun
	Uruk period, c. 4000–3000	Earliest known written documents: accounts using complex metrological systems Development of clay token accounting system, with sealed 'bullae'	*URBANISATION* *Beginnings of writing and bureaucracy* *Cylinder seals* *Monumental architecture* *Potter's wheel* *Bronze, gold and silver work*	
4000 BCE — NEOLITHIC AGE			*Irrigation agriculture* *Wide use of brick* *Temples* *Copper and pottery*	Megalithic cultures of western Europe First temple towers in South America
	Ubaid period, c. 5500–4000	Small, regularly-shaped clay 'tokens' apparently used as accounting devices	*Mesopotamia begins to be settled by farmers from the surrounding hills (the Fertile Crescent)*	
5000 BCE				Farming begins to reach Europe from the Near East

Time chart showing major political, societal, technological and mathematical developments in the ancient Near East.[12]

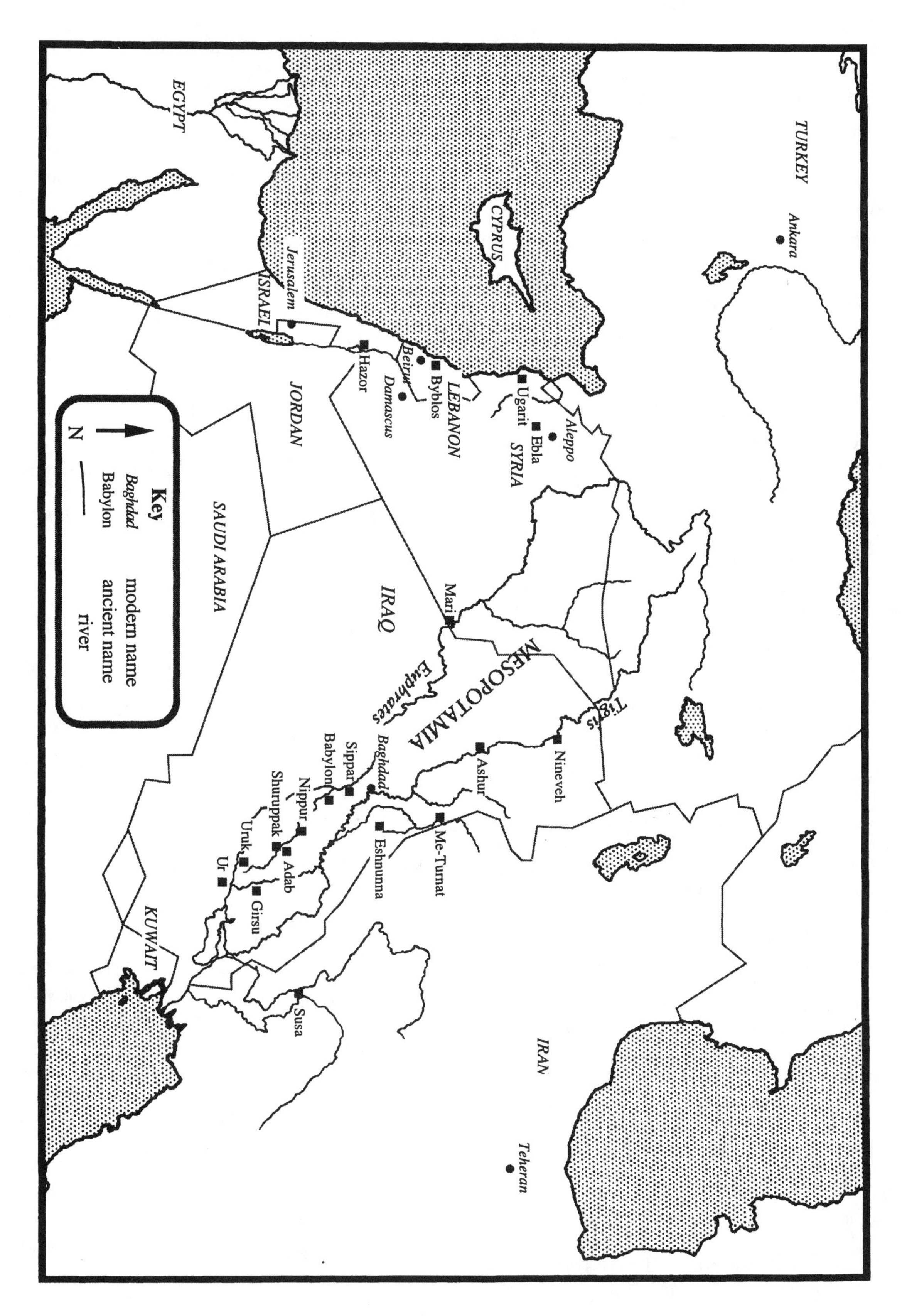

Map showing the principal modern cities of the Near East and all the ancient sites mentioned in the text.

Friberg, J.: 1986, "Three remarkable texts from ancient Ebla," *Vicino Oriente* 6, 3–25.

——: 1987–90, "Mathematik," in *Reallexikon der Assyriologie und vorderasiatische Archäologie* VII (ed. D. O. Edzard et al.), Berlin, 531–585.

Horowitz, W.: 1993, "The reverse of the Neo-Assyrian planisphere CT 33 11," in *Die Rolle der Astronomie in Kulturen Mesopotamiens* (Grazer Morgenländische Studien 3), (ed. H. D. Galter), Graz, 149–159.

——: 1997, "A combined multiplication table on a prism fragment from Hazor," *Israel Exploration Journal* 47, 190–197.

Høyrup, J.: 1990, "Algebra and naive geometry: an investigation of some basic aspects of Old Babylonian mathematical thought," *Altorientalische Forschungen* 17, 27–69; 262–354.

——: 1993, "Remarkable numbers' in Old Babylonian mathematical texts: a note on the psychology of numbers," *Journal of Near Eastern Studies* 52, 281–286.

Joseph, G. G.: 1991, *The crest of the peacock: non-European roots of mathematics,* Harmondsworth.

Katz, V.: 1993, *A history of mathematics: an introduction,* New York.

Kilmer, A. D.: 1960, "Two new lists of key numbers for mathematical operations," *Orientalia* 29, 273–308.

Nemet-Nejat, K. R.: 1993, *Cuneiform mathematical texts as a reflection of everyday life in Mesopotamia* (American Oriental Series 75), New Haven.

Neugebauer, O.: 1935–37, *Mathematische Keilschrifttexte* I–III, Berlin.

Neugebauer, O. and Sachs, A.: 1945, *Mathematical cuneiform texts* (American Oriental Series 29), New Haven.

Nissen, H. J., Damerow, P. and Englund, R.: 1993, *Archaic bookkeeping: early writing and techniques of economic administration in the ancient Near East,* Chicago.

Powell, M. A.: 1976, "The antecedents of Old Babylonian place notation and the early history of Babylonian mathematics," *Historia Mathematica* 3, 414–439.

——: 1987–90, "Masse und Gewichte," in *Reallexikon der Assyriologie und vorderasiatische Archäologie* VII (ed. D. O. Edzard et al.), Berlin, 457–530.

——: 1995, "Metrology and mathematics in ancient Mesopotamia," in *Civilizations of the ancient Near East* III (ed. J. M. Sasson), New York, 1941–1957.

Robson, E.: 1999, *Mesopotamian mathematics 2100–1600* BC*: technical constants in education and bureaucracy* (Oxford Editions of Cuneiform Texts 14), Oxford.

Stone, E.: 1987, *Nippur neighbourhoods* (Studies in Ancient Oriental Civilization 44), Chicago.

Thureau-Dangin, F.: 1938, *Textes mathématiques babyloniens* (Ex Oriente Lux 1), Leiden.

Westenholz, A.: 1977, "Old Akkadian school texts: some goals of Sargonic scribal education," *Archiv für Orientforschungen* 25, 95–110.

Whiting, R. M.: 1984, "More evidence for sexagesimal calculations in the third millennium," *Zeitschrift für Assyriologie* 74, 59–66.

Further reading on the history and culture of the ancient Near East

Black, J. A. and Green, A.: 1992, *Gods, demons and symbols of ancient Mesopotamia,* London.

Collon, D.: 1995, *Ancient Near Eastern art,* London.

Dalley, S.: 1989, *Myths from Mesopotamia* (Oxford University Press World's Classics), Oxford.

Foster, B. R.: 1993, *Before the muses: an anthology of Akkadian literature* I–II, Bethesda.

Freedman, D. N. (ed.): 1992, *The Anchor Bible Dictionary* I–V, New York.

Kuhrt, A.: 1995, *The ancient Near East: c. 3000–330* BC I–II, London.

Postgate, J. N.: 1992, *Early Mesopotamia: society and economy at the dawn of history,* London.

Roaf, M.: 1990, *Cultural atlas of Mesopotamia and the ancient Near East,* Oxford.

Roux, G.: 1992, *Ancient Iraq,* Harmondsworth.

Saggs, H. W. F.: 1995, *The Babylonians* (Peoples of the Past), London.

Sasson, J. M. (ed.): 1995, *Civilizations of the ancient Near East* I–IV, New York.

Walker, C. B. F.: 1987, *Cuneiform* (Reading the Past), London.

Endnotes

[1] From a hymn of self-praise to king Shulgi, 21st century BCE; cf. Castellino, 1972: 32.

[2] 1;24 51 10 ($\approx 1.4142129\ldots$) in YBC 7289 [Fowler and Robson, 1998]. In general I have tried to cite the most recent, reliable and easily accessible sources, rather than present an exhaustive bibliography for the topic.

[3] See, for instance, Joseph, 1991: 91–118; Katz, 1993: 6–7, 24–28.

[4] For general works on ancient Near Eastern history and culture, see the suggestions for further reading at the end.

[5] According to a recent theory, tokens could have been used like abacus counters for various arithmetical operations [Powell 1995].

[6] VAT 14942: see Nissen, Damerow and Englund, 1993: pl. 22.

[7] That is, in cuneiform signs which indicate both the absolute value of the number and the system of measurement used.

[8] The major publications of Old Babylonian mathematical texts are still Neugebauer, 1935–37; Thureau-Dangin, 1938; Neugebauer and Sachs, 1945; Bruins and Rutten, 1961. For an index of more recent publications, editions and commentaries, see Nemet-Nejat, 1993.

[9] BM 96954 + BM 102366 + SÉ 93, published in Robson, 1999: Appx. 3.

[10] TMS 14; Robson, 1999: ch. 7.

[11] HS 245, Sm 162, Sm 1113. See most recently Horowitz, 1993.

[12] Dates earlier than 911 BCE are not accurate, and vary from book to book and scholar to scholar, as do the names and dates of the periods into which Mesopotamian political history is conventionally divided.

An Excursion in Ancient Chinese Mathematics

Siu Man-Keung
University of Hong Kong

Nobody can hope to do justice, in one short article, to the two millennia of indigenous mathematical development in China up to the end of the 16th century.[1] This article attempts only to convey a general flavour of ancient Chinese mathematics and illustrate some of its characteristic features through a few examples. An annotated bibliography is provided at the end for the reader's convenience.[2]

Characteristic Features of Ancient Chinese Mathematics

The characteristic features of ancient Chinese mathematics can best be appreciated by looking at the work of the ancient Chinese mathematicians. Evidenced in their choice of topics is a strong social relevance and pragmatic orientation, and in their methods a primary emphasis on calculation and algorithms. However, contrary to the impression most people may have, ancient Chinese mathematics is not just a "cook-book" of applications of mathematics to mundane transactions. It is structured, though not in the Greek sense exemplified by Euclid's *Elements*. It includes explanations and proofs, though not in the Greek tradition of deductive logic. It contains theories which far exceed the necessity for mundane transactions.

We start with some ideograms (characters) related to mathematics. In ancient classics the term mathematics (數學) was often written as "the art of calculation" (算術) or "the study of calculation" (算學), indicating a deep-rooted basis in calculation. The ideogram for "number" and "to count" (數) appeared on oracle bones about 3000 years ago, in the form of a hand tying knots on a string (see Fig. 1a). The ideogram for "to calculate" (算) appeared in three forms, according to *Shuowen Jiezi* (*Analytic Dictionary of Characters*) by Xu Shen (AD 2nd century). The first is a noun, composed of two parts, "bamboo" on top and "to manipulate" in the bottom, with the bottom part itself in the form of two hands plus some (bamboo) sticks laid down on a board, some placed in a horizontal position and some placed in a vertical position (see

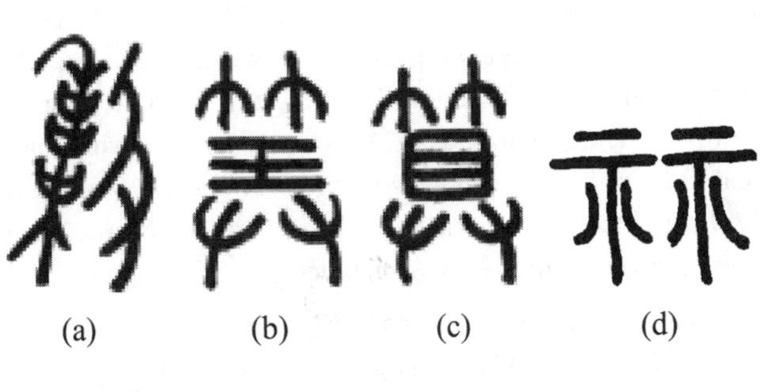

FIGURE 1

Fig. 1b). The second is a verb, also written with the parts of bamboo and hands (see Fig. 1c). The third, somewhat more puzzling, is in the form of a pair of ideogramatic parts pertaining to religious matters (see Fig. 1d). It is a tantalizing thought that the subject of mathematics in ancient China was not exactly the same subject as we understand it today. Indeed, in some ancient mathematical classics we find mention of "internal mathematics" and "external mathematics", the former being intimately tied up with *Yijing* (*Book of Changes*), the oldest written classic in China.[3]

Besides its appearance in these ideograms, the theme of calculation permeated the whole of ancient Chinese mathematics. This is best illustrated by the calculating device of the counting rods. Ample evidence confirms the common usage of counting rods as early as in the fifth century BCE, and these probably developed from sticks used for fortune-telling in even earlier days. The earliest relics from archaeological findings are dated to the second century BCE These were made of bamboo, wood and even metal, bone or ivory and were carried in a bag hung at the waist. The prescribed length in the literature (verified by the relics) was from 13.86 cm to 8.5 cm, which shortened as time went on. The cross-section changed with time, from circular (of 0.23 cm in diameter) to square so that the rods became harder to roll about. One mathematically extremely interesting feature is the occurrence of a red dot on a counting rod to denote a positive number, and a black dot to denote a negative number. These counting rods were placed on a board (or any flat surface) and moved about in performing various calculations.

The Chinese adopted very early in history a denary positional number system. This was already apparent in the numerals inscribed on oracle bones in the Shang Dynasty (c. 1500 BCE), and was definitely marked in the calculation using counting rods in which the positions of the rods were crucial. Ten symbols sufficed to represent all numbers when they were put in the correct positions. At first only nine symbols were used for the numerals 1 to 9, with the zero represented by an empty space, later by a square in printing, gradually changed to a circle, perhaps when the square was written by a pen-brush. To minimize error in reading a number, numerals were written alternatively in vertical form (for units, hundreds,...) and horizontal form (for tens, thousands,...). In a much later mathematical classic, *Xiahou Yang Suanjing* (*Mathematical Manual of Xiahou Yang*) of the fifth century, this method for writing counting rod numerals was recorded as:

> Units stand vertical, tens are horizontal, hundreds stand, thousands lie down. Thousands and tens look the same, ten thousands and hundreds look alike. Once bigger than six, five is on top; six does not accumulate, five does not stand alone.

For instance, 1996 would have been written as

Calculation using counting rods has several weak points: (1) The calculation may take up a large amount of space. (2) Disruption during the calculation causing disarray in the counting rods can be disastrous. (3) The calculating procedure is not recorded step by step so that intermediate calculations are lost. Counting rods evolved into the abacus in the twelfth-thirteenth centuries, and by the fifteenth century the abacus took the place of counting rods. The weak points (1) and (2) were removed by the use of the abacus, but (3) remained, until the European method of calculation using pen and paper was transmitted in the beginning of the seventeenth century. However, calculation using counting rods had its strong points. Not only did the positions of the counting rods display numerals conveniently, but also the positions in which these rods were placed on the board afforded a means to allow some implicit use of symbolic manipulation, giving rise to successful treatment of ratio and proportion, fractions, decimal fractions, very large or very small numbers, equations, and so on. Indeed, the use of counting rods was instrumental in the whole development of algorithmic mathematics in ancient China.

Even a casual reading of a few mathematical classics will disclose the unmistakable features of social relevance and pragmatic orientation. From the very beginning mathematical development was intimately related to studies of astronomical measurement and calendrical reckoning. The first written text containing serious mathematics, *Zhoubi Suanjing* (*Zhou Gnomon Classic of Calculation*) compiled at about 100 BCE—with its content dated to earlier times, was basically a text in astronomical study. In an ancient society based on agriculture, calendrical reckoning was always a major function of the government. Along with that, mathematics was performed mainly for bureaucratic needs. A sixth century mathematics classic actually carried the title *Wucao Suanjing* (*Mathematical Manual of the Five Government Departments*). The titles of the nine chapters of the most important mathematical classic *Jiuzhang Suanshu* (*Nine Chapters on the Mathematical Art*), which is believed to have been compiled some time between 100 BCE and 100 CE, speak for themselves. These are (1) survey of land, (2) millet and rice (percentage and proportion), (3) distribution by progression, (4) diminishing breadth (square root), (5) consultation on engineering works (volume of solid figures), (6) impartial taxation (allegation), (7) excess and deficiency (Chinese "Rule of Double False Positions"), (8) calculating by tabulation (simultaneous equations), (9) gou-gu (right triangles). The social relevance of the content of mathematical classics was so plentiful that historians have found in the texts a valuable source for tracing the economy, political system, social habits, and legal regulations of the time! The emphasis on social relevance and pragmatic orientation, in line with a basic tenet of traditional Chinese philosophy of life shared by the class of "shi" (intellectuals), viz. self-improvement and social interaction, was also exhibited in the education system in which training in mathematics at official schools was intended for government officials and clerks.[4]

Finally let us come to the issue of mathematical proofs. "If one means by a proof a deductive demonstration of a statement based on clearly formulated definitions and postulates, then it is true that one finds no proof in ancient Chinese mathematics, nor for that matter in other ancient oriental mathematical cultures.... But if one means by a proof any explanatory note which serves to convince and to enlighten, then one finds an abundance of proofs in ancient mathematical texts other than those of the Greeks."[5] The Chinese offered proofs through pictures, analogies, generic examples, and algorithmic calculations. These can be of pedagogical value to complement and supplement the teaching of mathematics with traditional emphasis on deductive logical thinking.

Jiuzhang Suanshu

Jiuzhang Suanshu is the most important of all mathematical classics in China. It is a collection of 246 mathematical problems grouped into nine chapters. There is good reason to believe that the content of *Jiuzhang Suanshu* was much older than its date of compilation, as substantiated by an exciting archaeological finding in 1983 when a book written on bamboo strips bearing the title *Suanshu Shu* (*Book on the Mathematical Art*) was excavated[6]. It is dated at around 200 BC and its content exhibits a marked resemblance to that of *Jiuzhang Suanshu*, including even some identical numerical data which appeared in the problems. The format of *Jiuzhang Suanshu* became a prototype for all Chinese mathematical classics in the subsequent one-and-a-half millennia. A few problems of the same category were given, along with answers, after which a general method (algorithm) followed. In the very early edition that was all and no further explanation was supplied—perhaps it was to be supplied by the teacher. Later editions were appended with commentaries which explained the methods, corrected mistakes handed down from the ancients, or expanded the original text. The most notable commentator of *Jiuzhang Suanshu* was Liu Hui (c. third century), some of whose works will be examined in the next section.

The format of *Jiuzhang Suanshu* may lead one to regard the book as a medley of recipes for solving problems of specific types. Indeed many who studied from the book in accordance with the official system in ancient China might have actually regarded the book as such and thus resorted to rote learning just like in recitation of other classics. This may explain why only a handful of mathematicians of some standing were produced from the tens of thousands of "mathocrats" who went through mathematical training in the official system during two millennia, while most noted mathematicians in history were either self-educated or studied at private academies[7].

However, upon closer scrutiny, the text reveals itself as quite different from a book of recipes. The body of knowledge contained in a classic such as *Jiuzhang Suanshu* is structured around several themes, the two main themes being the concept of "lu" (率, ratio) in arithmetic and the concept of "gou-gu" (勾股, right triangle) in geometry. A brief description on how ratio forms a backbone for most chapters will now be given, while right triangles will be left to the next section. In the commentary of Chapter 1, Liu Hui gave a definition: "a ratio is a relation between numbers."[8] He continued to offer a working definition of ratio by representing it as a reduced fraction. To reduce a fraction the rule of "reciprocal subtraction," known to Westerners as the Euclidean algorithm, was introduced.

> If both numerators and denominators are divisible by 2, then halve them both. If they are not both divisible by 2, then set up the numbers for numerator and denominator respectively continually and alternately subtracting the smaller from the larger, and seek their equality.

This is a good illustration of how the calculation itself is already a proof (or convincing argument), as can be seen from Problem 6 of Chapter 1:

> Reduce the fraction $\frac{49}{91}$.

$$(49,91) \to (49,42) \to (7,42) \to (7,35)$$
$$\to (7,28) \to (7,21) \to (7,14) \to (7,7).$$

Hence $49 = 7 \times 7$, $91 = 7 \times 13$, and $\frac{49}{91} = \frac{7}{13}$. At the beginning of Chapter 2 Liu Hui explained the so-called "Rule of Three" (also found in contemporary Indian manuscripts), which enables one to apply the concept of ratio to a number of situations, including distribution in direct proportions or in inverse proportions (Chapters 3, 6), formulation and treatment of problems in excess and deficiency, i.e., the method of "double false positions" (Chapter 7), and systems of simultaneous linear equations (Chapter 8). Although the Chinese terminology "fangcheng" (方程), which is the title of Chapter 8, was adopted as a translation for "equation" towards the end of the last century (and has become a standard term today) for a wrong but historically interesting reason,[9] the spirit of Chapter 8 lies rather in the direction of ratio than in the direction of equation. In the light of ratios, the technique amounting to the modern matrix method by Gaussian elimination arises naturally.

In ending this section consider an example after the style of *Jiuzhang Suanshu* which blends together social relevance, ratio and even an application in statistical sampling. It is Problem 6 of Book 12 of *Shushu Jiuzhang* (*Mathematical Treatise in Nine Sections*) by Qin Jiushao, published in 1247:

> When a peasant paid tax to the government granary in the form of 1534 shi of rice, it was found out on examination that a certain amount of rice with husks was present. A sample of 254 grains was taken for further examination. Of these 28 grains were with husks. How many genuine grains of rice were there, given that one shao contains 300 grains?

(In the mensuration system of the Song Dynasty, 1 shi = 10 dou = 100 sheng = 1000 he = 10000 shao. According to tradition recorded in *Jiuzhang Suanshu*, a grain of rice with husk was counted as half a grain of rice.) The answer was given to be 4,348,346,456 grains, out of the original 1534 × 10000 × 300 = 4,602,000,000 grains.

Some Examples and Their Solution Methods[10]

(1) Problem 14 of Chapter 9 of *Jiuzhang Suanshu* is a word problem on right triangles:

> Two persons A (Jia) and B (Yi) stood at the same spot. In the time when A walked 7 steps, B could walk 3 steps. B walked east and A walked south. After 10 steps south A turned to walk in a roughly northeast direction to meet B. How many steps had each walked (when they met)?

The rule that follows the problem essentially gives the ratio of the length a, b, c of the three sides of a right triangle with c as that of the hypotenuse, viz.

$$a : b : c = \frac{1}{2}(m^2 - n^2) : mn : \frac{1}{2}(m^2 + n^2),$$

where $m : n = (a + c) : b$. In this problem, $m = 7$, $n = 3$ and $a = 10$. Hence $a : b : c = 20 : 21 : 29$ and $b = 10\frac{1}{2}$, $c = 14\frac{1}{2}$. The mathematical meaning of this result goes much deeper than just an answer to the problem as it stands, for it offers a way to generate the so-called Pythagorean triplets, i.e., (positive) integers a, b, c with $a^2 + b^2 = c^2$. While no explicit formula for Pythagorean triplets was stated by the ancient Chinese, they were quite well-versed in these problems in which their Greek contemporaries were also interested, and in ancient Chinese mathematics arithmetic and geometry were intertwined through calculation. The achievement becomes all the more astounding if one notes that the ancient Greeks were aware of the notions of prime number and factorization while their Chinese contemporaries were not. Instead, the Chinese adopted a geometric viewpoint by looking for two quantities with suitable geometric interpretation in terms of which a, b, c can each be rationally expressed. In the case of Problem 14, the two quantities are the sum of the length of one side and the hypotenuse $(a + c)$ and the length of the third side (b). The explanation offered by Liu Hui can be illustrated as in Figure 2. In his commentary Liu Hui actually described in detail how to make use of colored pieces and to reassemble them for a convincing argument. If the original diagrams of the commentary were extant, they would make nice visual aids!

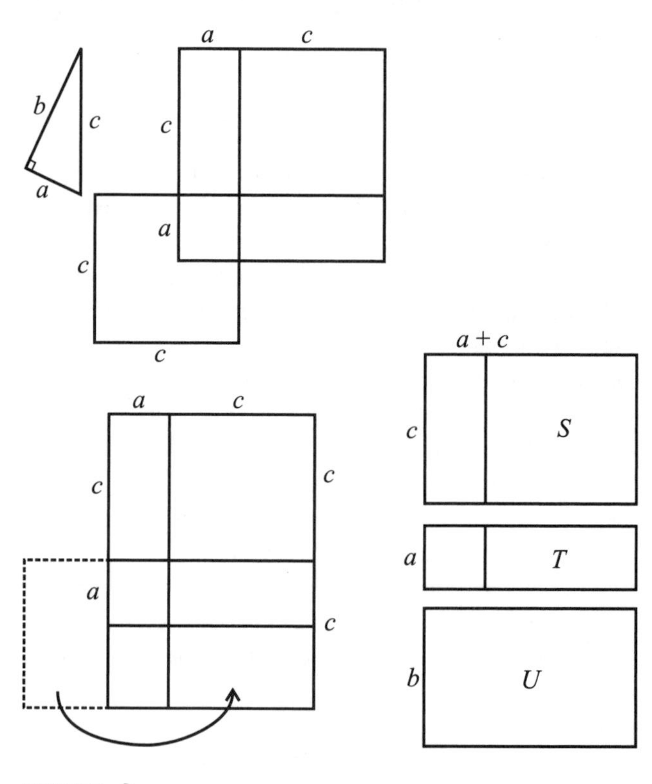

FIGURE 2

From Figure 2 we can see that

$$c : a : b = S : T : U$$
$$= \tfrac{1}{2}\left[(a+c)^2 + b^2\right]$$
$$: (a+c)^2 - \tfrac{1}{2}\left[(a+c)^2 + b^2\right] : (a+c)b.$$

Hence

$$a : b : c = \tfrac{1}{2}\left[(a+c)^2 - b^2\right] : (a+c)b : \tfrac{1}{2}\left[(a+c)^2 + b^2\right]$$
$$= \tfrac{1}{2}(m^2 - n^2) : mn : \tfrac{1}{2}(m^2 + n^2),$$

where $(a+c) : b = m : n$.

The influence of this prototype classic of *Jiuzhang Suanshu* can be found in later work, for example Problem 2 of Chapter 5 of *Shushu Jiuzhang* by Qin Jiushao, published more than a thousand years later:

> A triangular field has sides of length 13 miles, 14 miles and 15 miles. What is its area?

The solution was given in the book as (in modern day mathematical notations)

$$(\text{Area})^2 = \frac{1}{4}\left[A^2C^2 - \left(\frac{A^2 + C^2 - B^2}{2}\right)^2\right]$$

where A, B, C are the length of the three sides in decreasing magnitude. This is a rare gem in Chinese mathematics because this was perhaps the one occurrence of a triangle other than a right triangle in all Chinese mathematical texts before the transmission of Euclid's *Elements* into China.[11] A probable derivation of the formula by Qin Jiushao is as follows.[12] First note that, from our preceding example,

$$\frac{a}{b} = \frac{1}{2}[(a+c)^2 - b^2]/(a+c)b,$$

so that

$$a = \frac{1}{2}\left[(a+c) - \left(\frac{b^2}{a+c}\right)\right].$$

Construct a right triangle with sides of length a, b, c (c is the hypotenuse) where a, c are lengths as shown in Figure 3. Since $C^2 - a^2 = h^2 = B^2 - c^2$, we have $B^2 - C^2 = c^2 - a^2 = b^2$. Hence

$$a = \frac{1}{2}\left[(a+c) - \left(\frac{b^2}{a+c}\right)\right]$$
$$= \frac{1}{2}\left[A - \frac{B^2 - C^2}{A}\right] = \frac{1}{2}\left[\frac{A^2 + C^2 - B^2}{A}\right].$$

Finally,

$$(\text{Area})^2 = \frac{1}{4}h^2A^2 = \frac{1}{4}(C^2 - a^2)A^2 = \frac{1}{4}(A^2C^2 - a^2A^2)$$
$$= \frac{1}{4}\left[A^2C^2 - \left(\frac{A^2 + C^2 - B^2}{2}\right)^2\right].$$

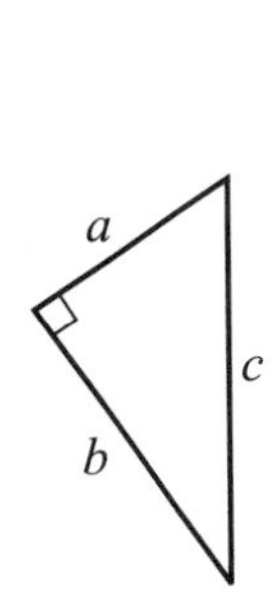

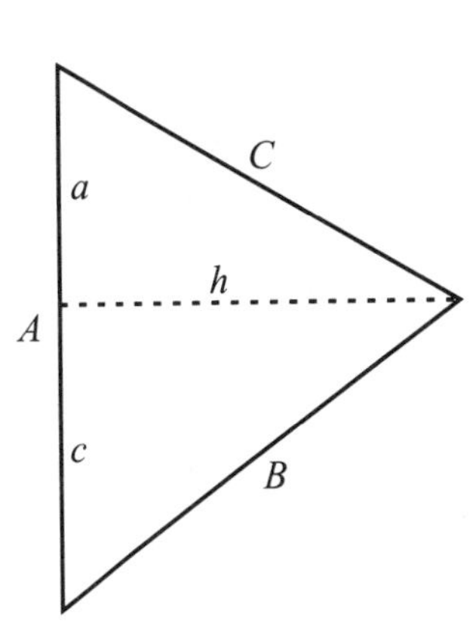

FIGURE 3

(2) Early Chinese calculation of π is given in Problem 32 of Chapter 1 of *Jiuzhang Suanshu*:

> A circular field has a perimeter of 181 steps and a diameter of 60 and 1/3 steps. What is its area?

The answer was given as "the area equals half the perimeter times half the diameter". This is a correct formula, as one can easily check that $A = (\frac{1}{2}C)(\frac{1}{2}d) = (\frac{1}{2}C)(r) = (\pi r)(r) = \pi r^2$. The data in this problem imply the formula $C = 3d$, which means π was then taken to be 3. In his commentary, Liu Hui explained why the formula is reasonable and pointed out how to obtain a more accurate value for π. He said:

> In our calculation we use a more accurate value for the ratio of the circumference to the diameter instead of the ratio that the circumference is 3 to the diameter's 1. The latter ratio is only that of the perimeter of the inscribed regular hexagon to the diameter. Comparing arc with the chord, just like the bow with the string, we see that the circumference exceeds the perimeter. However, those who transmit this method of calculation to the next generation never bother to examine it thoroughly but merely repeat what they learned from their predecessors, thus passing on the error. Without a clear explanation and definite justification it is very difficult to separate truth from falsity.

In this passage we see a truly first-rate mathematician at work, who probes into knowledge handed down and seeks understanding and clarification, thereby extending the frontier of knowledge. In modern day mathematical language Liu Hui's method is as follows. Put

A_n = area of an inscribed regular n-gon in a circle of radius r,
a_n = length of a side of the inscribed regular n-gon,
C_n = perimeter of the inscribed regular n-gon.

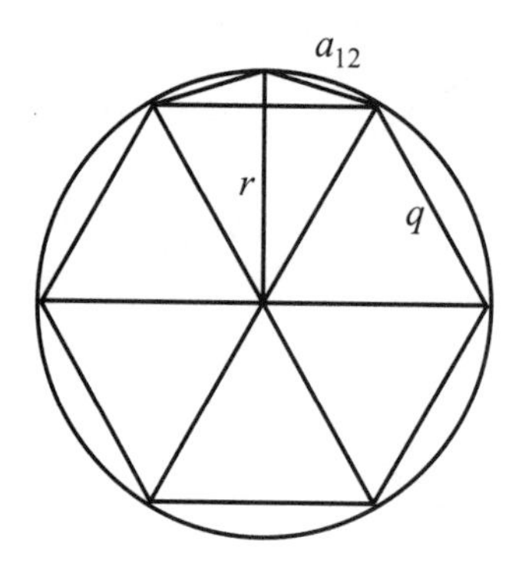

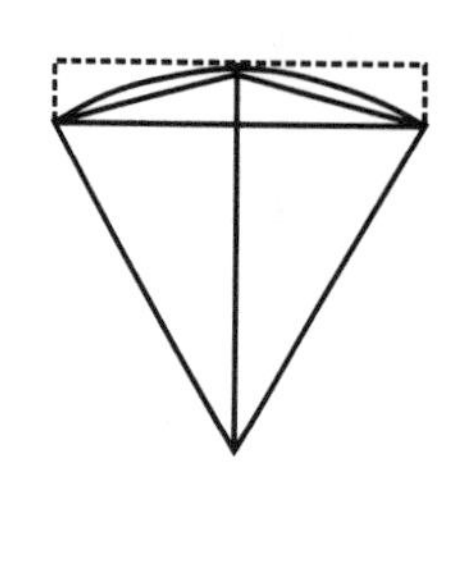

FIGURE 4

Starting with a regular hexagon ($n = 6$) and doubling the number of sides, Liu Hui enlarged it to a regular 12-gon, then a regular 24-gon, then a regular 48-gon, and so on, up to a regular 192-gon. He observed that $A_{12} = 3a_6r = \frac{1}{2}C_6r$, $A_{24} = 6a_{12}r = \frac{1}{2}C_{12}r$, $A_{48} = 12a_{24}r = \frac{1}{2}C_{24}r$, etc. He also knew that this was not the end but only the first few steps in an approximation process. He claimed, "the finer one cuts, the smaller the leftover; cut after cut until no more cut is possible, then it coincides with the circle and there is no leftover." We see here the budding concepts of infinitesimal and limit. He even gave an estimate, viz.

$$A_{2m} < A < A_{2m} + (A_{2m} - A_m),$$

as can be seen from Figure 4. With this he concluded that "ultimately" $A = \frac{1}{2}Cr$. He also carried out the computation for finding A_{192}. In doing that he first established the formula

$$a_{2n} = \sqrt{\left[r - \sqrt{r^2 - \left(\frac{a_n}{2}\right)^2}\right]^2 + (a_n/2)^2}.$$

A modern computer can obtain $A_{192} = 3.141032$ (with $r = 1$) with error term 0.001681. Imagine how Liu Hui did it with only the help of counting rods over 17 centuries ago, obtaining $A_{192} = 314\frac{64}{625}$ (with $r = 10$). Effectively he calculated π accurate to two decimal places[13].

(3) The algorithmic feature of ancient Chinese mathematics can best be illustrated by the method of solving simultaneous linear congruence equations. In abstract algebra there is a fundamental result known as the "Chinese Remainder Theorem". Its name comes from a concrete instance, viz. Problem 26 of Chapter 3 of *Sunzi Suanjing* (*Master Sun's Mathematical Manual*), (c. 4th century):

> There are an unknown number of things. Counting by threes we leave 2; counting by fives we leave 3; counting by sevens we leave 2. Find the number of things.

The problem became quite popular and appeared under different names. In a much later text *Suanfa Tongzong* (*Systematic Treatise on Arithmetic*) of Cheng Dawei, published in 1592, there appeared even a poem about it: "T'is hard to find one man of seventy out of three. There are twenty-one branches on five plum blossom trees. When seven persons meet, it is in the middle of the month. Discarding one hundred and five, the problem is done." The poem conceals the magic numbers 70 (for 3), 21 (for 5), 15 (for 7) of this specific problem, whose general answer is $2 \times 70 + 3 \times 21 + 2 \times 15$ plus or minus any multiple of $105 = 3 \times 5 \times 7$. In general, the problem is to solve a system of linear congruence equations

$$x \equiv a_i \bmod m_i, \quad i \in \{1, 2, \ldots, N\}.$$

Mathematicians were led to investigate linear congruences because of calendrical reckoning and had become quite deft in handling them. Already in this specific problem we can see a very significant step made, viz. reduction of the problem to solving $x \equiv 1 \bmod m_i$, $x \equiv 0 \bmod m_j$ for $j \neq i$ (the solution to the original problem being a suitable "linear combination" of the solutions of these different systems). The investigation was completed by Qin Jiushao who named his method the "Dayan art of searching for unity" in his *Shushu Jiuzhang* (1247). He showed how to find a set of magic numbers for making the "linear combination". Consider the case when the m_i's are mutually relatively prime, using modern notations. (Qin Jiushao also treated the general case.) It suffices to solve separately single linear congruence equations of the form $kb \equiv 1 \bmod m$ by putting $m = m_i$ and $b = (m_1 \cdots m_N)/m_i$. The key point in the method Qin Jiushao employed to find k is to find a sequence of ordered pairs (k_i, r_i) such that $k_i b \equiv (-1)^i r_i \bmod m$ and the r_i's are strictly decreasing. At some point $r_s = 1$ but $r_{s-1} > 1$. If s is even, then $k = k_s$ will be a solution. If s is odd, then $k = (r_{s-1} - 1)k_s + k_{s-1}$ will be a solution. This sequence of ordered pairs can be found by using "reciprocal subtraction" explained in *Jiuzhang Suanshu*, viz., $r_{i-1} = r_i q_{i+1} + r_{i+1}$ with $r_{i+1} < r_i$ (the process will stop before one reaches the case $r_{i+1} = 0$), and put $k_{i+1} = k_i q_{i+1} + k_{i-1}$. (Put $k_{-1} = 0$, $r_{-1} = m$, $k_0 = 1$, $r_0 = b$.) The way the ancient Chinese performed the calculation was even more streamlined and convenient, since they put consecutive pairs of numbers at the four corners of a board using counting rods, starting with

$$\boxed{\begin{array}{cc} 1 & b \\ 0 & m \end{array}}, \quad \text{going to} \quad \boxed{\begin{array}{cc} k & 1 \\ * & * \end{array}}.$$

The procedure was stopped when the upper right corner became a 1, hence the name "searching for unity". A typical intermediate step will look like

$$\boxed{\begin{matrix} k_i & r_i \\ k_{i-1} & r_{i-1} \end{matrix}} \longrightarrow \boxed{\begin{matrix} k_i & r_i \\ k_{i-1} & r_{i+1} \end{matrix}} \longrightarrow \boxed{\begin{matrix} k_i & r_i \\ k_{i+1} & r_{i+1} \end{matrix}}, \quad \text{if } i \text{ is even}$$

or

$$\boxed{\begin{matrix} k_{i-1} & r_{i-1} \\ k_i & r_i \end{matrix}} \longrightarrow \boxed{\begin{matrix} k_{i-1} & r_{i+1} \\ k_i & r_i \end{matrix}} \longrightarrow \boxed{\begin{matrix} k_{i+1} & r_{i+1} \\ k_i & r_i \end{matrix}}, \quad \text{if } i \text{ is odd.}$$

One can see how the positions on a board of counting rods help to fix ideas. In fact, the procedure outlined in *Shushu Jiuzhang* can be phrased word for word as a computer program!

(4) An example on the lighter side is Problem 34 of Chapter 3 of *Sunzi Suanjing*:

> One sees 9 embankments outside; each embankment has 9 trees; each tree has 9 branches; each branch has 9 nests; each nest has 9 birds; each bird has 9 young birds; each young bird has 9 feathers; each feather has 9 colours. How many are there of each?

The problem, an easy exercise in raising a number to certain powers, is not of much interest in itself. What is interesting is the frequent occurrence of such problems of a recreational nature in all mathematical civilizations. The medieval European mathematician, Leonardo Fibonacci posed a problem in his book "Liber Abaci" (1202)[14]:

> Seven old women went to Rome; each woman had seven mules; each mule carried seven sacks; each sack contained seven loaves; and with each loaf were seven knives; each knife was put up in seven sheaths. How many are there, people and things?

It reminds us of a children's rhyme: "As I was going to Saint Ives, I met a man with seven wives. Every wife had seven sacks. Every sack had seven cats. Every cat had seven kits. Kits, cats, sacks and wives, how many were there going to Saint Ives?" And then there was that similar Problem 79 in the oldest extant mathematical text, the Rhind Papyrus of ancient Egypt (c. 17 century BC)[15]:

Houses	7
Cats	49
Mice	343
Heads of wheat	2301
Hekat measures	16807
	19607

David Hilbert (1862–1943) once said, "Mathematics knows no races.... For mathematics, the whole cultural world is a single country."[16]

Bibliography

There is a wealth of general accounts and in-depth research works on ancient Chinese mathematics. Many of them are contained in a vast store of books and papers written in Chinese, most of which have not been translated and thus remain inaccessible or even unknown to the non-Chinese-speaking community. For lack of space and in line with the nature of an introductory article, the works of these authors cannot be documented in full here, although liberal use has been made of their scholarship, for which the author feels at the same time thankful and apologetic. For the convenience of readers who do not read Chinese but who wish to go further into the subject, a few helpful references available in English are cited below.

The most well-known and oft-cited standard reference, an immensely scholarly and well-documented work (with access to over 500 items in Chinese and Japanese) on the whole history of Chinese mathematics, is of course

J. Needham (with the collaboration of Wang Ling), *Science and Civilization in China*, vol. 3, Chapter 19, Cambridge University Press, Cambridge, 1959.

A more up to date reference, which also emphasizes the context of ancient Chinese mathematics besides its content, is

J. C. Martzloff, *A History of Chinese Mathematics* (translated from *Histoire des mathématiques chinoises*, Masson, Paris 1987), Springer-Verlag, Heidelberg, 1997.

For a first reading, two good choices are

Li Yan and Du Shiran, *Chinese Mathematics: A Concise History* (translated by J. N. Crossley and A. W. C. Lun from *Zhongguo Gudai Shuxue Jianshi*, Zhonghua, Beijing, 1963), Clarendon Press, Oxford, 1987;

G. G. Joseph, *The Crest of the Peacock: Non-European Roots of Mathematics*, Tauris, London, 1991.

For a quick introduction through several informative and interesting articles, one can read

F. J. Swetz, The evolution of mathematics in ancient China, *Math. Magazine*, 52 (1979), 10–19;

F. J. Swetz, The amazing Chiu Chang Suan Shu, *Math. Teacher*, 65 (1972), 423–430;

F. J. Swetz, The "piling up of squares" in ancient China, *Math. Teacher*, 70 (1977), 72–79.
(All three articles above are collected in *From Five Fingers to Infinity: A Journey Through the History of Mathematics*, edited by F. J. Swetz, Open Court, Chicago, 1994.)

A chronological outline of the development of Chinese mathematics with an accompanying extensive bibliography of references written in Western languages (up to 1984) can be found in

F. J. Swetz and Ang Tian Se, A brief chronological and bibliographic guide to the history of Chinese mathematics, *Historia Mathematica*, 11 (1984), 39–56.

Endnotes

[1] At the end of the sixteenth century the first wave of dissemination of European science in China began. What happened after the sixteenth century, the vicissitude of indigenous mathematical development and its integration with transmitted western mathematics, form a fascinating topic in itself, but will not be discussed in this article.

[2] This article is based on an introductory lecture scheduled for the conference on História e Educação Matemática (Braga, Portugal, 24–30 July 1996). Circumstances prevented the author from attending. The lecture was instead given by Mr. Chun-Ip Fung, to whom the author owes his thanks.

[3] A typical passage can be found in the preface to *Shushu Jiuzhang* by Qin Jiushao (1247). The discussion in this article will be confined to "external mathematics" owing to the author's ignorance of the aspect of "internal mathematics".

[4] For further discussion of mathematics education in ancient China, see: M. K. Siu, Mathematics education in ancient China: What lesson do we learn from it? *Historia Scientiarum*, 4–3 (1995), 223–232. See also Chapter 1 in: F. Swetz, *Mathematics Education in China: Its Growth and Development*, MIT Press, Cambridge 1974.

[5] This is quoted from: M. K. Siu, Proof and pedagogy in ancient China: Examples from Liu Hui's commentary on *Jiuzhang Suanshu*, *Educational Studies in Mathematics*, 24 (1993), 345–357. The paper contains a number of illustrative examples.

[6] It was reported in, for instance: Li Xueqin, A significant finding in the history of ancient Chinese mathematics: A glimpse at the Han bamboo strips excavated at Zhangjiashan in Jiangling (in Chinese), *Wenwu Tiandi*, 1 (1985), 46–47.

[7] Such a statement has to be taken with a grain of salt! A better perspective can only be gained when one views mathematical development against a broader socio-cultural background at the time. In particular, the community of "mathematicians" in ancient China was not a well-defined recognized group of scholars. The author is in the process of studying, in collaboration with A. Volkov, mathematical activities in ancient China in this wider context.

[8] Compare this with Definition 3 of Book 5 of Euclid's *Elements*: "A ratio is a sort of relation in respect of size between two magnitudes of the same kind".

[9] This might have to do with the promulgation of the thesis of "Chinese origin of Western knowledge" in the Qing Dynasty in an effort to reassert the role of indigenous mathematics. A detailed discussion is beyond the scope of this article.

[10] Specific reference to each problem is omitted. Most statements of these problems are translated by the author. But there do exist translated texts for *Sunzi Suanjing* (L. Y. Lam and T. S. Ang, *Fleeting Footsteps: Tracing the Conception of Arithmetic and Algebra in Ancient China*, World Scientific, Singapore, 1992), *Shushu Jiuzhang* (U. Libbrecht, "Chinese Mathematics in the Thirteenth Century: The Shu-shu Chiu-chang of Ch'in Chiu-shao, MIT Press, Cambridge, 1973) and *Jiuzhang Suanshu* (K. Vogel, "Neun Bücher Arithmetischer Technik", Friedr. Vieweg & Sohn, Braunschweig, 1968). A French translation and an English translation of *Jiuzhang Suanshu* are under preparation.

[11] One naturally calls to mind the formula by the Greek mathematician Heron of Alexandria (c. 1st century), viz.

$$(\text{Area})^2 = S(S-A)(S-B)(S-C)$$

where $S = (A+B+C)/2$. Indeed, the two formulas are equivalent.

[12] It is interesting to compare it with the proof of Heron by synthetic geometry, which can be found in, for instance: I. Thomas, *Greek Mathematical Works*, II, Harvard University Press, 1939; reprinted with additions and revisions, 1980, pp. 471–477.

[13] It is interesting to compare this computation of π with that by Archimedes, which can be found in, for instance: R. Calinger (ed), *Classics of Mathematics*, Moore Publishing, 1982; Prentice-Hall, 1995, pp. 137–141.

[14] See: H. Eves, *An Introduction to the History of Mathematics*, 4th edition, Holt, Rinehart & Winston, New York, 1976, pp. 43–44.

[15] See: A.B. Chace, *The Rhind Mathematical Papyrus*, Mathematical Association of America, Oberlin, 1927–29; reprinted by the National Council of Teachers of Mathematics, Reston, 1978, p. 59.

[16] See: Constance Reid, *Hilbert*, Springer-Verlag, Heidelberg, 1970, p.188.

The Value of Mathematics—A Medieval Islamic View

George W. Heine, III
Math and Maps

Mathematicians and teachers of mathematics are frequently asked by students, administrators, employers and society, "What is mathematics good for? Why should we (take this course, pay for this)?" We may be troubled by how often the question is asked, and we may find it difficult to give a satisfactory answer. Yet the question is clearly important, and its answers give insight into the relationship between our discipline and our culture.

The answers that we usually give today tend to fall into three categories: *Pragmatic* answers emphasize the essential role of mathematics in the natural sciences, engineering and computer science, economics, and other fields which are central to our technological society. *Pedagogical* answers state that the study of mathematics is useful for training the mind in abstract thought as a prelude to other subjects. *Aesthetic* answers stress that mathematical problems can be interesting, challenging, satisfying, and otherwise bring pleasure to those who work on them.

To put our answers in perspective, it is interesting to examine how another culture might answer these questions. In this paper, I will try to discern the view of mathematics in medieval Islam. My focus will be the lifetime and works of the Persian scholar Abu 'Ali al-Husayn ibn 'Abdullah ibn Sina (980–1037[1]), known in the west as Avicenna. Ibn Sina's liftetime was a creative period, when a large number of philosophers, mathematicians, natural scientists, and other creative thinkers were active. This culture shared many historical links with our own, and we might expect that it would assent to all three of the above justifications for mathematical activity, with perhaps a difference in emphasis. Yet the culture was also different from ours in many ways, and it is not surprising that we find some entirely alien elements.

Historical Background[2]

During the first century after the death of the prophet Mohammed, the community of Islam formed a kind of Arab military aristocracy. The rulers, or caliphs,[3] were chosen during this period from the Umayyad family.[4] They led one of the most successful military campaigns in history, conquering the Sasanian empire in Persia, the Byzantine/Hellenistic areas of Syria, Palestine, and Egypt, and, eventually, north Africa and most of Iberia. However, the Umayyad government, organized like an Arabian tribe, was ill equipped for the administration of its large new empire. Non-Arabs and Arabs not in the ruling clans, even those who were converts to Islam, were second-class citizens. This caused resentment, since the new religion taught that all were equal before the Creator. In 749, a revolutionary army overthrew the Umayyads and replaced them with a line of caliphs from the 'Abassid family.[5]

The succeeding three centuries of 'Abassid rule in Persia and Iraq were the golden age of Islamic science and humanities. Scholars familiar with the Hellenic, Persian, Babylonian and Indian traditions were united under one government and one common language. The society was very prosperous; many private libraries were founded, and rulers actively promoted learning and scholarship. The caliph Al Ma'mun (813–833) began a state project to translate texts from Greek, Syriac, Pahlavi, and Sanskrit into Arabic; many ancient works have survived to our time only in Arabic translation.

The culture exhibited an astonishing degree of tolerance, even in matters of theology.[6] For example, Mohammed ibn Zakariya al-Razi (865–925) wrote that

> Books on medicine, geometry, astronomy, and logic are more useful than the Bible and the Qur'an. The authors of these books have found the facts and truths by their own intelligence, without the help of prophets.[7]

Other writers argued vigorously against such heresies, but the man himself was unmolested and lived to a peaceful old age.[8]

The above historical synopsis may puzzle readers who are accustomed to think of Islamic society as conservative, especially in matters of religion, and suspicious of or even hostile to outside ideas. These tendencies did develop, but only later, beginning a century after ibn Sina's death. Perhaps this was partly a protective reaction to the waves of foreign invasions—the Crusades, the Spanish *reconquista,* and the Mongol sack of Baghdad:

> At the height of its power and glory Islam accepted and adapted much of the culture of the peoples it came in contact with. In the ages of its decline confidence was lost and the guardians of its tradition feared and resisted all foreign invasion.[9]

Another reason for the conservative reaction was surely the teachings of Abu Hamid al-Ghazali (1058–1111), whose career and influence roughly parallel that of Saint Augustine. Trained in philosophy and rational theology, he attained a respectable position at a fairly young age, but then suffered a crisis of conscience, became a Sufi mystic, and spent the rest of his life vigorously denouncing those who pursued "essential truth" through rational thought:

> On the contrary, faith in prophecy is to acknowledge the existence of a sphere beyond reason; into this sphere an eye penetrates whereby man apprehends special objects-of-apprehension. From these reason is excluded in the same way as the hearing is excluded from the apprehension of colors and sight from apprehending sounds. . . [10]

Al-Ghazali's influence was felt for centuries, and served to push scientists, mathematicians, and philosophers to the margins of Islamic society.

But let us return to the lifetime of Ibn Sina and see what we can discover about attitudes toward mathematics during the golden age.

Life and Works of Ibn Sina

By the time of ibn Sina's birth, the power of the 'Abassid caliphs was much diminished. Nominally, they still ruled Syria, Palestine, Mesopotamia, and Persia, but in fact, eastern and central Persia were controlled by various small independent princes, and the northeast region of the caliphate, where ibn Sina's family lived, was threatened by Turkish tribesmen. Despite the uncertain political situation, cultural life continued to flourish.[11]

We can be fairly certain about the facts in Ibn Sina's life; an autobiography survives[12] (with the narrative completed after his death by one of his pupils), along with material by at least five medieval historians.[13] He received his early education in Bukhara, at the far northeastern corner of the caliphate, in what is today Uzbekistan (See Figure 1.[14]) Like other scholars of the day, he made his livelihood at a series of courts as physician, astronomer/astrologer, coun-

FIGURE 1
Cities where ibn Sina lived, according to his autobiography. Bukhara, his birthplace, is the easternmost of these; the others represent the courts of various rulers whom he served.

sellor, and teacher. The first such position, which he took in his early twenties, was in the city of Gurganj, capital of the province of Khwarazm.[15] There, one of his fellow scholars was Abu'l-Rayhan Muhammad ibn Ahmad al-Biruni, who is remembered today for his contributions to mathematics, geodesy, history, and other fields.

Because of the instability of the times, Ibn Sina was forced to move frequently from one court to another. According to one story,[16] the Sultan Mahmud of Ghazna, on the verge of conquering Gurganj, had asked that the learned men of the town be sent to his court. Al-Biruni acquiesced, and went on to a relatively tranquil career, later accompanying his patron on a conquest of India and writing about the history and customs of that land. But Ibn Sina left Gurganj in the middle of the night, crossing the desert in an ordeal in which one of his companions died. Similar episodes occurred later in his life; on one occasion, his political enemies managed to get him sent to prison. The biography hints that over-indulgence in drinking and sex may have contributed to his death at age fifty-eight.

Ibn Sina was a prolific writer; depending on how we count duplicated material and uncertain attributions, the number of surviving works seems to be somewhere between seventy[17] and two hundred and fifty.[18] His *al-Qanun fi-Tibb,* an encyclopedia of medical knowledge, made the latinized form of his name, Avicenna, famous in Europe—Geoffrey Chaucer mentions him in the *Canterbury Tales.* His mathematical writing is less well known; much of it seems to consist of paraphrases and commentaries on works by earlier Greek, Arabic, and Indian writers, especially Euclid, Ptolemy, and Nicomachus of Gerasa. Al-Daffa and Stroyls[19] have partially catalogued his contributions to mathematics. Although they find his expository work shows understanding and "sometimes improves on its sources," they find only a handful of minor results which might be original. These seem to be mostly arithmetical; for example, "If the successive odd numbers are placed in a square table, the sum of the numbers lying on the diagonal will be equal to the cube of the side; the sum of the numbers filling the square will be the fourth power of the side." (See Figure 2.)

9	7	5	3	1
19	17	15	13	11
29	27	25	23	21
39	37	35	33	31
49	47	45	43	41

FIGURE 2
Illustration of Ibn Sina's result for a 5-by-5 array. In this case, the entries on each diagonal sum to $125 = 5 \cdot 5 \cdot 5$, and all the entries in array sum to $625 = 5 \cdot 5 \cdot 5 \cdot 5$.

Certainly Ibn Sina's original mathematical output does not compare with that of al-Biruni. Perhaps, like Blaise Pascal after age thirty, he did little in mathematics because his attention was elsewhere. This at any rate is the opinion of the twelfth-century historian Ibn Funduq:[20]

> a person who has tasted the sweetness of metaphysics is niggardly in spending his thoughts on mathematics, so he fancies it from time to time and then abandons it.

However, even the expository work shows that he was familiar with the mathematics and the attitudes toward mathematics in his culture. He carried on an extensive correspondence with al-Biruni, discussing what we today might call foundational questions.[21] He banished all traces of Pythagorean mysticism from number theory:

> It is customary, among practitioners of the arithmetic art, to appeal to explanations foreign to that art, and even more foreign to the custom of those who operate by deductive proof, and closer to the discourses of orators and poets. This must be renounced.[22]

He added the new sciences of algebra and Indian (decimal) calculation to the Greek quadrivium, thus contributing to the breakdown of the traditional wall of separation between commensurable and incommensurable magnitudes. Rashed[23] argues that the new legitimacy of irrational numbers—whose values can never be known exactly—was a significant influence on ibn Sina's metaphysics.

The Pragmatic Value of Mathematics

In medieval Islam, calculation was an everyday necessity in the business world. By the tenth century, the place-value method was widely known. Thus we read in Ibn Sina's autobiography that his father "sent me to a vegetable seller who used Indian calculation, and so I studied with him."[24]

What is interesting in this account is that, in contrast to the Greek attitude, ordinary calculation is considered a respectable part of one's education. (Ibn Sina reports on studying calculation after studying the Qur'an, and before studying logic and geometry.) Further evidence of this new respectability of calculation is that, in his mature writings, Indian calculation and algebra appear alongside the traditional disciplines of arithmetic and geometry.[25]

Although medieval Islam would agree with us that mathematics has practical applications, some of those applications had the additional stimulus of religious duty. The complicated rules of inheritance in the Qur'an[26] likely stimulated interest in algebra, diophantine analysis, and combinatorics. The importance of such problems is shown by the emphasis Al-Khwarizmi gives to them in the introduction to his major work on algebra:

> a short work on calculating by *al-jabr* and *al-muqabala,* confining it to what is easiest and most useful in arithmetic, such as men constantly require in cases of inheritance, legacies, partition, law-suits, and trade, and in all their dealings with one another, or where the measuring of lands, the digging of canals, geometrical computation, and other objects of various sorts and kinds are concerned.[28]

Similarly, Al-Biruni wrote

> I say that a civilized person is keen to possess worldly valuables; hence the need arose for sciences which enable people to check the amount and area of possessions, when transferred from one owner to another, either by exchanges, or by inheritance. The principles of these sciences are called mathematics and formulae; they are embodied in the science of geometry.[28]

Another Qur'anic prescription was that prayer (and thus, the orientation of the mosque) must always be in the direction of the "sacred House of worship" in Mecca.[29] Methods were needed for determining the *qibla,* or bearing to Mecca, from any point on earth. This stimulated an intensive study of geography, astronomy, and spherical trigonometry. Al-Biruni's treatise *al-Kitab Tahdid al-Amakin,*[30] the work whose introduction we just quoted, was written to solve the *qibla* problem for the city of Ghazna in present-day Afghanistan. He continues:

> If the investigation of distances between towns, and the mapping of the habitable world, so that the relative positions of towns became known, serve none of our needs except the need for correcting the direction of the *qibla,* we should find it our duty to pay all our attention and energy to that investigation. The faith of Islam has spread over most parts of the earth, and its kingdom has extended to the farthest west; and every Muslim has to perform his prayers and to propagate the call of Islam for prayer in the direction of the *qibla.*[31]

David King[32] suggests that determining the *qibla* (and similar astronomical problems, such as determining the times of prayer) gave rise to two types of solutions: approximate solutions, most commonly used in actual situations—for example, many mosques in the early centuries were simply built facing south—and difficult, more exact, solutions devised by the scholars of the period. No doubt, the scholars who set to work on these problems got the same satisfaction that we do from solving an intriguing and complex problem. But they must have also had the extra stimulus of performing a religious duty, and the extra reward of having benefited and contributed to the piety of the whole community.[33] To quote al-Biruni again:

> I do not think that my work on a correct determination, of my exposition of the methods for a correct determination, will not be rewarded in this world or in the hereafter.[34]

The Pedagogical Value of Mathematics

Plato said that arithmetic, properly taught, "strongly directs the soul upward and compels it to discourse about pure numbers"[35] and that geometry "tends to draw the soul to truth."[36] This attitude was adopted and expanded by Islamic writers. Abu Yusuf ibn Ishaq al-Kindi (c.801–c.873) wrote a treatise with the suggestive title *In that Philosophy cannot be Attained except by way of Mathematics.*[37] Ibn Nabata quotes him:

> The philosophical sciences are of three kinds: the *first in teaching* is mathematics which is intermediate in nature; the second is physics, which is the last in nature; the third is theology, which is the highest in nature.[38]

After the time of al-Kindi, mathematics was known as "the first study". One consequence seemed to be that no one could be taken seriously as a philosopher unless one actually did some mathematics. This attitude is implicit in Ibn Sina's autobiography, since he studies first literature, then "Indian calculation", then Euclid, then Ptolemy, before turning to the study of philosophy. It may have been for this reason that Ibn Sina felt compelled to introduce summaries of Nicomachus and Euclid into his writings.

At first glance, then, the thinkers at the time of Ibn Sina would have agreed that the study of mathematics is important to train the mind in abstract thought as a prelude for other subjects. Yet when we look deeper, the reasons for this attitude are much different from ours. In the modern world, we tend to think that studying mathematics is a good preparation for studying quantitative subjects such as engineering or economics. In medieval Islam, the reasons for studying mathematics were more profound. We can see evidence of this in the writings of the Ikhwan al-Safa ("Sincere Brethren" or "Brethren of Purity"). This was an underground sect that in tenth-century Persia was considered unorthodox or even heretical. Moreover, the Ikhwan were allied with the Isma'iliyya, the party in power in Egypt and inimical to the Baghdad caliphate; thus, they were viewed as dangerous foreign agents. We know that Ibn Sina was aware of their teachings, and that they were important in shaping his views on mathematics. In the autobiography, we read

> My father was one of those who responded to the propagandist of the Egyptians and was reckoned among the Isma'iliyya. From them he, as well as my brother, heard the account of the soul and the intellect in the special manner in which they speak about it and know it. Sometimes they used to discuss this among themselves while I was listening to them and understanding what they were saying, but my soul would not accept it, and so they began appealing to me to do it. And there was also talk of philosophy, geometry, and Indian calculation.[39]

Some authors[40] have speculated that Ibn Sina was a follower of the Ikhwan al- Safa and that it was for this reason that he had to spend so much of his life fleeing from strictly orthodox rulers like Mahmud of Ghazna.

The principal work of the Ikhwan al-Safa was a set of *Treatises* (*Rasa'il*) probably written in the mid-tenth century. Typical of their attitude toward mathematics is the statement that the final aim of geometry is to permit the faculties of the soul to reflect and meditate independently of the external world so that finally 'it wishes to separate itself from this world in order to join, thanks to its celestial ascension, the world of the spirits and eternal life.'[41] Karen Armstrong writes that the Ismai'iliyya

> had developed their own philosophy and science, which were not regarded as ends in themselves but as spiritual disciplines to enable them to perceive the inner meaning (*batin*) of the Qur'an. Contemplating the abstractions of science and mathematics purified their minds of sensual imagery and freed them from the limitations of their workaday consciousness. Instead of using science to gain an accurate and literal understanding of literal reality, as we do, the Ismailis used it to develop their imagination.[42]

Here is a *raison d'être* for mathematics which hints at a totally alien way of thought. "Purifying the mind" and "freeing the consciousness" have been, at most, marginal notions in Western scientific culture; we do not immediately associate them with mathematics.

Although the scholars of medieval Islam would assent to the pedagogical view of the importance of mathematics, they would assign it a deeper significance, and some of their reasons would appear strange to us. Mathematics was not meant merely to train the mind, but also to purify the soul.

Aesthetic Pleasures of Mathematics

We have already noted how Islamic scholars spent time and energy developing difficult and exact solutions to some of the problems posed by religious duty, even when approximations would have sufficed. They must have enjoyed solving mathematical problems. But we have more than indirect evidence; some of Ibn Sina's most evocative writing deals with the pleasures of the intellectual life.

There is no question that Ibn Sina viewed the exercise of intellect as the greatest pleasure available to the human spirit. Despite his religious training, he rejected the literal truth of those verses of the Qur'an which speak of the physical pleasures of Paradise. To him, Hell was literally the soul longing for intellectual fulfillment which can no longer be realized after the death of the body, and Heaven its opposite. Such beliefs caused him to be charged with heresy during his lifetime, after his death, and definitively a century later by al-Ghazali. We can do no better than quote Ibn Sina's own words from the *Kitab al-Najat,* still eminently readable a thousand years later:

> Now the peculiar perfection towards which the rational soul strives is that it should become as it were an intellectual microcosm, impressed with the form of the All.... So it will have become graven after its idea and pattern, and strung upon its thread as a pearl is strung upon a necklace... When this state is compared with those other perfections so ardently beloved of the other [sensual] faculties, it will be found to be of an order so exalted as to make it seem monstrous to describe it as more complete or more excellent than they; indeed, there is no relationship between it and them whatsoever, whether it be of excellence, completeness, abundance, or any other of the respects wherein delight in sensual attainment is consummated.[43]

Ibn Sina and the intellectuals of his period took pleasure in solving problems, and they would assent to our "aesthetic" reasons for studying mathematics. But their answer had a deeper dimension. Intellectual pleasure became the basis for a whole doctrine of higher and lower pleasures, and was conceived as a means of purifying the soul, and reaching contemplation of the All. This leads us into a view of mathematics, and intellectual study in general, which is not familiar to our culture. In order to better understand this, in the next section we briefly examine the Theory of the Intellect and the special place occupied by mathematics in medieval Islamic cosmology.

The Theory of the Intellect[44]

In the fifth book of *De Anima,* Aristotle wrote:

> Since in every class of things, as in nature as a whole, we find two factors involved, (1) a *matter* which is potentially all the particulars included in the class, (2) a *cause* which is productive in the sense that it makes them all (the latter standing to the former, as, e.g., an art to its material), these distinct elements must likewise be found within the soul.[45]

Whatever Aristotle himself may have had in mind, later Greek philosophers interpreted the "matter" to be the human mind, and the "cause" to be an independent nonhuman entity which induces the mind to produce thoughts. Alexander of Aphrodisias[46] called this external entity "Active Intellect," and the human mind, "Material Intellect". Plotinus, in the *Enneads,* specified the relationship between the two:

> when 'a soul is able to receive,' Intellect 'gives' it clear principles, and then 'it combines' those principles until it reaches perfect intellect.[47]

Lest anyone wonder what those principles might be, Themistius, in a paraphrase of Aristotle available in Arabic translation, made the explicit connection with mathematics, stating as evidence for the existence of the Active Intellect that all persons knew, without being taught, the same "common notions" (κοιναι εννοιαι), "first definitions" (πρωτοι 'οροι), and "first axioms" (πρωτα αχιοματα).[48]

In the century before Ibn Sina, Abu Nasr al-Farabi (c. 870–950) made the notion of Active and Material Intellect part of his cosmology.[49] Like many ancient and medieval writers, al-Farabi envisions the terrestrial globe surrounded by nine concentric spheres (one each for the moon, sun, five planets, the fixed stars, and an outer "diurnal" sphere). The "Necessary One", i.e., God, emanates a first intellect, which has two thoughts, of its Creator and of its own essence. From the first thought, the existence of a second intellect "proceeds necessarily", while the existence of the first (outermost) sphere "proceeds necessarily" from the second thought. The process is repeated nine times, resulting in nine spheres and nine intelligences. From the ninth intelligence emanates the lunar sphere and a tenth intelligence, which al-Farabi identifies with the Active Intelligence, an incorporeal entity radiating principles to those human intellects ready to receive them. A much-used analogy compares this to the role of light on the human sense of vision; light does not create the eye, but it makes vision, and the perception of colors and forms possible.

In his treatise "On the Different Meanings of the Intellect," al-Farabi proposed that the intellect begins by dealing only with perceptions of the senses. Training it to comprehend abstractions is a step in the process which leads eventually to "communion, ecstasy, and inspiration." Following Plotinus, al-Farabi believes that whenever the human soul has

> to a certain degree attained freedom and separation from matter, the active intellect tries to purify it from matter . . . so that it arrives at a degree close to the active intellect.[50]

Ibn Sina developed his own cosmology, resembling al-Farabi's in many respects. In contrast to the mystic, who seeks direct knowledge of God through rejection of rational thought, Ibn Sina describes the attainment of the same goal by the embrace of rational thought and in fact by building a chain of syllogisms.

One interesting feature of Ibn Sina's work is his application of al-Farabi's theory of the intellect to explain how the human mind constructs logical proofs, or in his own terms, "finds the middle term of the syllogism."[51] The human mind must formulate the original question. When it is properly prepared, it receives an "emanation" (what we might call an inspiration or a flash of insight) from the Active Intellect. It is then up to the mind to act upon and make sense of this emanation. If it correctly finds the meaning of the emanation, it can go on to formulate the middle term. The result can then be remembered or written down; the mind can then go on to repeat the process for other syllogisms and so build a chain of reasoning.

With this in mind, the following sentence in Ibn Sina's autobiography becomes more comprehensible:

> And because of those problems which used to baffle me, not being able to solve the middle term of the syllogism, I used to visit the mosque frequently and worship, praying humbly to the All-Creating until He opened the mystery of it to me and made the difficult seem easy.[52]

Although any student, or any research mathematician, can sympathize with Ibn Sina's frustration, it is unlikely that such a sentence would appear in a modern biography. We are accustomed to think of mathematics and religion as quite irrelevant to one another.

Conclusion

It appears that medieval Islam would agree with all three of the reasons that we give for studying mathematics, but their answers would have a different significance. Mathematics was practical, but among its practical applications was the fulfillment of religious duty. Mathematics was good for training the mind, but not merely to study engineering and economics; the grasp of abstractions was an essential step in the process leading the soul to heavenly consciousness. Mathematics was enjoyable, but this enjoyment had a deeper significance because it led the soul to appreciate higher, rather than merely sensual, pleasures. In brief summary, the medieval Islamic culture differed from ours in the ever-present consciousness of a metaphysical dimension.

It is easy to assume, unless we are careful, that a historical person was motivated by the same things that motivate us today. Perhaps the most difficult task for the historian is the attempt to lift the veils of our own cultural prejudice. But it is a valuable study, and one that seems worth pursuing. Not only does it give us greater understanding of history, it prepares us to empathize with the different world-views of our colleagues, clients, and students today, and perhaps can lead to a more thoughtful personal answer to the question "Why is mathematics important?"

Acknowledgements

The author thanks Victor Katz, Janet Barnett, and the referee for many helpful comments and suggestions.

References

Afnan, S. M.: 1958, *Avicenna—His Life and Works,* Unwin Brothers, London.

Ali, Jamil: 1967, *The Determination of the Coordinates of Positions for the Correction of Distances between Cities—a Translation from the Arabic of al-Biruni's Kitab Tahdid Nihayat al-Amakin Litashih Masafat al-Masakin,* American University of Beirut.

Al-Daffa, A. A. and Stroyls, J. J.: 1984a, "Ibn Sina as a mathematician," In *Studies in the Exact Sciences in Medieval Islam,* John Wiley and Sons, New York, pp. 60–118.

——: 1984b, "Some myths about logarithms in near eastern mathematics," In *Studies in the Exact Sciences in Medieval Islam,* John Wiley and Sons, New York, pp. 26–30.

Arberry, A. J.: 1951, *Avicenna on Theology,* The Wisdom of the East, John Murray, London.

Armstrong, K.: 1993, *A History of God,* Alfred A. Knopf, New York.

Badawi, A.: 1963, "Muhammad ibn Zakariya al-Razi," In [Sharif, 1963], chapter XXII, pp. 421–433.

Berggren, J. L.: 1986, *Episodes in the Mathematics of Medieval Islam,* Springer-Verlag.

Bosworth, C. E.: 1968, "The political and dynastic history of the Iranian world, AD 1000–1217," In [Boyle, 1968], chapter 4, pp. 303–421.

Boyle, J. A., ed.: 1968, *The Cambridge History of Iran, Volume 5, the Saljuq and Mongol Periods,* Cambridge University Press.

Davidson, H. A.: 1992, *Alfarabi, Avicenna, and Averroes on Intellect,* Oxford University Press.

El-Ehwany, A. F.: 1963, "Al-Kindi," In [Sharif, 1963], chapter XXI, pp. 421–433.

Fakhry, B.: 1968, "A tenth-century Arabic interpretation of Plato's cosmology," Reprinted as Chapter VI in [Fakhry, 1994].

——: 1994, *Philosophy, Dogma, and the Impact of Greek Thought in Islam,* Variorum Collected Studies, Ashgate Publishing Company, Brookfield, Vermont, USA.

Frye, R., ed.: 1975, *The Cambridge History of Iran, Volume 4, from the Arab Invasion to the Saljuqs,* Cambridge University Press.

Gohlman, W. E.: 1974, *The Life of Ibn Sina—A Critical Edition,* Studies in Islamic Philosopy and Science, State University of New York Press.

Heinze, R.: 1899, "Themistii de anima," In *Commentaria in Aristotelia Graeca,* vol V, part III, Royal Academy, Berlin.

Kennedy, E.S.: 1968, "The exact sciences in Iran under the Saljuqs and Mongols," In [Boyle, 1968], chapter 10, pp. 659–680.

——: 1973, *A Commentary on Biruni's Kitab Tahdid al-Amakin,* American University of Beirut.

——: 1975, "The exact sciences in Iran under the Saljuqs and Mongols," In [Frye, 1975], chapter 11, pp. 378–395.

Kennedy, H.: 1981, *The Early Abassid Caliphate,* Longman, New York.

——: 1986, *The Prophet and the Age of the Caliphates,* Barnes and Noble.

King, D. A.: 1990, "Science in the service of religion: the case of Islam," Reprinted as chapter I in King, D. A. (1993), *Astronomy in the Service of Islam,* Variorum Collected Studies, Ashgate Publishing, Brookfield, Vermont, USA.

Madkour, I.: 1963, "Al-Farabi," In [Sharif, 1963], chapter XXIII, pp. 450–468.

McKeon, R., ed.: 1941, *The Basic Works of Aristotle,* Random House, New York.

Mottahedeh, R.: 1975, "The 'Abassid caliphate in Iran," In [Frye, 1975], chapter 2, pp. 57–89.

Nasr, S. H.: 1964, *An Introduction to Islamic Cosmological Doctrines,* Belknap Press, Cambridge, MA.

Rashed, R.: 1984, "Mathématiques et philosophe chez Avicenne," In Jolivet, J. and Rashed, R., *Etudes sur Avicenne,* Société d'Edition les Belles Lettres, Collection Sciences et Philosophie Arabes, Paris, 1984, pp. 29–39.

Rosen, F., editor and translator: 1831, *The Algebra of Muhammed ben Musa,* Oriental Translation Fund, London, Reprinted in 1986 by Georg Olms Verlag (Hildesheim, Zürich, New York).

Schroeder, F. M. and Todd, R. B.: 1990, *Two Greek Aristotelian Commentators on the Intellect—Introduction, Translation, Commentary, and Notes,* vol 33 of *Mediæval Sources in Translation,* Pontifical Institute of Mediæval Studies.

Sharif, M. M., editor: 1963, *A History of Muslim Philosophy,* Otto Harrassowitz, Wiesbaden.

Tibawi, A. L.: 1972, *Islamic Education—Its Traditions and Modernization into the Arab National Systems,* Crane, Russak, and Company, Inc., New York.

Watt, W. M.: 1963, *Muslim Intellectual—A Study of al-Ghazali,* Edinburgh University Press.

——: 1967, tr. "Deliverance from Error," in Watt, W.M., ed. (1967), *The Faith and Practice of al-Ghazali,* Allen and Unwin, Ltd., London.

Zarrinkub, 'Abd al-Husain: 1975, "The Arab conquest of Iran and its aftermath," In [Frye, 1975], ch. 1, pp. 1–56.

Endnotes

[1] Works on Islamic history customarily cite dates in both the traditional Islamic lunar calendar (AH) and the western calendar (CE). In this paper, we give only the CE date.

[2] For detailed political history, we relied on [Bosworth, 1968], [H. Kennedy,1981], [H. Kennedy, 1986], [Mottahedeh, 1975], and [Zarrinkub, 1975]. For shorter historical summaries with special attention to intellectual life, see [Tibawi, 1972] and [Watt, 1963].

[3] *Caliph* comes from an Arabic phrase meaning "successor to the Prophet." Each of the three caliphates (Umayyad, 'Abassid, Isama'ili) claimed descent from the family of the prophet Muhammad.

[4] A dispute between the Umayyads and supporters of the Prophet's son-in-law, 'Ali, led eventually to the modern schism between Shi'ite and Sunni Islam.

[5] The revolutionary army that won the 'Abassid caliphate never reached Spain or north Africa. A branch of the Umayyads continued to rule these areas for two centuries. The Isma'ili, a Shi'ite group, ruled in Tunis and then in Egypt from the mid-tenth century.

[6] At this epoch, the bulk of the population was not Muslim; in fact, because converts to Islam were exempt from most taxes, government officials seem to have often discouraged conversion. The Qur'an ordained that Christians and Jews be given special consideration as "People of the Book," and members of these religions often attained high positions at court.

[7] [Badawi, 1963], pp. 445–6.

[8] We do not mean to suggest that the society was open in the modern sense. There were sporadic, usually unsuccessful, attempts to enforce orthodoxy by such means as the withholding of government appointments; there were also instances of harsh repression, but apparently only when the regime perceived an actual military threat.

[9] [Tibawi, 1972], p. 45

[10] [Watt, 1967], p. 78

[11] Some even argue that cultural life prospered *because* of the instability; see [Nasr, 1964], pp. 14–17.

[12] In preparing this paper, we used the English translations of [Arberry,1951] and [Gohlman, 1974].

[13] [Rashed, 1984], p. 37, note 7.

[14] This region, the valley of the Oxus or Amu Dharya river, was much less isolated in the Middle Ages than it is today. An agricultural society prospered from at least 1300 BCE, and the region lay astride the ancient "Silk Road" to China. In addition, it was only a few days ride to the Persian province of Khorasan; the latter was of great political importance during the 'Abassid period.

[15] The great mathematician Muhammad ibn Musa al-Khwarizmi (d.850) also came from Khwarazm, as attested by his name.

[16] [Afnan, 1958], pp. 63–64.

[17] [Gohlman, 1974], pp. 13–15.

[18] [Nasr, 1964], p.180.

[19] [Al-Daffa and Stroyls, 1984a]

[20] [Gohlman, 1974], p. 122, note 27.

[21] Samples of this correspondence can be found in Chapter 10 of [Nasr, 1964].

[22] Ibn Sina's *Al-Shifa,* quoted by [Rashed, 1984], p. 32. "Il est d'usage, chez ceux qui traitent de l'art arithmetique, de faire appel ... à des développements étrangers à cet art, et plus encore étrangers à l'usage de ceux qui procèdent par démonstration, et plus proches des propos des rhéteurs et des poètes. Il faut y renoncer."

[23] [Rashed,1984]

[24] [Gohlman, 1974], p. 21. A marginal note in some texts of the biography suggests that he also learned algebra from a man known as "The Mathematician".

[25] The respectability of ordinary computation may have helped lead, at this time, to the invention of the method of prosthaphaeresis to do multiplication via addition and trigonometric tables. There is some controversy about whether prosthaphaeresis was known to medieval Islam. See [Al-Daffa and Stroyls, 1984b].

[26] Chapter iv, verses 7–12 and verse 176.

[27] [Rosen, 1831], p. 3.

[28] [Ali, 1967], pp. 4–5.

[29] Chapter ii, verse 150.

[30] English translation, [Ali, 1967]. A useful commentary is [E. Kennedy, 1973].

[31] [Ali, 1967], p. 13.

[32] [King, 1990].

[33] For further information on mathematical problems with a special Islamic character, see each of the sections entitled "The Islamic Dimension" in [Berggren, 1986], chapters 2, 3, 4, and 5.

[34] [Ali, 1967], p. 14.

[35] *Republic,* book vii, 525d.

[36] *Republic,* book vii, 527b.

[37] See [Afnan, 1958], p. 22.

[38] [El-Ehwany, 1963], p. 424. Emphasis added.

[39] [Gohlman, 1974], p. 19.

[40] [Afnan, 1958], pp. 63–65.

[41] [Nasr, 1964], p. 49

[42] [Armstrong, 1993], p. 178.

[43] [Arberry, 1951], p. 67.

[44] The relationship between Greek and Arab cosmology has been the object of much research. Sources that have been helpful here include [Armstrong, 1993], [Davidson, 1992], [Fakhry, 1994], and numerous articles in [Sharif, 1963].

[45] [Mckeon, 1941], p. 591. Emphasis added.

[46] [Davidson, 1992], pp. 20–24.

[47] [Davidson, 1992], p. 24.

[48] Greek edition, [Heinze, 1899], pp. 103–104; English translation, [Schroeder and Todd, 1990], p. 105.

[49] For al-Farabi's theory of the intellect I have relied principally on [Davidson, 1992], chapter 3 and [Madkour, 1963].

[50] [Davidson, 1992], p. 49.

[51] [Davidson, 1992], pp. 110–111.

[52] [Gohlman, 1974], p. 29.

Mathematics, Humanism, and Urban Planning in Renaissance Italy

Uwe Gellert
Freie Universität Berlin

Introduction: The Renaissance and mathematics

The period of the Renaissance is commonly held to constitute the entrance into modern scientific thinking—or at least its basis. Taking classical Greek and Hellenistic achievements into account and following the idea of inquiry by systematic research, the citizens of the Renaissance towns questioned the medieval clerical world. They tried to shape their modern society by the means of new machines and a reformed interpretation of the system of government in accordance with a humanistic body of thought.

Humanism could be termed an intellectual movement within western history. This movement tried to revive classical culture in order to use it for the benefit of general education. It attaches prime importance to man and human values, and is often regarded as the central theme of Renaissance civilization. Philosophically, humanism made human interests the measure of all things.

For this interest, new disciplines arose, such as astronomy and architecture, which unified scientific outcomes with a new imagination concerning life in society. But, what about mathematics, later (about 1800) called the queen of all sciences? Had a specified body of knowledge, a structured conception of mathematics, already existed? Are there new results in geometry widening Euclid's 'Elements' or algebraical and arithmetical insights going beyond the mathematics which the Arabs imported to Spain? What new mathematics was taught in the universities of Padova, Pisa and Bologna, enlarging the manuscripts of Toledo and Cordoba? Are we able to name famous Renaissance mathematicians of the 15th century?

An initial answer could be 'no'. The main task in the fifteenth century was to distribute the translations of Arab and Greek mathematics to the emerging scientific community. But, in a broader sense, including the influence and applications of mathematics on architecture, art, warfare, music, astronomy, geography, and navigation, Leonardo da Vinci, Alberti, Brunelleschi must be considered key figures in initiating the mathematization of Italian and, a few years later, of European society.

Urban planning in Renaissance Italy

The changes in the method of town planning in Italy in the second half of the fifteenth century serve as a significant example which clearly shows the rapidity of the new mathematization. It was a time when town planning entered a close relationship with intense realization and symbolism of religious or mystic images and, consequently, opened up to a rational mathematical counterpart of the humanistic world view. According to humanistic thought, the shape of the town gives an expression of the related State.

This can be demonstrated in the modification of the Tuscan village Corsignano into the model town Pienza by Pope Pius II around 1460. The realized concept of the central place (or 'Piazza') and the new cathedral of Pienza curiously shows the difficulties of the step from the town symbolizing the church authority of the sovereign, to the town intending to give evidence of the sovereign's new identity as a humanist. The unchanged existing village of Pienza bears testimony to a unique mixture of irrationality, as well as the then newly found rationality in the Quattrocento concept of an ideal town and society. But without Enea Silvio Piccolomini (1405– 1464), the son of an impoverished nobleman of Siena, Pienza, today, would not have existed as a town. Piccolomini was a universally educated humanist; Kaiser Friedrich III appointed him 'Poeta Laurentus' because of his extensive literary masterpiece; and his 'Cosmographia', the first scientific geography, influenced Christopher Columbus. Piccolomini made his way as the Bishop of Trieste and later of Siena to the Vatican in the city of Rome. While politically engaged in a crusade against the Turks, he concentrated his tendencies toward Humanism on the modification of his native town which, then, he called Pienza—city of Pius.

This modification focused on the central place, including the construction of a cathedral and palaces for the Pope, the bishop, and the town council, and should be interpreted as a demonstration of a cultural vision, rather than a vision of concrete modernization or of renovation. The central structural characteristics of this realized vision are:

- a church hall built on natural caves, with three naves flooded with light,
- a church façade decorated by columns alluding to classical temple architecture,
- the town palace of the Pope, the 'Palazzo Piccolomini', integrating the style of a villa by opening the rear to the landscape over a hanging garden as an epitome of artifically arranged nature,
- the connecting centerpiece, the 'Piazza', divided into nine sections by travertine lines, linking the heterogeneous structures of the façades and walls of the adjoining buildings.

The Palazzo Piccolomini could be considered as a very modern conception of official and private living, and of living in a town as well as with nature. The cathedral shows the link of the main influences of sacred architecture: the caves as an Etruscan place of worship, the three naves driving into the Romanesque period, the Gothic order of the big windows, and the Renaissance façade.

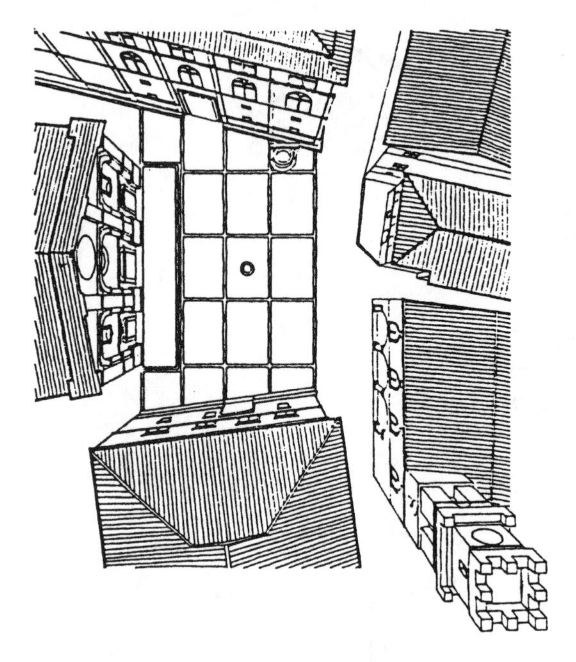

FIGURE 1
The relation between the travertine lines and the surrounding buildings [Pieper, 1986, p. 1729]

Concerning the level of mathematization, the main point of interest is the geometrical structure of the Piazza and, until 1978, its hidden connections with the geographical position and size of the cathedral. A first hint for the discovery of these subtle interrelations is the unusual orientation of the cathedral. Normally, the chancel is directed to the East, but in Pienza it is oriented towards the south, opening the view to the Monte Amiata, which dominates southern Tuscany. A more sophisticated reason for this "incorrect" orientation is only observable in the equinoxes, the dates when the nights and the days have the same length in spring and autumn. On these days, the construction of the building is intended to give a shadow fitting into the geometrical order of the Piazza.

The calculation with angles and related distances of the shadow is used to paraphrase an Orphic-Christian number symbolism. The division of the Piazza into nine sections alludes to traditional views about the underworld: according to Christian tradition, Jesus Christ, after his crucifixion, went into an underworld consisting of nine hells. Dante Alighieri's "Inferno" also reported a ninefold hell (Hopper, 1938). The calculated shadow play on the Piazza leaves room for interpretations in the context of light, death, and underworld, based on numerology and exactly constructed by means of mathematics. The involved rationality, how-

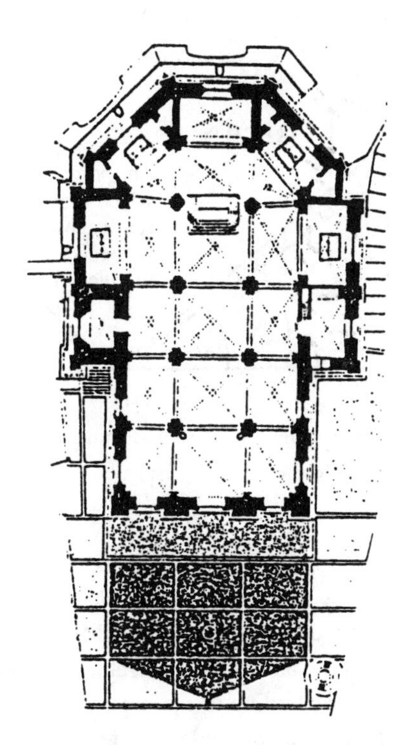

FIGURE 2
Ground plan and shadow of the cathedral on equinoxes [Pieper, 1986, p. 1714]

ever, is not easy to discover, and the astronomy of the calendar architecture of Pienza seems to be difficult:

(1) The theoretical examination with the help of magnetism and with the mathematical-technical tool, the compass, did not allow Pius' architect Bernardo Rossellino (1409–1464) to correctly determine the north-south direction. Although the magnetic declination of the compass was discovered around 1460, it was not known in Italy at that time. Therefore, Rossellino's determination differs $12°30'$ from the exact south;[1] a sundial situated on the Piazza shows this deviation, too. Consequently, nowadays, we observe the congruence between the shadow and the pattern in the Piazza $(12.5°/360°) \times 24h = (5/6)h = 50$ min after noon. As a result of this delay, Rossellino's plan had to take into account that the sun meanwhile drops about $1°26'$.

(2) The calendar architecture of Pienza also illustrates a basic astronomical problem of the fifteenth century, the incorrectness of the Julian calendar. The Julian calendar, established by Julius Caesar in 46 BCE, is based on a year of $365\frac{1}{4}$ days. Therefore, the calendar becomes one day out every 129 years, and in the year of the construction (1459) it was eleven days wrong. By about 1440, Nicolaus von Kues (well known as Cusanus, 1401–1464), an important scientist, philosopher, and intimate friend of Pius, had drawn up a calendar reform which was not realized until 1582 by Pope Gregory XIII. Various reasons relating to church policy prohibited the reform from taking place earlier. The decision to fix the spring equinox on the 21st of March was carried out in 325 CE by the Council of Niceae, and determines the dates of Easter, the most important celebration in the whole ecclesiastical year. This holiday takes place on the first Sunday after the first full moon after the spring equinox. Consequently, changing the calendar meant changing the dates of Easter—a problematic question among the different Christian churches.

A result of this disagreement between church and science was that Pius and his counsellors, Cusanus and Rossellino, were confronted with a matter of conscience difficult to tackle. According to exact measurement, the spring equinox in 1460 was not on March 21 but on March 10/11 (in the Julian calendar). The corresponding angles of incidence of the sun at noon were of size $\alpha = 49°22'$ (March 21, 1460, in the Julian calendar) and $\beta = 45°55'$ (March 10/11, 1460, in the Julian calendar).

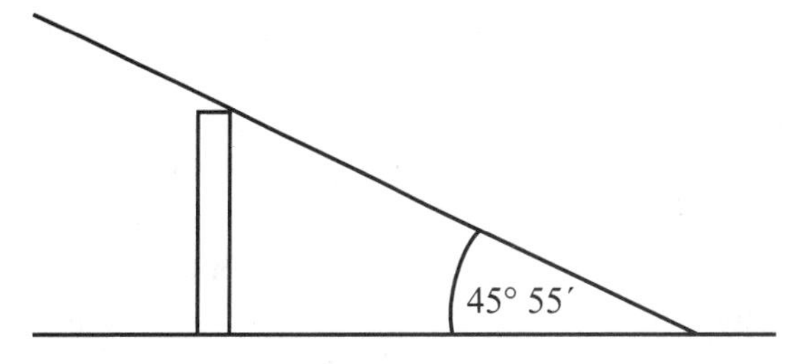

FIGURE 3
The 'true' angle of incidence of the sun at noon on the equinoxes in Pienza

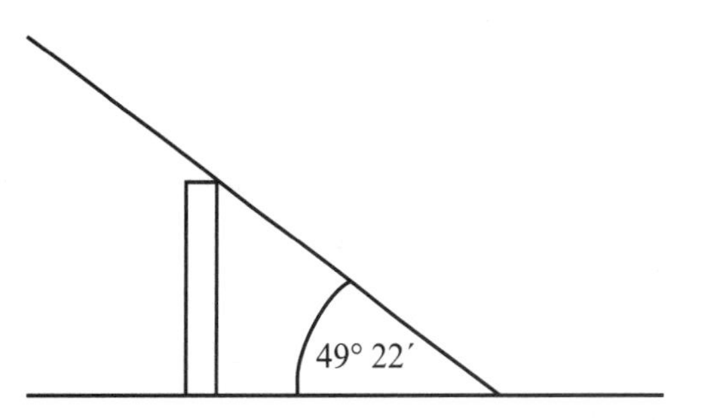

FIGURE 4
The angle of incidence of the sun at noon on the spring equinox in Pienza around 1460 according to the Julian calendar (on an exaggerated scale)

[1] The magnetic declination varies irregularly. The average variation is about $9'$ per year. The maximum declination for the geographical position of Pienza is $24°45'$ East or West.

But as Pope Pius II, he was aware of the implied disputes of a strong insistence on his scientific and humanistic views. On the other hand, a modern architecture and model town planning based on the ignorance of science in favor of an old and incorrect calendar must have been an intellectual sacrifice for the humanists, Pius and Cusanus.

Fortunately, these mental battles are still reconstructible from the details of the cathedral:

(a) One curious occurrence inside the building is the very strange-looking, artificially raised capitals of the columns which take the weight of the vault of the hall with three naves. Obviously, these capitals break the outline of the whole cathedral. Pius also wrote, in his "Comments" on the building process, that when the columns had already been finished and the capitals had been laid, the architect noticed a gap of seven feet in height, and interpreted the belated heightening as a fortunate and enriching error for the beauty of the cathedral.

Shall we follow Pius in his "Comments"? Is it probable that one of the most famous and experienced Renaissance architects 'suddenly noticed' a space of seven feet in height? Another interpretation is the following: The height of the artificial heightening of the capitals is about 2.41 m. Bearing in mind the difference in size between the 'Gregorian' and the 'Julian' angle of incidence of the sun on the spring equinox at noon of $3°27'$, we discover that the 2.41 m mentioned compensate for the higher position of the sun (the calculated difference based on the incorrect calendar is 2.44 m), as is easily explainable. The total height of the façade responsible for the shadow is measured by theodolite to $a = 21.41\ m$. The length of the shadow in the place is $s = 18.38\ m$. The calculation of a by using s and the angle of incidence of the sun α on March 21, 1460, at noon, in the "wrong" Julian calendar proves:

$$a = s \text{ multiplied by } \tan\alpha$$

$$a = 18.38\,\text{m multiplied by } \tan 49°22'$$

$$a = 21.42\,\text{m (within the scope of precision of the measurement).}$$

Calculation towards the height of the façade on the true equinox b under the angle of incidence at noon of the sun $\beta = 45°55'$ shows:

$$b = s \text{ multiplied by } \tan\beta$$

$$b = 18.38\,\text{m multiplied by } \tan 45°55'$$

$$b = 18.98\,\text{m}.$$

Consequently, the difference $a - b = 21.42\text{ m} - 18.98\text{ m} = 2.44\text{ m}$ has been compensated by heightening the capitals inside the church.

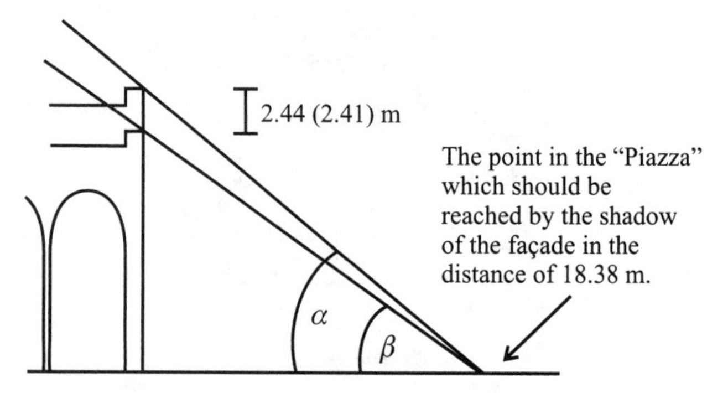

FIGURE 5
Belated heightening of the façade (on an exaggerated scale)

Apparently, Pius' or Rossellino's first plan was founded on the true—scientific—date of the equinox. Not until the building was nearly finished did Pius II at last decide to use the church as treasurer of truth. He consequently dated the equinox according to the old calendar bearing in mind that his mathematical model of a new town as a symbol for a rational society now became blurred by the rootedness in out-dated clerical definitions. Hence, Rossellino's task was to inconspicuously change the conception of the upper part of the building; and in the interior he succeeded partly through the curious shape of the capitals. On the other hand, he was forced to give up any proportional system on the façade.

(b) Among the Renaissance architects, theorists, and mathematicians, the best way to arrive at art and beauty was controversial. All parties involved agreed that proportion had to be regarded as mother, and queen, of all arts. The use of special numbers to describe and construct these proportions mathematically was regarded as the only method to reach beauty and art. Also, the source of the numbers ought to be from the classical period:

- Vitruvius' (84 BCE–?) '*De architectura libri decem*', the only architectural treatise preserved which was written before the Common Era, was considered by Petrarca and Boccaccio. It was then well known among architects and artists, who favored the Pythagorean numbers as the standard for proportion.
- Rossellino, and later Andrea Palladio (1508–1580) referred to Plato's expositions about the square doubling to gain "perfetti" proportions (Palladio, 1570) by using the square root of two.
- The mathematician and monk Luca Pacioli's (1445–1514) '*Divina proportione*' picked up Euclid's expression for the golden section and tried to link geometry with divine harmony.

• Leon Battista Alberti (1404–1472), humanist, architect, and mathematician, developed aesthetics of building based on "numerus" (number), "finitio" (relation), and "collocatio" (arrangement) in his treatise, "*De re aedificatoria*" (Alberti, 1485). By claiming the proportion between diameter and height of Doric, Ionic, and Corinthian columns to be 1:7, 1:8, and 1:9, he emphasized the "normative demand" of Renaissance architecture (Thoenes, 1995), and called for the use of integers for the construction of proportions.

In the façade of the cathedral it is not possible to discover any common proportion. If, however, we reduce the total height of the building by 2.41 m, the standard proportion according to Rossellino becomes obvious: The mathematical relation of the height of the basement to the top floor is $1 : \sqrt{2}$.

Conclusion: Mathematics is not 'universal'

The process of construction of the central area of Pienza serves as a significant example of the transition to a rational, mathematical society.

Pius II and his humanistic counsellors viewed mass education as a *conditio-sine-qua-non* for the implementation of humanism in society. Their humanism made clear by the concept of a cathedral integrating different epochs of architecture, put education at the top of the agenda. This focus on education aimed at accelerating the circulation of information. In addition, the buildings in Pienza show myths and mysticism contrasted with scientific methods and rationality and they symbolize the change in the consciousness of societies.

Although Pius II intended to give evidence of the sovereign's new identity as a humanist, he was forced to act as a church authority and to give up the realization of the claim of scientific thought. Engaged in leaving the old conception of town planning, of directing all views towards the church and, instead, showing Pienza as a symbol and an expression of modern states, he experienced the old constraints of working against a perfect mathematical idea.

The concepts of humanism, social order, and scientific thought have changed since the fifteenth century. Education, however, in schools and universities has become an unquestioned, state-run cornerstone in all Western societies. In the educational process of enculturation, the role of mathematics is often regarded as crucial, as well as doubtful, and remains under discussion.

Mathematics and science are the only school subjects thought of by many people to be relatively unaffected by the society in which learning takes place. As Bishop (1993, p. 225) points out, "whereas for the teaching of the language(s) of that society, or its history and geography, its arts and crafts, its literature and music, its moral and social customs, which all would probably agree should be considered specific to that society, mathematics and science are usually considered as 'universal'." On the contrary, the example of Pienza shows how intimately mathematics, religion, history, society, and politics are intertwined. However, the teaching of mathematics has the duty of divesting it of its 'universality'. Mathematics should no longer be considered as being free of any values.

Taking into account that mathematics is 'pregnant with values', the National Council of Teachers of Mathematics (1989) identified "New Goals for Students" in the learning of mathematics. It describes under the heading, the needs of society and the needs of students, how "learning to value mathematics", "becoming a mathematical problem-solver", "learning to reason mathematically", and "learning to communicate mathematically" can fulfill social as well as individual demands.

In order to follow these arguments, the beginning mathematization of town planning activities in Pienza could be interpreted as a demonstration of how the development of mathematics and science influenced a change in social concepts.

References

Alberti, L.B.: 1485, *Leonis Baptiste Alberti de re aedificatoria incipit,* Florence: Nicolaus Laurentius.

Bishop, A.J.: 1993, "On Determining New Goals for Mathematical Education," In Keitel, C. and Ruthven, K. (Eds.) *Learning from Computers: Mathematics Education and Technology,* Berlin: Springer-Verlag, p. 222-242.

Hopper, V.F.: 1938, *Medieval Number Symbolism,* New York.

National Council of Teachers of Mathematics (Ed.): 1989, *Curriculum and Evaluation Standards for School Mathematics,* Reston, VA: NCTM.

Pacioli, L.: 1509, *Divina proportione,* Venice: Paganino de' Paganini.

Palladio, A.: 1570, *I quattro libri dell' architettura,* Venice: Dominico de Franceschi.

Pieper, J.: 1986, "Pienza," In: *Bauwelt* 1986, Heft 45, pp. 1711–1732.

Thoenes, C.: 1995, "Anmerkungen zur Architekturtheorie," (Comments on the theory of architecture) In: Evers, B. (Ed.) *Architekturmodelle der Renaissance, Die Harmonie des Bauens von Alberti bis Michelangelo,* München: Prestel, pp. 28–39.

Vitruvius, M.P.: 1497, *L. Vitruvii Pollionis de architectura libri decem,* Venice: per Simonem Papiensem dictum Bevilaquam.

An Epistemological History of Number and Variation

Luis E. Moreno-Armella and Guillermina Waldegg C.
Centro de Investigación y de Estudios Avanzados del Instituto Politécnico Nacional, Mexico

In this paper we map out a thread of connected developments in the epistemology of number and variable. Instead of following a detailed historical path, we present some "snapshots" of these developments: from the initial binary Greek categories of discrete number and continuous magnitude, through Stevin's unification of these notions in the wake of the introduction of decimals; from Oresme's use of forms and Galileo's study of curves as a geometric representation for variation, together with Vieta's idea of algebraic variable quantity, to Descartes' conjunction of these concepts in the invention of analytic geometry. Our intention here is to stress the epistemic shift in the concepts of number and magnitude and the evolution of the idea of a variable quantity as being of principal significance to the entire sweep of the history of mathematics.

Introduction

Two developments were instrumental in the history of number and variation: (i) the identification of number with the geometrical continuum, and (ii) the geometrical representation of variation. With regard to number, the drastic change that this concept underwent in evolving from the Euclidean conception into the unified concept of number introduced by Stevin in 1585 was significant. In the development of the concept of variable quantity, the keystone was the possibility of interpreting those quantities as continuous magnitudes and representing them in the language of algebra.

It has been reported [see for instance Klein, 1968 and Jones, 1978] that the Euclidean number concept came from the works of Pythagoras and Aristotle. There are two elements of this history that, from our present viewpoint, are surprising: (i) *one* is not considered a number, and (ii) *number* can only be applied to the study of discrete collections; in other words, there was no notion of continuity associated with the concept of number. It is because of these features that we wish to identify the point in time that number and (continuous) magnitude become integrated into the same concept.

We must recall that Aristotle dismissed the Arrow Paradox by saying that time could not be made of moments, and lines could not be made of points—which was a way of saying that the category of *quantity* had two disjoint components: the discrete (number) and the continuous (magnitude). These components were reflected in mathematics as the study of magnitudes and numbers, i.e., as the study of geometry and arithmetic. For Aristotle, continuity was characterized as "never ending" divisibility, from which it was possible to conclude that the line could not be composed of points. Lines and points belonged to different mathematical realms.

The geometric continuum appeared as an abstraction of the physical continuum. Because of the characterization of continuity as never ending divisibility, it was possible to conclude that the continuum was not made of indivisibles. On the other hand, number was the prototype of discreteness; number was a collection of units and the unit was not a number.

This scenario changed radically with the work of Simon Stevin. The Greek concept of number had developed as a result of an abstraction process applied to the material world. Stevin challenged the Greek viewpoint to accommodate the utilitarian matter of measurement in the real world, a perspective which will be elaborated below. Nevertheless, Euclidean conceptions were so deeply rooted that Stevin found it necessary to argue against this tradition, both from a practical viewpoint and from what we consider to be an epistemological viewpoint. It was not just a matter of saying "one is a number" but of producing a substantiated argument to justify that assertion.

Let us turn now to a more systematic account of Greek mathematics necessary to understand the conceptual atmosphere in which Stevin developed his work and, also, the construction of algebraic language and the notion of variation.

The Greek Concept of Quantity

Aristotle introduced *quantity* as one of the eight essential categories of thought [Aristotle, *Categories* 1b, 25] and divisibility as the fundamental operation that renders the classification of quantities possible. A magnitude was a quantity divisible *ad infinitum*; this defining characteristic was called continuity. A number was a quantity only divisible in a finite number of steps; this defining characteristic was called discreteness. The *quantum,* as an element of the class of quantities, was

> that which is divisible into two or more constituent parts of which each is by nature a 'one' and a 'this'. A 'quantum' is a plurality if it is numerable, a magnitude if it is measurable. 'Plurality' means that which is divisible potentially into non-continuous parts; 'magnitude', that which is divisible into continuous parts; ... limited plurality is number, limited length is a line... [Aristotle, *Metaphysics,* 1020a, 5].

The following quotation illustrates why geometry and arithmetic were unconnected except in the particular case of commensurable magnitudes. In this case arithmetical results could be applied since these magnitudes can be thought of as numbers:

> The axioms that are premises of demonstration may be identical in two or more sciences: but in the case of two different `genera' such as arithmetic and geometry, you cannot apply arithmetical demonstrations to the properties of magnitudes unless the magnitudes in question are numbers. [Aristotle, *Posterior Analytics* 75b, 5.]

The word *genera* (plural of genus) in this quotation refers to the class of objects that share certain characteristics of "birth" or generation and reveals how the differences in the nature or the origin of mathematical objects determined Aristotle's mode of operation with them. Although the propositions in the two theories are analogous and the rules of logic the same, they could not be indiscriminately applied to objects with a different genesis. Euclid maintained a separation between arithmetic and geometry analogous to that of Aristotle [Jones 1978, p. 377]. Books I to VI and XI to XIII of the *Elements* are geometric, while Books VII to IX are arithmetic. With the exception of Book X, the terms *number* and *magnitude* never appear together in the same book. The proposition X-5, however, makes use of the license granted by Aristotle as follows:

> Commensurable magnitudes have to one another the ratio which a number has to a number.

Aristotelian Epistemology

For Aristotle, concepts had their origins in the physical world and inherited their properties and relationships from Nature. To acquire knowledge, Nature should be closely observed and scientific concepts abstracted. In the same way, number and magnitude were concepts extracted from the material world, as can be appreciated by the following quotation:

> All these objections, then, and others of the sort make it evident that number and spatial magnitudes cannot exist apart from things. [Aristotle, *Metaphysics,* 1085b, 30]

For the Greeks what could be known was a reality external to the knower and independent from him. [Piaget 1950, Chapter III, on "Mathematical knowledge and reality".] This epistemological position permitted the construction of a satisfactory theoretical and methodological structure but imposed a series of restrictions on mathematical objects. For our purposes we will look at the restrictions that can be detected in the concepts of number and magnitude and the relations between them.

Number and Magnitude

In the *Elements,* the concept of number was based on multiplicities or collections of concrete, individual and indivisible objects, each one of them identified with an abstract unit:

> A unit is that by virtue of which each of the things that exist is called one. [*The Elements,* VII, Def 1]
>
> A number is a multitude composed of units. [*The Elements,* VII Def 2]

These definitions, like Aristotle's definition of quantum quoted earlier, lead us to concepts of unit and of number linked to the process of counting as an activity of "reading figures" in things. The Greek numerical domain had a basic structure derived not only from how the numerical series was generated but also from the operation with its elements.

Once the nature of the concepts was determined—as a reflection of an external physical reality—the methodology required to operate (manipulate) with these concepts was defined. The unit, the principle of generation of number, was the concept that resulted from abstracting the features that were exclusively concerned with the singularity of each "thing". Thus, from the Greek point of view, to think of the division of the unit was meaningless, since it would lose its essence in the same way that a thing (a man, a horse) could not be divided without losing its essence. The impossibility of dividing the unit illustrates the essence of discrete quantity as defined by Aristotle. Subdivisions of a discrete quantity could not continue beyond the unit and, as a result, only a finite number of divisions could be carried out. On the other hand, geometric magnitudes—the prototype of continuous quantities—formed a heterogeneous domain. Length, area, and volume belonged to this domain. They shared the property of being divisible, potentially *ad infinitum,* and, if they were of the same species, one could establish ratios and proportions among them. There was no structure similar to that found in the numerical domain. Line segments, for example, formed an aggregate of abstract elements isolated from and independent of each other. Given two segments, it was possible to operate with them, to compare them, and to order them, but this did not mean that they lost their individuality. In Greek geometry the straight line was just a segment that could be extended indefinitely, but there was no concept analogous to our current "real line."

The comparison between magnitudes in terms of their ratios, which is the first qualitative level of the measuring process, led to the notion of "relative (not absolute) commensurability" between magnitudes. However, this was not yet a classification—in the sense of an equivalence relation—since only two elements were taken at a time. Divisibility was at the root of the definition of commensurability and incommensurability. If it was possible to find a finite magnitude that "measured" (divided exactly) two given magnitudes simultaneously, then these two magnitudes were said to be commensurable. If the search for this "unit-magnitude" led us to take smaller and smaller magnitudes in a process of potentially infinite division, the magnitudes were said to be incommensurable. In the first case, the ratios behaved in a similar way to numerical ratios and, as a result, could be treated like the latter, according to proposition X-5 of the *Elements.* If this were not the case, then there was no possible link between magnitude and number. Greek mathematics considered the existence of a unit magnitude not only necessary for the practice of measurement but also theoretically indispensable for the commensurable-incommensurable characterization.

According to Greek realist epistemology, a metric unit could not have the status of "absolute" that the arithmetic unit had, since the former did not have an "external" reality. To select a unit magnitude would be the equivalent of endowing the magnitude with attributes imposed by the subject and not "read" from reality itself; thus the election of a unit of magnitude was not independent of the activity of the subject. However, the unit of magnitude that resulted from the comparison of two commensurable magnitudes (their "greatest common divisor") was a unit intrinsic to the two magnitudes compared and consequently independent of the subject that carried out the comparison. This unit had the theoretical characteristics imposed by Greek mathematics, but it also had the drawback of not always being the same, since it varied as the pair of compared magnitudes changed.

Accepting that the arithmetic unit and the geometric unit corresponded to different "genera", let us summarize their behavior. The arithmetic unit was not a number; the geometric unit was a magnitude. The arithmetic unit was absolute; the geometric unit depended on the magnitudes that were being measured. The arithmetic unit generated numbers; the geometric unit did not generate magnitudes. The differences between the two domains that we have highlighted make an identification between numbers and magnitudes in Euclidean mathematics impossible, even at an operational level.

Variation in a Geometrical Context

Now let us turn to Oresme's work, *Treatise on the Configurations of Qualities and Motions* (see Clagett, 1968 for a splendid English translation). There we find the study of variation in a geometrical context, in particular, the idea of a *flowing quantity.* Oresme introduced geometric figures to represent the behavior of a *quality.* According to him, the study of a body could be realized from two viewpoints: from an *extensional* one—for instance, the weight of the body—and from an *intensional* viewpoint—for instance, the body's temperature. In the latter case, the measurement was made point-by-point. Reading through the chapter "On the continuity of intensity" [Oresme, Part I. i, pp. 165–169], we find the Euclidean conception of number where Oresme said that any measurable thing *except number* could be represented by a magnitude; that is, "in the manner of continuous quantity". This also explains why, when talking about intensities, he said that "the points of a line" were a *necessary fiction* used to represent a place on the studied body. Each *intensional measurement* made on a body, was represented by means of a segment. A segment also represented the body itself. Oresme considered the whole set of intensities as the gateway to the study of variation. This set was called a *surface latitude,* and it contained all the information about the variation of the intensity. But here a shift in his viewpoint appears as the surface latitude was used to study the variation of *forms,* rendering his work a qualitative one.

Let us consider Oresme's study of velocity. Velocity was imagined as a quality acquired by a moving body. The terms "uniform" and "uniformly difform" were introduced to name a quality that did not change with time in the first case, and a variable latitude with a constant rate of change in the second case. Oresme also considered "difformly difform" surface latitudes.

Although this classification procedure was, in general, independent of physical considerations, Oresme's way of considering variation endowed his representational model with the capability of studying motion, in particular, geometrically. This viewpoint is also found in Galileo's *Two New Sciences.* However, in the latter case the conceptual framework is very different. Let us consider Galileo's

> Theorem I (Proposition I)
> The time in which any space is traversed by a body starting from rest and uniformly accelerated is equal to the time in which that same space would be traversed by the same body moving at a uniform speed whose value is the mean of the highest speed and the speed just before acceleration began. [Galileo in Struik, 1969, pp. 208–209]

An analogous version of this theorem, known as the Merton Rule, can be found in Oresme's *Treatise on the Configurations of Qualities and Motions.* The wording there is the following:

> Every quality, if it is uniformly difform, is of the same quantity as would be the quality of the same or equal subject that is uniform according to the degree of the middle point of the same subject. [Oresme 1350, Part III. vii, p. 409]

We find it enlightening to quote Clagett on this matter. He says,

> [Oresme] does not, however, apply the Merton Rule of the measure of uniform acceleration of velocity by its mean speed, discovered at Oxford in the 1330's, to the problem of the free fall, as did Galileo almost three hundred years later, although Oresme of course knew the Merton Theorem and in fact gave the first geometric proof of it in another work, the De Configurationibus, but as *applied to uniform acceleration in the abstract rather than directly to the natural acceleration of falling bodies* (our emphasis) [Clagett, Op. Cit., pp. 13–14]

This clearly establishes that the conceptual framework within which Galileo worked was not the same as Oresme's. The latter was interested in the *general* study of intensities and configurations. The former was interested in the very special case of free falling bodies.

A New Concept of Number

In 1585 Simon Stevin published his book *L'Arithmetique* [Stevin 1585, in Girard (ed.), 1634]. which produced an epistemic shift in mathematical knowledge. It was a treatise about the theoretical and practical aspects of arithmetic. In chapter X Stevin presented his (new) concept of number. For him, "number is that through which the quantitative aspects of each thing are revealed" (our translation).

In Greek mathematics the category of quantity had been separated into disjoint classes: discrete and continuous. Stevin did not take this separation into account. Number as an isolated entity was for him "continuous" in the Aristotelian sense, since it could be divided indefinitely and, in any case, it inherited the characteristics of continuity or discreteness of the thing that it was quantifying. For example, referring to *one horse,* the number "one" was discrete; however, attached to *one yard,* the number "one" was continuous. With this formulation, continuous and discrete ceased to be ontological categories; they now became merely circumstantial properties of the objects quantified. The decimal notation Stevin introduced could solve the problems posed by the tension between form and content. In fact, as there was no distinction made within this newly revised category of quantity, there could be no distinction between the object of study of arithmetic and that of geometry. The representational tool needed to cope with this new situation had to be flexible enough to deal with the problems of discrete quantity and, simultaneously, with the problems of divisibility. To talk about parts of unity, decimal notation was instrumental. This representation was deeply linked to the new concept of number. This is one of the best examples of how an adequate symbolic representation becomes an instrument with which to explore mathematical concepts.

While Euclidean mathematics produced a concept of number by means of an *empirical abstraction*—in the sense of Piaget's genetic epistemology—Stevin's concept was the result of *reflective abstraction*. Stevin constructed his concept of number from generalizing the practice of measurement. Here, the boundary between theoretical mathematics and applied mathematics faded and practical needs came to determine the kind of mathematics that should be developed. Stevin, as Klein says,

> puts his 'practical' commercial, financial, and engineering experience into the service of his 'theoretical' preoccupation—as, conversely, his 'theory' is put to use in his 'practical activity.' [Klein 1968, p. 186]

Stevin worked out the operational aspect in a work entitled *La Disme*, published shortly before *L'Arithmetique*. In that work he presented a systematization with some innovations of the decimal notation already known by this time but far from being in general use. He attacked the theoretical problems in *L'Arithmetique*. There, he identified magnitude and number, attributing numerical properties to continuous quantities and continuity to numbers. From this point on, it was not possible for mathematicians to separate the concept of quantity from its symbolic representation. This should make clear the more abstract nature of the new concept of number. In fact, it is interesting to compare once more the Greek concept of number in which unity is the fundamental principle, and the one introduced by Stevin.

We interpret the reflective abstraction process that led Stevin to his concept of number as a process by means of which the arithmetic operations "revealed"—in the sense that Stevin gave to this word in his definition of number—the actions or transformations that were carried out on material objects insofar as these objects could be regarded as quantities. For Stevin number revealed the quantity of each thing. Therefore, arithmetic operations were sustained by the transformations that were carried out on quantities.

A very important step towards the widening of the numerical domain was to consider that results of arithmetic operations, carried out with numbers, were numbers. Stevin made this point in several ways. For instance, the following argument was used repeatedly by Stevin, for which he was harshly criticized at the time. It is first found at the point where he asserted that the unit was a number:

> The part is 'of the same material' as the whole. The unit is a part of a multitude of units. Therefore the unit is 'of the same material' as the multitude of units; but the material of a multitude of units is number. Therefore the material of the unit is number. [Stevin 1585, p. 1] [1]

This argument presented a noticeable ambivalence regarding its level of abstraction. The first premise, "the part is of the same matter as the whole", referred to material objects, while the second, "the unit is part of the multitude of units", referred to mathematical, and thus abstract objects. The unrestricted step from one level to another showed the normative underpinning that physical reality gave to the concept of number for Stevin. He sustained his concept of number on the basis of the arithmetical operations that could be carried out with it, and the operations themselves on the transformations that could be carried out on matter (to the extent it could be quantified). These transformations whose results did not alter the total quantity of the matter that intervened in the process were reflected as operations characterizing, in turn, the closure of the mathematical domain.

The Choice of the Unit

The divisibility of the unit has its roots in the context of measurement processes through which a number is associated with a magnitude. Stevin proposed a theoretical structure based on the systematization of this practice. He suggested the popularization of the use of decimal subdivisions:

> all measurements will be divided, be they measurements of length, liquids, dry things, money, etc., by the precedent progressions of tenths and

[1] QUE L'UNITÉ EST NOMBRE
La partie est de mesme matiere que son entier,
La unité est partie de multitude d'unitez,
Ergo l'unité est de mesme matiere que la multitude d'unitez; ,
Mais la matiere de multitude d'unitez est nombre,
Doncques la matiere d'unité est nombre.

each one of these famous species will be called Commencement; for example, the Mark, Commencement of weights, by which gold and silver are weighed, the Pound, Commencement of other common weights; the Flemish Pound, the Pound Sterling of England, the Spanish Ducat, and so with every Commencement of coins. [*La Disme,* in Stevin 1585, p. 221]

Stevin called each one of the units of measurement *Commencement* (beginning), independently of their nature or extension. He associated the numerical unit with this *Commencement.* Throughout *La Disme,* Stevin clearly showed how he established the equivalence between each one (and all) of the metric units with a single natural numerical unit by means of the practice of measurement. Stevin's unit was thus not only the result of an abstraction from objects as quantities, but mainly of an abstraction from the coordinated actions that were carried out in the process of measuring these objects. This last argument shows why his number concept was the result of a reflective abstraction process.

Some comments on the decimal expansions are now pertinent. Stevin did not deny the theoretical existence of infinite decimal expansions. Indeed, he suggested a method, by means of the division algorithm, for approximating an infinite decimal as closely as we want [*La Disme* in Stevin, 1585, p. 210]. However, there was a predominant concern for those objects that could be obtained (constructed) by means of arithmetic operations. The following quotation, referring to incommensurable magnitudes, describes the epistemological position of Stevin:

we can never know the incommensurability of two given magnitudes through this experience; firstly because of the errors of our eyes and hands (which cannot see or divide perfectly), we would finally conclude that all magnitudes, both commensurable and incommensurable, are commensurable. And, in the second place, although it be possible for us to subtract the smaller quantity from the greater one hundreds of thousands of times and to continue in this way for hundreds of thousands of years, (if the given numbers are incommensurable) we would always work eternally, remaining ignorant of what could finally be found. Therefore, this form of cognition is not legitimate; it puts us in a position of impossibility if we wish to declare of what Nature really consists [Stevin, p. 215]

Now we turn to a study of some aspects of the transition from arithmetical to algebraic thought.

Viète and the Analytic Art

The last quotation shows the importance of the arithmetical operation in Stevin's conceptions. In Stevin's work the concept of number is justified not only because it could accommodate all the needed computation, but also because the symbolic character of his work was in line with the development of algebra. Stevin talked about "arithmetical numbers" and also about "geometrical numbers". According to him, if the numerical value of a geometrical number was unknown, it entered algebraic computations as an *indeterminate quantity.* This was not yet a genuine algebraic variable insofar as Stevin required the homogeneity of the quantities involved. That is, two geometrical quantities could be added only if both belonged to the same category: length, area, or volume. Moreover this indeterminate quantity *was a number,* although it was unknown to us.

The algebraic concepts underwent a very subtle development in the early seventeenth century. This first conception of algebra was as a kind of generalized arithmetic. Another viewpoint was presented in Viète's *Artem Analyticem Isagoge* (1591) [the Appendix in Klein 1968 is an English translation of Viète's work: *Introduction to the Analytic Art.*] In this book, one of the cornerstones of the new algebraic thinking, we read:

The supreme and everlasting law of equations or proportions, which is called the law of homogeneity because it is conceived with respect to homogeneous magnitudes, is this:

I. Only homogeneous magnitudes are to be compared with one another.

For it is impossible to know how heterogeneous magnitudes may be conjoined. And so, if a magnitude is added to a magnitude, it is homogeneous with it. If a magnitude is multiplied by a magnitude, the product is heterogeneous in relation to both. [Viète 1591, Chapter III, p. 324]

In Viète's work the term "magnitude" was used in a general sense, not just geometrically. The magnitude one was looking for when solving an equation, for instance, could be a number. In this respect, we quote J. Klein:

What is characteristic of this 'general magnitude' is its indeterminateness, of which, as such, a concept can be formed only within the realm of symbolic procedure... [the Euclidean presentation] does not do two things which constitute the heart of symbolic procedure: It does not identify the object represented with the means of its representations, and it does not replace the real

> determinateness of an object with a possibility of making it determinate, such as would be expressed by a sign, which instead of illustrating a determinate object, would signify possible determinacy [Klein 1968, p.123]

This was a large step towards the constitution of a symbolic mathematics; nevertheless, Viète still required the law of homogeneity. Furthermore, the symbolic character of Viète's work was (as in Stevin's case) the result of a process of reflective abstraction. In a sense the concepts of the "new science" were obtained by a reflection on the total context of that concept or, in other words, by a process that can be termed "symbol-generating abstraction". Viète's work can be seen as the first work of this new discipline.

In his book *Rules for the Direction of the Mind* [Descartes 1628 in Kolak 1994], Descartes considers the problem of multiplying the product ab of the magnitudes a and b by a third magnitude c. He says in Rule XVIII, that for this to be possible, "we ought to conceive (the product) *ab as a line*" [Descartes, p. 81]

This detachment of quantities from the geometrical constraints was possible because of the resultant abstract symbolism. The quantities involved in the operational activity were abstractions of the geometrical figures not the figures themselves. The identification of a number with the symbol used to represent it led to a conceptualization of number as a mental entity, not anymore as the Greek *arithmos* used to count material things. In this way Viète's symbolic algebra, which still was arithmetical and geometric, became fully symbolic in Descartes' hands through the loss of the dimensionality of the symbols used. Before Viète, the main activity was the search for a formula (a set of procedural rules) to compute the roots of an equation with numerical coefficients. This activity can be seen as a generalized arithmetic. All the operations involved were performed on the numerical coefficients, and, eventually, the unknown also became involved in the operations. With Viète's work this changed radically. Now, the *operation* was the new object of study. In terms of genetic epistemology we have entered an *inter-objectal* stage of development [Piaget and García 1989, pp. 142–143.] As we have already noted, the passage to this new stage was possible because a process of reflective abstraction had taken place.

Variable and Variation

The symbolic language of algebra enabled the construction of models at a higher level of representation. Within this operational field, symbols could be manipulated like given quantities and related through symbolic expressions. This was instrumental for the conscious study of functions as models designed to study sequences of stages. The *table of values* is a good example of this kind of model. Perhaps what is more important in this context is the use of literal symbols as objects with the capacity to take numerical values and thus to *vary* on a numerical substratum.

When analytic geometry was created, the possibility of moving from a symbolic representation (the algebraic equation) to a visual one (the graph) was established. The interplay between these two forms of representation led to the construction of deeper meanings for the conceptions involved. Connections controlled meaning; so in the encounter between the symbolic/algebraic and the visual/geometric conceptions, the new space for doing geometry—the Cartesian plane— transformed the geometric study of movement previously advanced by Oresme and the Scholastics, who had begun a serious study of movement with non-Cartesian representations. Nevertheless, it was the new conception of space—a geometrized space—which proved to be instrumental for the development of infinitesimal calculus.

Let us remark on the connection between the development of algebraic language and the study of motion in the geometrical context. In the Merton Rule it was important that the *distance* traversed by a body was represented as an *area*. There is a link between this fact and Descartes' Rule XVIII that identified *all* magnitudes with line segments. Viète's homogeneity law stands between these two perspectives and reflects a former period of algebraic development. The homogeneity law was an obstacle to the development of algebraic language. Descartes overcame it through the linearization of magnitudes. In a sense, placing all magnitudes in the same category made possible the emergence of geometrical/dynamical models in elementary calculus.

We can summarize this development in the diagram on the next page. The Cartesian plane supplemented by the operational field of the new "non-dimensional" algebra, furnished the necessary representational tools that allowed the construction of models for the study of variation.

Concluding Remarks

The mathematical ideas which have been studied in this paper constitute one of the cornerstones of calculus, as it was developed during the second half of the seventeenth century, mainly in the hands of Newton and Lebniz. This mathematical age characterized by the study of variation was possible only within a conceptual framework that included:

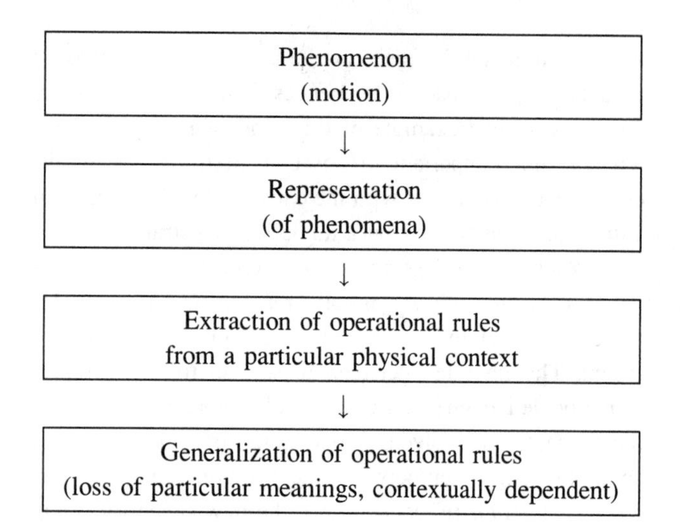

1. A unified concept of number capable of dealing with discrete quantities and continuous magnitudes.
2. A form of geometrical representation that enabled the variable quantities to be considered as (geometrical) magnitudes and thus as never ending divisible ones.
3. A new algebraic language that emphasized the study of mathematical processes.

References

Aristotle: 1978, *The Great Books of the Western World,* vol. VIII, Encyclopædia Britannica, Chicago.

Clagett, M.: 1968, *Nicole Oresme and the Medieval Geometry of Qualities and Motion,* University of Wisconsin Press, Madison, Wisconsin.

Descartes, R.: 1994, *Rules for the Direction of the Mind* (1628), in *From Plato to Wittgenstein. The Historical Foundations of Mind,* Kolak, D. (ed.), Wadsworth Publishing Company, Belmont CA., pp. 41–82.

Euclid: 1956, *The Elements,* in Sir Thomas Heath, *Euclid, The Thirteen Books of the Elements,* Dover Publications, Inc., New York.

Jones, C.V.: 1978, *One as a Number,* unpublished doctoral dissertation, Toronto University.

Kaput, J.: 1991, "Notations and representations as mediators of constructive processes," in E. von Glasersfeld (ed.), *Constructivism and Mathematical Education,* Kluwer, Dordrecht, Netherlands, pp. 53–74.

Klein, J.: 1968, *Greek Mathematical Thought and the Origins of Algebra,* MIT Press, Cambridge.

Oresme, N.: 1350, "Treatise on the Configurations of Qualities and Motions" in Clagett, 1968, Chapters I and II.

Piaget, J. and Garcia, R.: 1989, *Psychogenesis and the History of Science,* Columbia University Press, New York.

Piaget, J.: 1950, *Introduction à L'epistémologie Génétique,* vol. I: *La Pensée Mathématique,* Presses Universitaires de France, Paris.

——: 1978, *The Development of Thought: Equilibration of Cognitive Structures,* Blackwell, Oxford.

Stevin, S.: 1585, *L'Arithmetique et la Practique d'Arithmetique,* in *Les Oeuvres Mathématiques de Simon Stevin,* A. Girard, (ed.), Leyden, 1634.

Struik, D. J. (ed.): 1969, *A Source Book In Mathematics, 1200–1800,* Harvard University Press, Cambridge, Mass.

Vieta, F. : 1591, "Introduction to the Analytical Art" in Klein, 1968, Appendix, pp. 313–353.

Combinatorics: a Historical and Pedagogical Approach

Robin Wilson
The Open University, UK

In his 1666 *Dissertatio de arte combinatoria,* Gottfried Wilhelm Leibniz described combinatorics as the study of placing, ordering and choosing a number of objects. Since then, the scope of the subject has widened significantly and the subject is now usually taken to include the whole of finite or discrete mathematics, as well as having substantial overlaps with set theory, computer science, and many other areas.

Broadly speaking, combinatorial problems fit into one or more of the following categories:

- existence problems: does ... exist? Is it possible to ... ?
- construction problems: if ... exists, how can we construct it?
- enumeration problems: how many ... are there? Can we list them all?
- optimization problems: if there are several ..., which is the 'best'?

Because of their generality and wide applicability such problems occur throughout mathematics, in all disciplines and at all levels. In particular, while combinatorics is a fruitful area for research activity, it also yields a large number of teaching problems for students with varying abilities. In this article, we consider four areas of combinatorics, outline their history, and list some student activities; their solutions appear towards the end of the article.

Permutations and combinations

There are four types of selection problems in which r objects are to be selected from a set of n objects:

- if the selections are ordered and repetition is allowed, then the number of possible choices is n^r;
- if the selections are ordered without repetition (permutations), then the number of choices is

$$(n)_r = n(n-1)\cdots(n-r+1);$$

- if the selections are unordered without repetition (combinations), then the number of choices is

$$\binom{n}{r} = \frac{n!}{r!(n-r)!}$$

- if the selections are unordered with repetition, then the number of choices is

$$\binom{n+r-1}{r} = \frac{(n-r+1)!}{r!(n-1)!}$$

An early example of ordered selections with repetition occurred in the seventh century BCE *I ching* (Book of changes), where the symbols for the yin and yang were

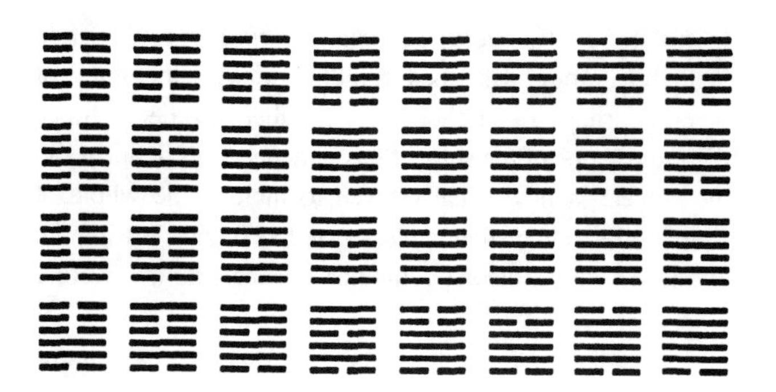

FIGURE 1

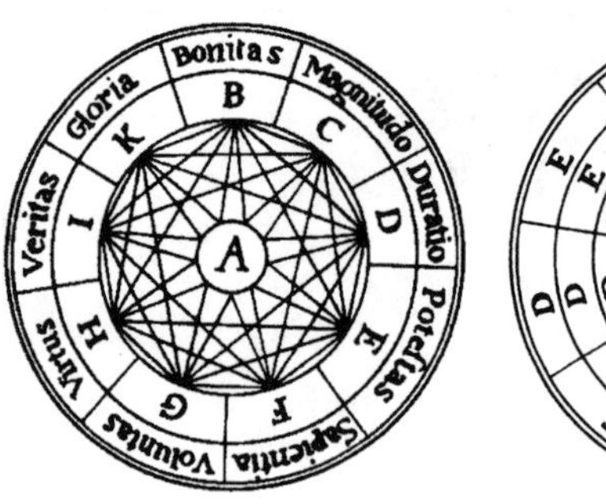

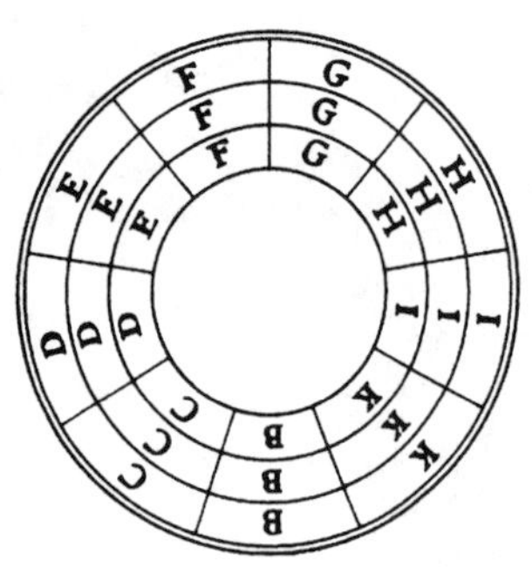

FIGURE 2

combined into hexagrams (systems of six symbols) in $2^6 = 64$ ways; half of these are illustrated in Figure 1.

Problem 1. *Draw the rest of the 64 hexagrams. It has been claimed (probably incorrectly) that Leibniz used these hexagrams as evidence that the Chinese invented a binary system of numbers. How might this have been done?*

There are also early examples from India. In a medical treatise of Susruta (sixth century BCE), there is a discussion of the combinations of tastes that can be made from the six basic qualities *sweet, acid, saline, pungent, bitter,* and *astringent.* A systematic list was given of all the possible combinations, giving the number of possible combinations when taken singly, in twos, in threes, in fours, in fives, and all together.

Problem 2. *How many combinations are there of these six qualities when taken three at a time, and four at a time?*

Other Hindu examples occur around 200 BCE in the work of the Jainas, where there are discussions involving combinations of males, females and eunuchs, and in the writings of Pingala, concerning the metrical rhythms that can be constructed from a given number of short and long syllables (such as – ⌣ ⌣ – – and ⌣ ⌣ – – –). Later, in the sixth-century *Brhatsamhita* of Varahamihira, it is stated that there are exactly 1820 ways of choosing four out of sixteen ingredients to make perfume. In the eleventh century, Bhaskara's *Lilavati* asks for the number of variations of the god Sambhu by the exchange of the ten attributes held in his several hands—the rope, elephant's hook, serpent, tabor, skull, trident, bedstead, dagger, arrow, and bow; the answer is correctly given as 3,628,800.

Problem 3. *Explain how Varahamihira and Bhaskara might have obtained these numbers.*

Much early interest in permutations and combinations arises from various theological concerns—both Jewish and Christian. A Jewish text *Sefer yetsirah* (Book of Creation), possibly from the eighth century, calculates the number of ways of choosing two distinct letters from the 22 letters of the Hebrew alphabet; such calculations were considered of importance because various combinations of letters were thought to have power over nature. A later Jewish writer Rabbi ibn Ezra used combinations in an astrological context to calculate the number of possible conjunctions of the planets. The thirteenth-century Catalan mystic Ramon Lull believed that all knowledge arises from a number of basic categories, and by moving through all possible combinations of these categories one can thereby discover everything. He used combinatorial diagrams (Figure 2) to present the active manifestations of the divine attributes (goodness, power, etc.), and constructed models to demonstrate combinations of these; the inner wheels in the right-hand diagram rotate independently to reveal all ternary combinations of the letters.

Lull's work was taken up by many Renaissance religious teachers, including Marin Mersenne (who used combinations in a musical context and exhibited all the factorials up to 64!, a 90-digit number) and Athanasius Kircher (who exhibited a table of the 324 ordered pairs of the 18 attributes and listed all permutations of the letters in the words *ORA, AMEN* and *PATER*). Leibniz was also an enthusiast for Lull's work in his earlier years (although not later), and Lull's influence is clearly visible in the youthful work mentioned at the beginning of this article. Further details of Lull's followers may be found in Knobloch [6] or Fauvel and Wilson [7].

Problem 4. *List all the permutations of the letters in the word AMEN.*

The importance of combinations arises not only from their combinatorial interest, but also from their appearance as binomial coefficients. In particular, they appear in the so-called 'Pascal triangle', named after Blaise Pascal who discussed this triangle in his 1665 *Traité du triangle arithmétique*, although it had appeared many years earlier in a Chinese text of 1303, in Jordanus de Nemore's *De arithmetica* (c. 1225), in al-Tusi's *Handbook of arithmetic using chalk and dust* (1265), and in earlier Chinese and Arabic works.

$$
\begin{array}{ccccccccccc}
 & & & & & 1 & & & & & \\
 & & & & 1 & & 1 & & & & \\
 & & & 1 & & 2 & & 1 & & & \\
 & & 1 & & 3 & & 3 & & 1 & & \\
 & 1 & & 4 & & 6 & & 4 & & 1 & \\
1 & & 5 & & 10 & & 10 & & 5 & & 1 \\
\cdot & \cdot & \cdot & \cdot & \cdot & \cdot & \cdot & \cdot & \cdot & \cdot & \cdot
\end{array}
$$

A full discussion of the 'arithmetical triangle' and its history appears in Edwards [8].

Problem 5. *What are the next three rows of this triangle? What patterns do you notice among the numbers?*

Partitions

A long-standing problem is to find the number $p(n)$ of ways of partitioning the positive integer n into positive integers; the order in which the numbers appear is irrelevant. For example, there are five partitions of the number 4:

$$4,\ 3+1,\ 2+2,\ 2+1+1 \text{ and } 1+1+1+1;$$

thus $p(4) = 5$.

It is a simple matter to write down the values of $p(n)$ for small values of n; for example,

$$p(1) = 1,\ p(2) = 2,\ p(3) = 3,\ p(4) = 5,\ p(5) = 7,$$

and so on, and a natural question is to ask for a general formula for $p(n)$.

Problem 6. *Find the values of $p(6)$ and $p(7)$.*

Similarly, one can ask for the number of partitions of a given number n into parts of certain types, such as odd parts, even parts, or distinct parts; when $n = 4$, these are respectively

odd parts:	$3+1$ and $1+1+1+1$
even parts:	4 and $2+2$
distinct parts:	4 and $3+1$

In his 1748 *Introductio in analysin infinitorum*, Euler noted many theorems on such partitions, including the following results:

- the number of partitions of n into odd parts is equal to the number of partitions into distinct parts.
- the number of partitions of n into an odd number of unequal parts is equal to the number of partitions into an even number of unequal parts, except when n is a number of the form $k(3k \pm 1)/2$, for $k = 1, 2, \ldots$, in which case these numbers differ by 1; they are often called *Euler's pentagonal numbers*.

Problem 7. *By writing down the appropriate partitions, verify these results when $n = 6$ and $n = 7$.*

Euler also obtained his fundamental partition-generation function

$$1 + p(1)x + p(2)x^2 + p(3)x^3 + \cdots$$
$$= (1+x+x^2+\cdots)(1+x^2+x^4+\cdots)(1+x^3+x^6+\cdots)\cdots,$$

from which one can successively calculate $p(1), p(2), p(3), \ldots$. To prove this result, notice that each term in the product on the right contributes exactly 1 to the corresponding term on the left; for example, the term $p(3)x^3$ arises from the terms $(x^3)(1)(1)$, $(x)(x^2)(1)$, and $(1)(1)(x^3)$ on the right. By writing the right-hand side as

$$(1-x)^{-1}(1-x^2)^{-1}(1-x^3)^{-1}\cdots,$$

we can then subject the partition-generation function on the left to the tools of mathematical analysis.

Problem 8. *Use this formula to calculate the first few values of $p(n)$.*

A useful consequence of Euler's work on the partition-generating function is his recurrence relation for $p(n)$:

$$p(n) = [p(n-1) + p(n-2)] - [p(n-5) + p(n-7)]$$
$$+ [p(n-12) + p(n-15)] - \cdots,$$

where the numbers $1, 2, 5, 7, 12, \ldots$ appearing in the brackets are the above-mentioned pentagonal numbers of the form $k(3k \pm 1)/2$, for $k = 1, 2, \ldots$.

Problem 9. *Use this recurrence relation to find the smallest value of n for which $p(n) > 100$.*

The values of $p(n)$ increase very rapidly with n. Euler calculated the values of $p(n)$ up to $n = 69$, and in the 1890s Percy MacMahon constructed a table of partition numbers up to $p(200) = 3{,}972{,}999{,}029{,}388$, but these results pale into insignificance in comparison with a 1918 paper of Hardy and Ramanujan. In this truly remarkable paper, the authors obtained an almost unbelievable exact formula for $p(n)$ as the nearest integer to an expression involving square roots, derivatives, exponentials and 24th roots of unity; further details may be found in Shiu [9]. Studying MacMahon's table of values, Ramanujan was able to prove several other results, such as: $p(5k+4)$ is always divisible by 5, and $p(7k+5)$ is always divisible by 7.

Problem 10. *Find three numbers n such that $p(n)$ is divisible by 35.*

Designs

In 1839 the geometer Julius Plücker observed that a system $S(n)$ of n points, arranged into triples in such a way that any two points lie in just one triple, is possible only when $n \equiv 1$ or $3 \pmod 6$; for example, when $n = 7$, we obtain the following seven triples:

1	2	3	4	5	6	7
1	2	3	4	5	6	7
2	3	4	5	6	7	1
4	5	6	7	1	2	3

Problem 11. *Consider such systems with 9, 13, 15, 19 and 21 points; how many triples are there in each case?*

Plücker's discovery was communicated, possibly by J. J. Sylvester, to William Woolhouse, editor of the *Lady's and Gentleman's diary,* who proposed a 'prize question' for his readers; Plücker's system $S(n)$ corresponds to the case $p = 3$, $q = 2$:

> Determine the number of combinations that can be made out of n symbols, p symbols in each; with this limitation, that no combination of q symbols which may appear in any one of them shall be repeated in any other.

In 1847, the Rev. Thomas Penyngton Kirkman showed how to construct such a system $S(n)$ for any number n of this form, and three years later he described a system $S(15)$ with 35 triples partitioned into seven sets of 5 triples in such a way that each symbol occurs just once in each set of 5; this system gives a solution of the *Kirkman schoolgirls problem,* stated in 1850, in which

> fifteen young ladies in a school walk out three abreast for seven days in succession; it is required to arrange them daily, so that no two shall walk twice abreast.

Problem 12. *By completing the following table, solve the '9 schoolgirls problem' of arranging for nine schoolgirls to walk in groups of 3 for four days so that no two girls walk together more than once:*

Monday			*Tuesday*			*Wednesday*			*Thursday*		
1	4	7	1	–	–	1	–	–	1	–	–
2	5	8	4	–	–	6	–	–	8	–	–
3	6	9	5	–	–	7	–	–	9	–	–

The system $S(n)$ is now known as a *Steiner triple system,* after Jakob Steiner who studied them in 1853, several years after Kirkman had done so. In 1971, D. K. Ray-Chaudhuri and R. M. Wilson resolved a long-standing problem by proving that whenever $n \equiv 3 \pmod 6$, there is a Steiner triple system $S(n)$ that can be partitioned as in Kirkman's schoolgirls problem.

Kirkman also constructed geometric systems with $r^2 + r + 1$ points and $r^2 + r + 1$ lines, with $r + 1$ points on each line and $r + 1$ lines passing through each point, where r is 4 or 8 or a prime number. Such systems are now called *finite projective planes* of order r, and are finite analogues of the real and complex projective planes in which exactly one line passes through each pair of points, and exactly one point lies on each pair of lines. Finite projective planes are known to exist whenever r is a power of a prime number.

In 1900 Gaston Tarry proved that there is no finite projective plane of order 6, and after a long computer search in the 1980s, Clement Lam proved the non-existence of a projective plane of order 10. It is not known whether there exists a projective plane of order r for any other composite value of r.

Problem 13. *Show how $S(7)$ can be regarded as a projective plane of order 2. Try to draw it. (The resulting picture is called the Fano plane.)*

A seemingly unrelated topic, which developed independently, is that of *orthogonal latin squares.* A question posed by J. Ozanam in his *Récreations mathématiques et physiques* (1725) is to arrange the sixteen court cards in a deck of cards in a 4×4 array, so that each value and each suit appears in each row and each column. A solution of this problem is as follows:

J ♢	Q ♡	K ♠	A ♣
Q ♠	J ♣	A ♢	K ♡
K ♣	A ♠	J ♡	Q ♢
A ♡	K ♢	Q ♣	J ♠

A square $n \times n$ array in which each of n symbols appears just once in each row and column is often called a *latin square.* Two or more latin squares are said to be *orthogonal* if, when they are superimposed, each of the n^2 possible ordered pairs of symbols appears just once. For example, the J, Q, K, A and ♢, ♡, ♠, ♣ latin squares above are orthogonal, as are the following two latin squares:

A	B	C	D	A	B	C	D	AA	BB	CC	DD
B	A	D	C	C	D	A	B	BC	AD	DA	CB
C	D	A	B	D	C	B	A	CD	DC	AB	BA
D	C	B	A	B	A	D	C	DB	CA	BD	AC
latin square 1				**latin square 2**				**1 and 2 superimposed**			

Problem 14. *Construct a* 4×4 *latin square with first row* $a\ b\ c\ d$ *which is orthogonal to both of the latin squares* 1 *and* 2.

In 1782, Leonhard Euler posed the following problem:

If there are 36 officers, one of each of six ranks from each of six different regiments, can they be arranged in a square in such a way that each row and column contains exactly one officer of each rank and one from each regiment?

A solution of this problem would correspond to a pair of 6×6 orthogonal latin squares and Euler, unable to find such a pair, conjectured that no such pair exists. More generally, he conjectured that there is no pair of orthogonal latin squares of order n, for any n of the form $4k+2$. For $n = 6$, Euler's conjecture was proved by G. Tarry in 1900, but in 1959–60, R. C. Bose, S. S. Shrikhande and E. T. Parker astounded everyone by constructing a pair of $n \times n$ orthogonal latin squares for *all* n of the form $4k+2$ greater than 6, thereby disproving all other cases of Euler's conjecture. In Problem 14, we observed that there exist 3 mutually orthogonal latin squares of order 4. In the 1930s, Bose and others showed that a finite projective plane of order n corresponds to a set of $n-1$ mutually orthogonal latin squares of order n, thereby linking these seemingly unrelated topics.

Problem 15. *By cyclically permuting the first row* $A\ B\ C\ D\ E$ *in various ways, construct a set of four mutually orthogonal latin squares of order* 5. *(Note that this set corresponds to a finite projective plane of order* 5, *with* 31 *points and* 31 *lines.)*

Much of the recent interest in such systems arises from the design of agricultural experiments in which a field may need to be planted with varieties of wheat that are to be compared in pairs. Pioneering work in this area was carried out in the 1930s by Ronald Fisher and Frank Yates, and led to the general study of balanced incomplete block designs; these have been extensively investigated in the past few years.

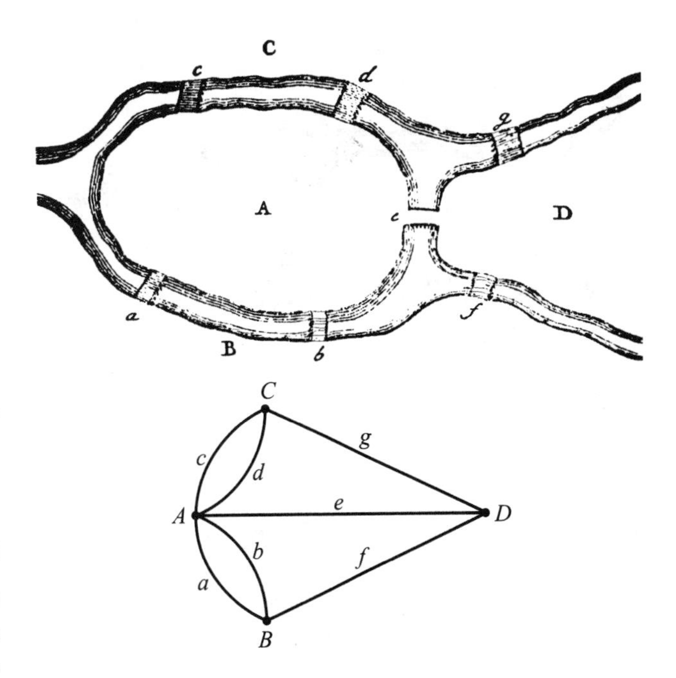

FIGURE 3

Graph theory

The subject of graph theory originated in the 1730s with Leonhard Euler's solution of the *Königsberg bridges problem,* in which four land areas are connected by seven bridges and it is required to find a route that crosses each of these bridges just once. Euler proved that there can be no such route, and showed how his method can be extended to any arrangement of islands and bridges. Although his approach was essentially graph-theoretic, he did not use graphs as such, and the above four-vertex graph (Figure 3) usually drawn to represent the problem did not appear until the end of the nineteenth century when W. W. Rouse Ball included the Königsberg problem in his book on recreational mathematics [12].

Problem 16. *The map in Figure 4 is taken from Euler's paper of 1736. Find a route that crosses each bridge just once.*

Sir William Rowan Hamilton, whose studies on noncommutative algebras led him to consider cycles on a dodecahedron passing just once through each vertex, discussed another type of traversal problem. Kirkman, who investigated which polyhedra have such a cycle, had already discussed problems of this type. Unlike the Eulerian problem, which has a simple solution, traversal problems of this kind are hard to solve.

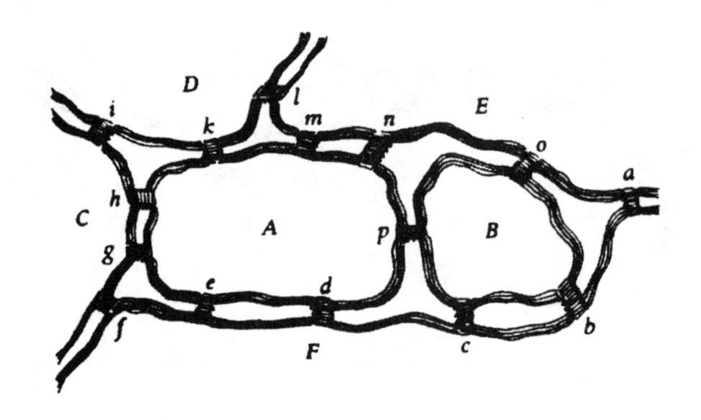

FIGURE 4

Problem 17. *Find as many cycles as you can on the dodecahedron in Figure 5, visiting each letter just once, and ending at the starting point. How many can you find starting with BCPNM or with JVTSR? (Hamilton asked these questions in 1859.)*

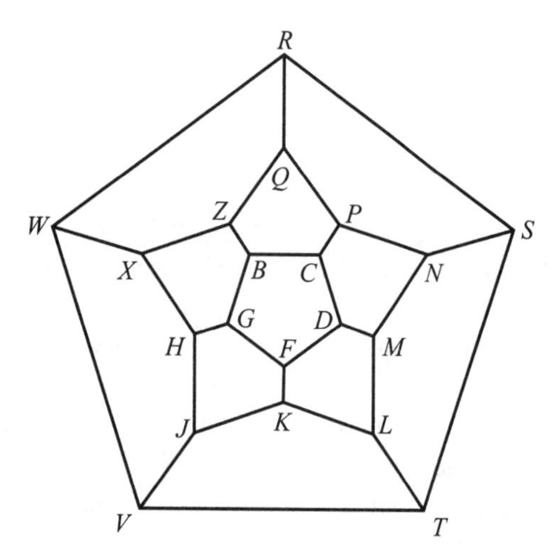

FIGURE 5

Problem 18. *In each of the pictures in Figure 6, can you visit each letter just once and return to your starting point?*

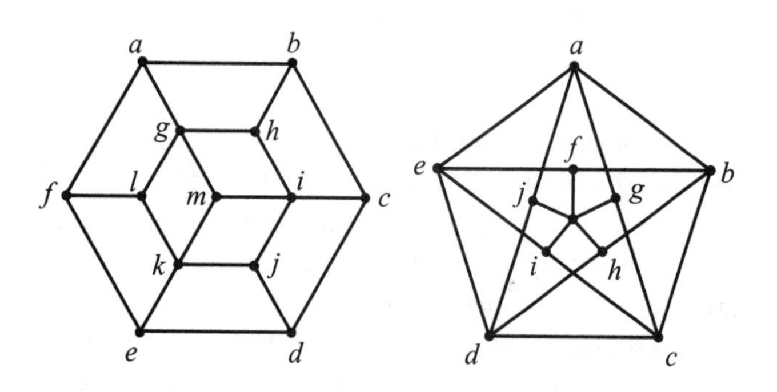

FIGURE 6

FIGURE 7

Whereas the Eulerian and Hamiltonian problems were essentially recreational in nature, the early study of trees was closely linked with the enumeration of chemical molecules. A *tree* is a connected graph without cycles; Figure 7 illustrates the two different trees with four vertices.

Problem 19. *Draw the three different trees with five vertices. How many different trees are there with six vertices?*

If we now regard the four vertices as carbon atoms (with valency 4), we can construct two chemical molecules with formula $\mathbf{C_4H_{10}}$ by adding on enough hydrogen atoms (with valency 1) to bring the valency of each carbon atom up to 4.

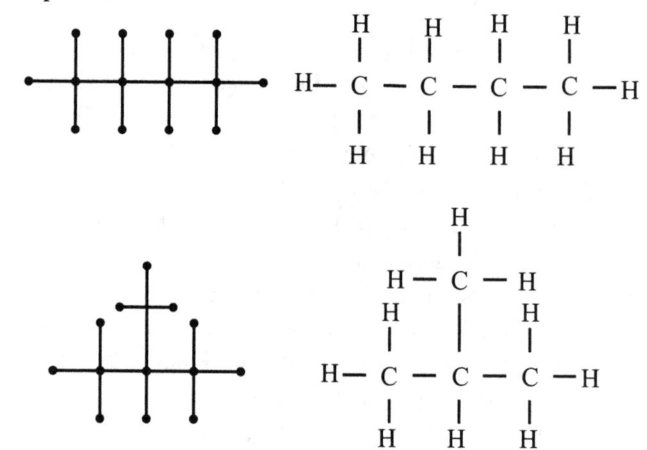

FIGURE 8

By using tree-diagrams of this kind, and building up the molecules step by step from their 'centres', Arthur Cayley was able to determine the number of chemical molecules (alkanes) with formula $\mathbf{C_nH_{2n+2}}$; further details are given in Biggs, Lloyd and Wilson [10].

Problem 20. *Draw the chemical molecules with formula* $\mathbf{C_5H_{12}}$ *that arise from the 5-vertex trees in Problem 19. How many molecules with formula* $\mathbf{C_6H_{14}}$ *arise from the 6-vertex trees?*

The most famous problem in graph theory is the *four colour problem,* which asks whether every map can be coloured with just four colours in such a way that neighbouring countries are differently coloured. This problem, due to Francis Guthrie in 1852, was communicated to Augustus De Morgan, who in turn communicated it to other mathematicians. In 1879, Alfred Kempe proved that every map can indeed be so coloured, but his proof was later found to be fallacious by Percy Heawood (1890). A correct proof was not produced until 1976 when Kenneth Appel and Wolfgang Haken produced an argument that involved the detailed analysis of almost 2000 different configurations of countries and hundreds of hours of computer time.

Problem 21. *The countries of the map in Figure 9 are to coloured red, yellow, green and blue. If three countries are coloured as shown, show that country A must be coloured red, and colour the rest of the map.*

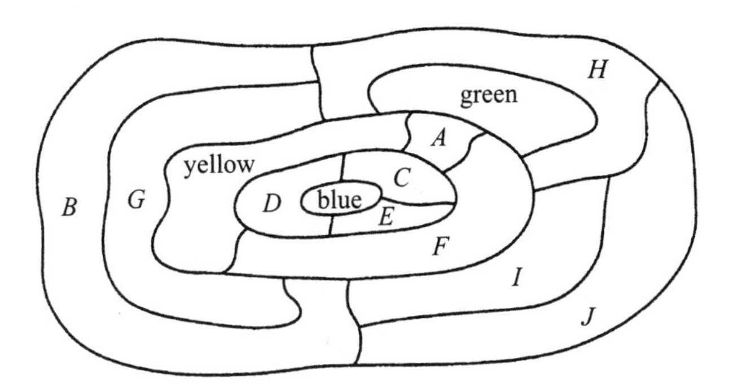

FIGURE 9

Heawood also gave a formula for the number of colours needed for maps on surfaces other than the plane or sphere—namely, the integer part of

$$\frac{7+\sqrt{1+48g}}{2},$$

where g (≥ 1) is the number of holes in the surface; for example, any map on a torus (ring doughnut), where $g = 1$, can be coloured with seven colours. Unfortunately, Heawood's proof was deficient and the gap was not filled until 1968 by Gerhard Ringel and Ted Youngs.

Problem 22. (a) *Find a map on the surface of a torus that needs 7 colours.*

(b) *How many colours are needed for a map on a double-torus (a surface with exactly two holes in it)?*

Solutions to the problems

1. The remaining hexagrams are shown in Figure 10. Replacing the two symbols in the 64 hexagrams by 0 and 1 gives the 64 binary words from 000000 to 111111.

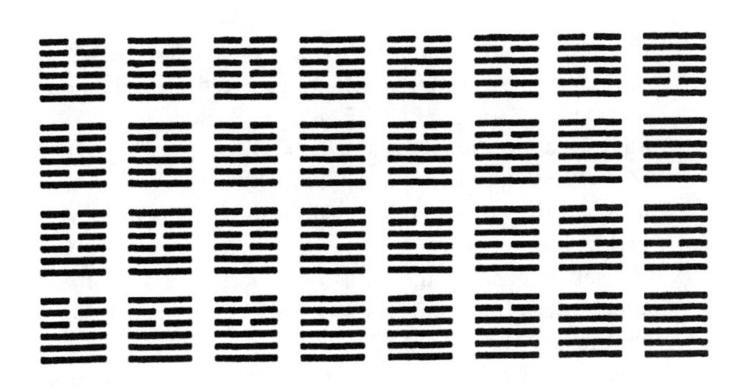

FIGURE 10

2. There are 20 combinations when taken three at a time; for example, *sweet-acid-pungent, acid-saline-pungent,* and *saline-bitter-astringent.*
   There are 15 combinations when taken four at a time; for example, *sweet-acid-saline-pungent, acid-saline-pungent-bitter,* and *saline-pungent-bitter-astringent.*
3. Varahamihira might have said: choose the first ingredient in 16 ways, the second in the remaining 15 ways, the third in 14 ways and the fourth in 13 ways. This gives $16\times15\times14\times13$ possibilities, but since order is irrelevant, we obtain each selection $4\times3\times2\times1$ ways; thus, the number of choices is equal to $(16\times15\times14\times13)/(4\times3\times2\times1) = 1820$.
   Bhaskara might have said: there are 10 choices for the first attribute, 9 choices for the second, 8 for the third,..., and 1 for the tenth, giving a total of $10\times9\times8\times7\times6\times5\times4\times3\times2\times1 = 3{,}628{,}800$.
4. *AMEN, AMNE, AEMN, AENM, ANME, ANEM, MAEN, MANE, MEAN, MENA, MNAE, MNEA, EAMN, EANM, EMAN, EMNA, ENAM, ENMA, NAME, NAEM, NMAE, NMEA, NEAM, NEMA.*
5. The next three rows are:

   1–6–15–20–15–6–1, 1–7–21–35–35–21–7–1,

   and

   1–8–28–56–70–56–28–8–1.

   There are many patterns; for example, each row reads the same forwards and backwards, each number (other than 1) is the sum of the two above it, each row adds up to a power of 2, etc.
6. $p(6) = 11$; $p(7) = 15$.
7. *First result:* $n = 6$—there are 4 partitions of each type:
   odd parts: $5+1$, $3+3$, $3+1+1+1$, $1+1+1+1+1+1$;
   distinct parts: 6, $5+1$, $4+2$, $3+2+1$.

   $n = 7$—there are 5 partitions of each type:
   odd parts: 7, $5+1+1$, $3+3+1$, $3+1+1+1+1$, $1+1+1+1+1+1+1$;
   distinct parts: 7, $6+1$, $5+2$, $4+3$, $4+2+1$.

   Second result: $n = 6$—there are 2 partitions of each type:
   odd number of unequal parts: 6, $3+2+1$;
   even number of unequal parts: $5+1$, $4+2$.

   $n = 7$—there are 2 partitions of the first type, and 3 partitions of the second type:
   odd number of unequal parts: 7, $4+2+1$;
   even number of unequal parts: $6+1$, $5+2$, $4+3$.

8. Multiplying out the brackets gives $1 + x + 2x^2 + 3x^3 + 5x^4 + 7x^5 + \cdots$, from which we can read off $p(1) = 1$, $p(2) = 2$, $p(3) = 3$, $p(4) = 5$, $p(5) = 7$, etc.
9. We deduce successively that $p(6) = 11$, $p(7) = 15$, $p(8) = 22$, $p(9) = 30$, $p(10) = 42$, $p(11) = 56$, $p(12) = 77$, and $p(13) = 101$, so 13 is the smallest value of n for which $p(n) > 100$.
10. We seek numbers of the forms $5k + 4$ and $7k + 5$—these numbers have the form $35k + 19$; the three smallest such numbers are 19, 54, and 89.
11. If there are n points, the number of triples is easily seen to be $n(n-1)/6$; thus the required numbers of triples are 12, 26, 35, 57, and 70.
12. Monday: 1–2–3, 4–5–6, 7–8–9;
    Tuesday: 1–4–7, 2–5–8, 3–6–9;
    Wednesday: 1–5–9, 2–6–7, 3–4–8;
    Thursday: 1–6–8, 2–4–9, 3–5–7.
13. We obtain a finite projective plane with 7 points and 7 lines, with 3 points on each line and 3 lines through each point; the seven triples of $S(7)$ give us the points on the seven lines. (In any plane drawing of it, at least one line must be curved as in Figure 11.)

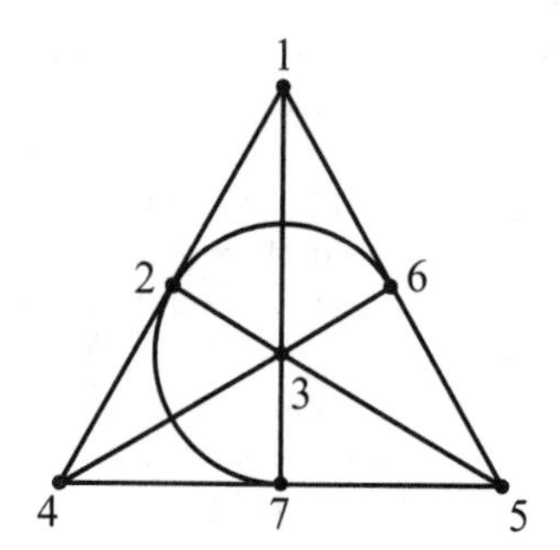

FIGURE 11

14.

a	b	c	d
d	c	b	a
b	a	d	c
c	d	a	b

15.

$A\ B\ C\ D\ E$	$A\ B\ C\ D\ E$	$A\ B\ C\ D\ E$	$A\ B\ C\ D\ E$
$B\ C\ D\ E\ A$	$C\ D\ E\ A\ B$	$D\ E\ A\ B\ C$	$E\ A\ B\ C\ D$
$C\ D\ E\ A\ B$	$E\ A\ B\ C\ D$	$B\ C\ D\ E\ A$	$D\ C\ B\ A\ E$
$D\ E\ A\ B\ C$	$B\ C\ D\ E\ A$	$E\ A\ B\ C\ D$	$C\ D\ E\ A\ B$
$E\ A\ B\ C\ D$	$D\ E\ A\ B\ C$	$C\ D\ E\ A\ B$	$B\ C\ D\ E\ A$

16. There are several possibilities, as long as we start in region D or E—for example, we can start in region E and cross the bridges in the order a–b–c–d–e–f–g–h–i–k–m–n–p–o–l, ending in region D. It is not possible to cross each bridge exactly once and return to the starting point.
17. There are many possibilities—for example,

    B–C–D–F–G–H–J–K–L–M–N–P–Q–R–S–T–V–W–X–Y–Z–B.

    Starting with B–C–P–N–M, there are two possibilities:

    B–C–P–N–M–D–F–K–L–T–S–R–Q–Z–X–W–V–J–H–G–B

    B–C–P–N–M–D–F–G–H–X–W–V–J–K–L–T–S–R–Q–Z–B.

    Starting with L–T–S–R–Q, there are four possibilities—for example,

    L–T–S–R–Q–Z–X–W–V–J–H–G–B–C–P–N–M–D–F–K–L.
18. *First picture*: there is no such route;
    second picture: there are several possible routes—for example, a–b–f–e–i–c–g–k–h–d–j–a.
19. There are three trees with five vertices and six trees with six vertices (Figure 12).

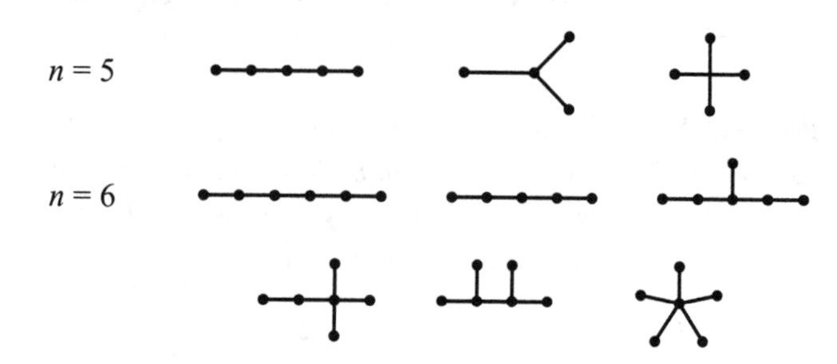

FIGURE 12

20. The chemical molecules arising from the 5-vertex trees in Problem 19 are pictured in Figure 13. There are only five molecules arising from the six 6-vertex trees; the last tree cannot give rise to a chemical molecule.

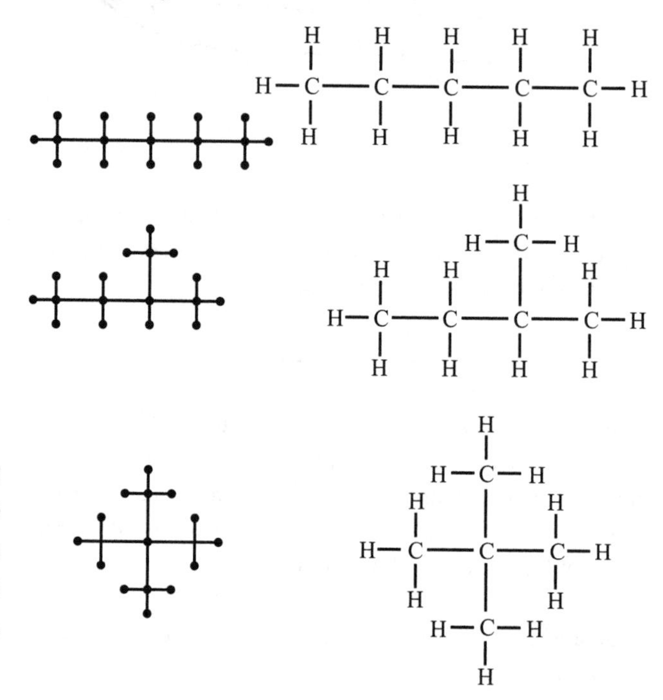

FIGURE 13

21. Country A must be coloured either *red* or *blue*.
    If A is coloured blue, then F must be *red*, D must be *green*, and it is then impossible to colour C; so A must be *red*. It follows that F is *blue*, H is *red*, G is *green*, B is *yellow*, C is *green*, D is *red*, E is yellow, I is *green*, and J is *blue*.

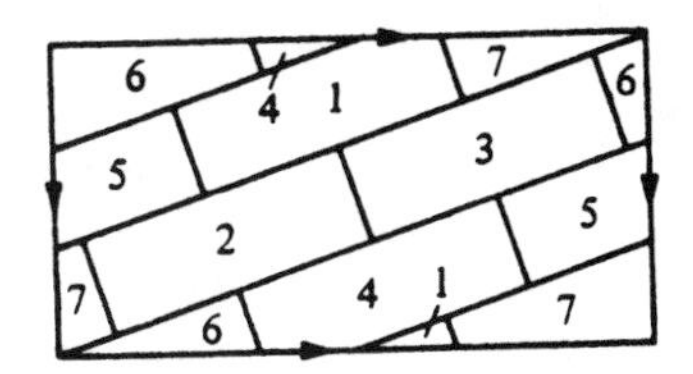

FIGURE 14

22. (a) A possible map is as in Figure 14; the torus is shown here in its flattened form, with opposite edges identified.
(b) For a double-torus ($g = 2$), Heawood's formula gives the integer part of $(7 + \sqrt{97})/2 = 8.424\ldots$, which is 8; thus, any map on the double-torus can be coloured with eight colours.

References

Extended accounts of the historical material in this article may be found in Biggs, Lloyd and Wilson [1] and Wilson and Lloyd [2], and a shorter version of the article appears in Wilson [3]. Early work on combinatorics is discussed more fully in Biggs [4] and Katz [5]. Further information on Renaissance combinatorics is given in Knobloch [6] and in Fauvel and Wilson [7], a fuller treatment of Pascal's triangle is given in Edwards [8], and Hardy and Ramanujan's work on partitions is discussed by Shiu [9]. A more detailed account of the history of graph theory is given in Biggs, Lloyd and Wilson [10], which includes extracts from some of the works mentioned here. Several of the Problems in this article are taken from the Open University Course MT365, *Graphs, Networks and Design* [11].

1. N. L. Biggs, E. K. Lloyd and R. J. Wilson: 1995, "The history of combinatorics," *Handbook of Combinatorics* (ed. R. Graham et al.), Elsevier Science B.V., pp. 2163–2198.
2. R. J. Wilson and E. K. Lloyd: 1994, "Combinatorics," *Companion Encyclopaedia of the History and Philosophy of the Mathematical Sciences* (ed. I. Grattan-Guinness), Routledge, pp. 952–965.
3. R. J. Wilson: 1996, "History of combinatorics," *Historia e Educaçao Matematica* (ed. E. Veloso), Braga, Portugal, pp. 95–100.
4. N. L. Biggs: 1979, "The roots of combinatorics," *Historia Mathematica* 6, 109–136.
5. V. J. Katz: 1996, "Combinatorics and induction in medieval Hebrew and Islamic manuscripts," *Vita Mathematica: Historical Research and Integration with Teaching* (ed. R. Calinger), Mathematical Association of America, pp. 99–106.
6. E. Knobloch: 1979, "Musurgia universalis: Unknown combinatorial studies in the age of Baroque absolutism," *History of Science* 17, 258–275.
7. J. Fauvel and R. J. Wilson: 1994, "The Lull before the storm: combinatorics and religion in the Renaissance," *Bull. Inst. Comb. Appl.* 11, 49–58.
8. A. W. F. Edwards: 1987, *Pascal's Arithmetical Triangle*, Griffin.
9. P. Shiu: 1997, "Computations of the partition function," *Mathematical Gazette* 81, 45–52.
10. N. L. Biggs, E. K. Lloyd and R. J. Wilson: 1998, *Graph Theory 1736–1936* (paperback reissue), Clarendon Press, Oxford.
11. *MT365, Graphs, Networks and Design*: 1995, Open University Course, The Open University, Milton Keynes, England.
12. W. W. Rouse Ball: 1892, *Mathematical Recreations and Problems of Past and Present Time* (later entitled *Mathematical Recreations and Essays*), Macmillan, London.

The History of Non-Euclidean Geometry

Torkil Heiede
Royal Danish School of Educational Studies Copenhagen, Denmark

The sum of the angles in a triangle and parallelism

If you ask somebody who is not a mathematician to give you a geometric fact, it is very probable that you will get the answer that the sum of the angles in a triangle is 180 degrees. But if you put the same question to a mathematician it is possible that you will get another question in return: What do you mean by a geometric fact? This is an interesting difference in attitude: The non-mathematician is ready to supply you with a mathematical fact, while the mathematician puts the whole issue of mathematical facts in doubt. This in turn must have something to do with the way the non-mathematician learned his or her mathematics, and since this mostly took place in school it could be informative to see how this "geometrical fact" is presented in different textbooks for schools.

In many elementary textbooks each of the pupils is asked to cut out a triangle of colored paper, tear off the corners, put them together, and see that they (seemingly) fit along a ruler; since this happens for everyone in the whole class, the truth of the statement is evident. There is here no borderline between geometry and physics—the truth of a geometrical assertion is decided by experiment. Sometimes the same idea is expressed a little differently. Instead of tearing off the corners, one folds them together so that they meet in a point of the base of the triangle. More subtly, the pupils may not perform this folding on real paper triangles, but are only asked to think of doing it, or to draw pictures of the unfolded and the folded triangle. The experiment has become a thought experiment one can reason about. It is, of course, an unspoken assumption that such a folding is possible for all triangles—also in a geometry which takes place in the minds of the pupils. Behind this there are assumptions about parallelism; the triangle has been folded into a quadrangle with four right angles, so it is a (hidden) assumption that such rectangles exist, and therefore also parallel lines.

Another argument is also very often met. A small line segment on one of the sides of a triangle is moved about; it is translated along the sides of the triangle and rotated around its corners, and it returns to its original position, seemingly having in all been rotated through 360 degrees, and the statement follows. But consider a large triangle on the surface of a sphere, the sides being a quarter of the "equator" of the sphere and two quarter-meridians stretching up to the "North Pole". Moving a small arc segment in the same way as before along the sides and around the corners of this triangle back to its original position would seem to show that the angle sum of this triangle is also 180 degrees. However, it is evident that it is 270 degrees, so the argument cannot stand alone. It is only valid together with an (maybe unspoken) assumption on parallelism, namely that parallel translations are at all possible. In fact parallel translations are not possible on the surface of a sphere, since there are no parallel lines. Two lines—that is: two great circles—always meet, even twice. By the way, on the surface of a sphere no rectangles exist so that we see once more that the argument about folding a triangle contains hidden assumptions on parallelism.

One can also meet more sophisticated versions of the argument—or experiment—on tearing off the corners of a triangle. One can use a tessellation of the plane with congruent triangles where angles from different triangles meet in a point and add up to 180 degrees. But also here parallelism plays an important role without being mentioned. From where does one really know that such parallel strips crossing each other in three directions are possible?

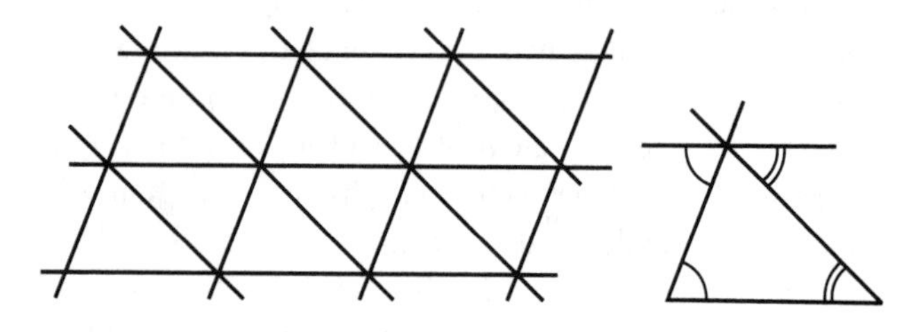

FIGURE 1

Then there is the classical argument which can still be seen in not quite so elementary textbooks. It involves explicitly a line which is parallel to one of the sides of the triangle and goes through the opposite corner. Using a theorem on angles at parallel lines one sees that the angle at that corner together with two angles congruent to the other two angles in the triangle add up to 180 degrees. Of course this theorem on angles at parallel lines must then be proved first, maybe from other theorems, but finally it must rest on some assumption about parallel lines.

This argument is really classical; it goes straight back to the oldest known source for all this, the treatment given by Euclid (c. 300 BCE) himself as proposition 32 of Book I of his *Elements*. The most authentic version which exists is still the one constructed more than 100 years ago by Johan Ludvig Heiberg (1854–1928, Danish) from all existing manuscripts. Here is the proposition in the translation by Thomas L. Heath (1861–1940, English) of Heiberg's text:

> **Proposition 32.** In any triangle, if one of the sides be produced, the exterior angle is equal to the two interior and opposite angles, and the three interior angles of the triangle are equal to two right angles.

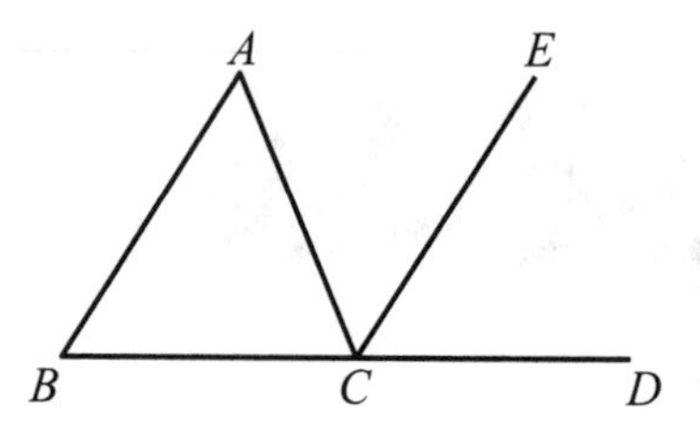

FIGURE 2

The figure is a little different from the one above, but the idea is the same, and the basic assumption about parallel lines on which the proof rests is Euclid's famous fifth postulate, or axiom, from the beginning of Book I. Here it is, also in Heath's translation from Heiberg:

> 5. That, if a straight line falling on two straight lines make the interior angles on the same side less than two right angles, the two straight lines, if produced indefinitely, meet on that side on which are the angles less than the two right angles.

It is evident that this has much to do with the sum of the angles in a triangle. For if the two lines Euclid is talking about meet somewhere then the three lines form a triangle, and if the sum of the three angles in a triangle is always 180 degrees, then the sum of the first two angles is necessarily less than 180 degrees. What the axiom says is that this is not only necessary but also sufficient for the two lines eventually to meet. One can add that if the sum of the first two angles is precisely 180 degrees, and if the sum of the three angles in a triangle is always 180 degrees, then there is no room for a third angle, and the two lines cannot meet.

One might think that everything was in order and there was nothing to worry about. However, right from the beginning, Euclid's parallel axiom was seen as questionable. All his other axioms talked of something which could be seen as taking place inside the borders of a sheet of paper:

> Postulates
>
> Let the following be postulated:
>
> 1. To draw a straight line from any point to any point.
> 2. To produce a finite straight line continuously in a straight line.
> 3. To describe a circle with any centre and distance.
> 4. That all right angles are equal to one another.

In contrast to this, the situation described in the parallel axiom might require one to go very far out to one side to find the intersection point of which it talks. How far? To the wall of Euclid's room in the Museion in Alexandria where he lived and worked, or to the end of Africa, or to the Moon, or further? How could one really be so sure of the truth of such an assertion? And these axioms were all meant to express something which was self-evident!

Attempts to prove the parallel axiom

Euclid himself seems to have considered the parallel axiom as something special. While he used the other four axioms freely right from the beginning, the first time he used the parallel axiom was in the proof of Proposition 29, just before he reached Proposition 32 on the sum of the angles in a triangle. (Book I contains in all 48 propositions, the last two being the theorem of Pythagoras and its converse.)

Already in antiquity many mathematicians tried to improve Euclid's *Elements* by proving the parallel axiom from the other axioms and thereby changing it into a theorem. They had no luck, or, if they thought so, others were always able to show that they had inadvertently built upon something else which in its turn they could not prove without using the parallel axiom.

There were also some who tried to resolve the problem by altering Euclid's definition of parallel lines:

> 23. Parallel straight lines are straight lines which, being in the same plane and being produced indefinitely in both directions, do not meet one another in either direction.

They exchanged the last sentence with this one:

> have the same distance between them in both directions

But if you take all the points equidistant to a line on one side of it, can you then really be sure that they constitute a line? One should think so, but it seemed that in any attempt to prove it one had to use the parallel axiom. From time to time one still encounters this definition in school textbooks whose authors probably do not know how bad an idea it is, since it clouds the whole issue.

One of the Greeks who tried to prove the pararallel axiom was the great astronomer of antiquity, Claudius Ptolemy (or Klaudios Ptolemaios) (c.85–c.165). Another one was the historian of mathematics (one of the first we know) who wrote an elaborate commentary to Book I of Euclid's *Elements,* Proclus (or Proklos) (410–485). He showed this theorem (which is the contrapositive of Euclid I, 30 and therefore equivalent to it):

> If a line meets one of two parallel lines then it also meets the other one.

From this he proved the parallel axiom. But to prove his claim he used this assumption:

> The distance between two parallel lines is bounded above.

One should think that this was evident, but it is not; it really turns out to be equivalent to the parallel axiom. Many other statements have from time to time been tried as substitutes for the parallel axiom. Here are some of them; they have all on closer examination proved to be equivalent to it (the second of them is often called Playfair's axiom):

> Two lines parallel to the same line are parallel (Euclid I, 30).
> Through a point outside a line there is at most one line parallel to it (Euclid I, 31).
> Triangles can be similar without being congruent.
> Similar triangles of different size exist.
> Triangles of the same shape but different size exist.
> Through a point outside two intersecting lines there exists a line meeting both.
> Every triangle has a circumcircle (that is a circle through its vertices).
> Through three different points go either a line or a circle.

The third, fourth, and fifth of these statements are evidently equivalent to each other. They are not true on the surface of a Euclidean sphere, since it is a theorem of Euclidean spherical geometry that if two triangles have equal angles they are congruent. The last three statements are true on the surface of a Euclidean sphere, so they can only be equivalent to the parallel axiom under some additional condition which fails there, as for instance that the distance between points on a line is unbounded above. The same can be said of the first two of the statements—they are true on the surface of a Euclidean sphere since on that there are no parallels! The last two of the eight statements are clearly equivalent to each other; the first of them was used by Farkas Bolyai (see later) in his attempts to prove the parallel axiom.

It would lead us too far afield to show here that all these statements are equivalent to the parallel axiom, but as an example let us look at the second one, Playfair's axiom. It was mentioned above that it is the same as Euclid I, 31, so already in Euclid's *Elements* one will find a proof that it follows from the parallel axiom and Euclid's other axioms and theorems derived from these. Let us prove that, conversely, the parallel axiom follows from Playfair's axiom

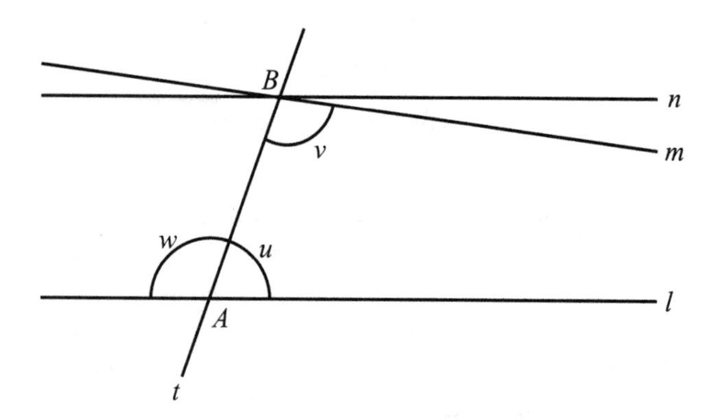

FIGURE 3

and Euclid's first four axioms and theorems derived from them.

Let the lines l and m be cut by a transversal t in points A and B such that the interior angles u and v on one of the sides of t have a sum less than $180°$. Now let w be the supplement of u at A on the other side of t; since $u + w = 180°$ we have $v < w$. Let n be the line through B making an interior angle as large as w with t (on the same side of t as u and v); n exists because of Euclid I, 23, and by Euclid I, 27, n is parallel to l. Moreover, since w is different from v, n is different from m. From Playfair's axiom it then follows that m cannot be parallel to l, so it intersects l in some point C. If C were on the side of t opposite to u and v, then v would be an exterior angle for $\triangle ABC$, and from Euclid I, 16 it would then follow that $v > w$, a contradiction. So C is on the same side of t as u and v.

Among the many mathematicians since the time of Ptolemy and Proclus who proved the parallel axiom from statements such as the ones listed above can be mentioned:

Alhazen (965–1041, Arabic)
Omar Khayyam (1048–1131, Persian, the great poet)
Christopher Clavius (1537–1612, German)
Pietro Antoni Cataldi (1548–1626, Italian)
John Wallis (1616–1703, English)
Girolamo Saccheri (1667–1733, Italian)
Johann Heinrich Lambert (1728–1777, German)
Louis Bertrand (1731–1812, Swiss)
John Playfair (1748–1819, Scottish)
Adrian Marie Legendre (1752–1833, French)

There were others, but these seem to be the most important, and among them two stand out: Saccheri and Lambert. They both tried to give an indirect proof of the parallel axiom, and they both believed they had succeeded in this, but they had not. Nevertheless, it was precisely the failure of an indirect proof which later on was so eminently fruitful.

Saccheri and Lambert

Girolamo Saccheri was a logician, and strongly interested in the logic of indirect proofs. Since all direct proofs of the parallel axiom seemed to him not to have worked he tried an indirect one. He began with a quadrangle with two opposite sides of equal length, both orthogonal to the side between them. Omar Khayyam had also worked with such quadrangles, but not to the same extent, so today they are most often called Saccheri quadrangles. Saccheri showed that the two remaining angles are equal. He also showed that if they are both right angles in one such quadrangle the same will be the case in all such quadrangles, and similarly if they are both obtuse or both acute. Now it was clear that if the parallel axiom were true one would have the first of these three cases. On the other hand, Saccheri could show that in the first of the three cases, or, as Saccheri put it, under the hypothesis of the right angle, one could prove the parallel axiom.

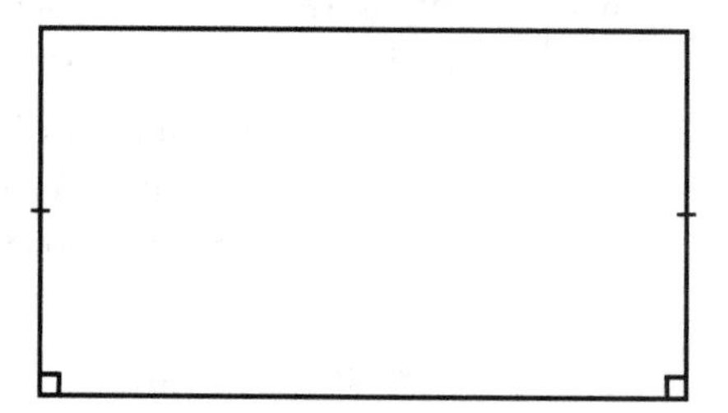

FIGURE 4

Saccheri went on to show that in the second of the three cases—the hypothesis of the obtuse angle—one could also prove the parallel axiom from which, as already mentioned, the hypothesis of the right angle would follow. But in his proof of the parallel axiom Saccheri used Euclid I, 16 which is dependent on the unstated assumption—which we have already met—that the distance between points on a line is unbounded above. This assumption does not hold under the hypothesis of the obtuse angle. However, Saccheri had in this way apparently reached an obvious contradiction: if the two angles were both obtuse, they were both right. Therefore, if he could also deduce a contradiction in the third case—the hypothesis of the acute angle—he would have given an indirect proof of the parallel axiom.

This is what Saccheri then started to do: to draw consequence after consequence from the hypothesis of the acute angle to see if he could end up with a contradiction. He found more strange consequences than anyone earlier who had had similar ideas. But even if he found them strange he could not honestly say that they contradicted each other or the axioms (other than the parallel axiom) or other theorems deduced from these, so he continued to draw consequences until he arrived at this one:

> The distance between two lines which do not meet can decrease indefinitely and they will then have a common perpendicular in a point infinitely far away in which they touch each other.

This he deemed to be "repugnant to the nature of the straight line" and took it to be the contradiction he had set out to find. He was, however, not quite content with it and tried to find a better one using the curve consisting of the points equidistant to a line on one side of it. Unlike his predecessors, he was quite aware that under the hypothesis of the acute angle such a curve is not a straight line. (By the way, this is also the case under the hypothesis of the obtuse angle: on the surface of a Euclidean sphere an equidistant curve to a great circle is a parallel circle and not a great circle.) Saccheri now computed (using infinitesimals) the length of an arc of the equidistant curve connecting the "upper" vertices of a Saccheri quadrangle. This arc had to be longer than the line segment connecting these vertices (the "fourth side" of the quadrangle), and he had already shown this to be longer than the base of the quadrangle (all this of course under the hypothesis of the acute angle). But he made an error in his computations and found that this arc had the same length as the base, and because of this contradiction he again rejected the hypothesis of the acute angle. He was still not quite satisfied, so he finished his book *Euclides ab omni naevo vindicatus* (*Euclid vindicated of all blemish*) on a note of doubt, comparing the clear contradiction he had reached under the hypothesis of the obtuse angle with the more obscure ones he had reached under the hypothesis of the acute angle. Perhaps he wanted to come to a conclusion; the book was in fact published in the year of his death.

Half a century after Saccheri a similar approach was used by Johann Heinrich Lambert, the main difference being that he started from quadrangles with three right angles; the question was then if the fourth angle was right, obtuse or acute. Alhazen had worked with such quadrangles, but today they are mostly known as Lambert quadrangles. Since a Lambert quadrangle can be thought of as half a Saccheri quadrangle it is not surprising that Lambert was able to show first that the hypothesis of the right angle gives the parallel axiom, next that the hypothesis of the obtuse angle gives a contradiction, and finally that the hypothesis of the acute angle gives a long list of strange consequences.

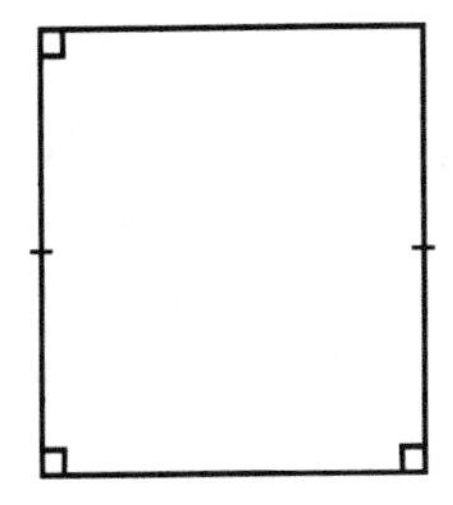

FIGURE 5

He found even more of these strange consequences than had Saccheri. In particular, he found that in every triangle the sum of the angles is less that 180 degrees. However, just like Saccheri, he ended up with a contradiction, again something about all points equidistant from a line. Lambert, though, did not publish anything, so maybe he was not quite convinced after all. His book *Theorie der Parallellinien* (*Theory of parallel lines*) came out nine years after his death, twenty years after he had written it.

One of Lambert's most remarkable statements was that one might almost draw the conclusion that all the strange consequences of the hypothesis of the acute angle were true on an imaginary sphere, just as the consequences of the hypothesis of the obtuse angle are true on a usual sphere. How such an imaginary sphere would look he did not say explicitly.

One hundred years after Saccheri and fifty years after Lambert three very different mathematicians became convinced (at nearly the same time) that no contradiction would ever appear:

Carl Friedrich Gauss (1777–1855, German, the prince of mathematicians)
Nikolai Ivanovich Lobachevsky (1792–1856, Russian)
János Bolyai (1802–1860, Hungarian)

All three said that one could have a geometry different from the Euclidean geometry whose uniqueness and truth had never been doubted in all the long history from Proclus to Legendre, a geometry in which all the strange consequences of denying the parallel axiom were valid geometric theorems.

Gauss, Bolyai and Lobachevsky

The first to arrive at the conviction that such a non-Euclidean geometry existed was Carl Friedrich Gauss. It probably happened around 1820, but he did not publish anything about it throughout his long life. He only wrote of it for his own pleasure and to a very few correspondents. He was the most famous mathematician of his time; he was quite sure that publication of these thoughts would create controversy; and he would not (as he said) expose himself to the yellings of the Boïotians. (In antiquity the people from Boïotia were considered by all other Greeks to be very coarse and not very intelligent.) For this reason, Gauss's work on this subject is known only from his notes and from letters published after his death.

János Bolyai was the son of an old friend and fellow student of Gauss, Farkas Bolyai (1775– 1856), who had himself worked in vain on the parallel axiom and who had warned his son against having anything to do with it. But in 1823 the son wrote to his father that he had created a strange new world out of nothing. In 1832 he published his discoveries as an appendix in Latin to a large book by his father. This book, also in Latin, was an extensive survey of endeavors to prove the parallel axiom; it is usually called *Tentamen* (*Attempt*) after the first word in its very long title. The appendix itself also has a long title of which this is only the beginning: *Appendix scientiam spatii absolute veram exhibens* (*Supplement containing the absolutely true science of space*). It is usual to call it *Science of space* or simply Bolyai's *Appendix*.

Farkas Bolyai sent an advance copy of his son's appendix to Gauss who wrote back that he could not praise the work of his old friend's son since that would be self-praise—because he had himself had the same thoughts many years ago. János Bolyai was of course furious, and he never wrote on the subject again. But his father did; in 1851 he published a book, *Kurzes Grundriss eines Versuches ...* (*A short sketch of an attempt...*) in which he believed he had proved the parallel postulate. He assumed the truth of the statement we have already mentioned that through three points not on a line passes a circle, which is in fact equivalent to the parallel axiom. Sad to say, he had not understood his son's discovery.

Nikolai Ivanovich Lobachevsky spent most of his life in Kazan, 720 km directly east of Moscow, on the Volga. He grew up there; he was a student at the University of Kazan (which was founded in 1804 and was regarded as the easternmost university in the world); he became a professor at the same university; and finally for many years he was its vice-chancellor.

One of Lobachevsky's teachers at the university had been J.M.C.Bartels (1769–1836, German), who much earlier had been Gauss's teacher in Brunswick. It seems that in 1815–17 Lobachevsky was trying to prove the parallel axiom, but that between 1823 and 1825 he became convinced that such a proof was not possible. He gave his first lecture on his discoveries in 1826 and published his first treatise on them in 1829: *O nachalah geometrii* (*On the principles of geometry*), in the journal of the Kazan University. If Lobachevsky was not the first discoverer of non-Euclidean geometry, he was the first to publish on it, but in Russian, and at a place very far away from the mathematical centers of Europe.

Over the years, and in between his many professional and administrative duties, Lobachevsky wrote many articles and books on his imaginary geometry, as he called it. He wrote articles not only in Russian but also in French and German, without ever attracting the attention his ideas merited. Evidently, the mathematical world did not care much about what someone in Kazan might think. It would surely have been different if Gauss had published anything on the subject. However, Lobachevsky sent his book *Geometrische Untersuchungen zur Theorie der Parallellinien* (*Geometrical researches on the theory of parallel lines*) from 1840 to Gauss who replied appreciatively and saw to it that Lobachevsky was elected a member of the Scientific Academy of Göttingen.

Three men were in possession of an epoch-making discovery, but nobody really noticed or understood what had happened before an Italian in 1868 proved what these three had only conjectured: no contradiction would ever occur in this new non-Euclidean geometry (provided that no contradictions were hidden in the old Euclidean geometry, which was considered unthinkable). By then all three discoverers were dead.

Beltrami, Riemann, Klein and Poincaré

Four mathematicians are especially important in the next period of the history of non-Euclidean geometry:

Bernhard Riemann (1826–1866, German)
Eugenio Beltrami (1835–1900, Italian)
Felix Klein (1849–1925, German)
Henri Poincaré (1854–1912, French)

In 1868 Eugenio Beltrami showed that in Euclidean space geometry one could build a model of the plane geometry of Gauss, Bolyai and Lobachevsky. In other words, in Euclidean space one can find a surface whose "intrinsic" or "inner" geometry is non-Euclidean. On, or rather in, this surface one can find curves which—taken as the lines of a plane geometry—satisfy all the Euclidean axioms except the parallel axiom, and instead of that its negation. If one accepts Euclidean geometry, in the sense that one takes for granted that no contradiction will ever appear in it, then the existence of such a model forces one to acknowledge that no contradiction will ever appear in non-Euclidean geometry either (since the model is embedded in Euclidean space)—and one is therefore forced to accept also non-Euclidean geometry.

Beltrami's proof of the existence of such a model was a decisive turn, for until then everybody had (if they had taken an interest in it at all) believed that either Euclidean geometry was true, or non-Euclidean. Since one was convinced of the truth of Euclidean geometry, one rejected non-Euclidean geometry. But Beltrami's work showed definitively that if Euclidean geometry was true then non-Euclidean geometry was also true.

On the other hand Lobachevsky had known—and maybe a few people before him—that if one imagines a non-Euclidean space, then in it one can find a type of surface whose "intrinsic" geometry is Euclidean, so Euclidean and non-Euclidean geometry are equally true. With this the word 'truth' changes its meaning in mathematics. Until this moment mathematics and especially geometry had been considered as a system of true statements about the world around us, deduced from self-evident truths, but from now on one was forced to regard mathematics quite differently, namely as something one imagines. Also, Euclidean space was now no more real than non-Euclidean space—they were both something we (only) imagine—and with this mathematics became an independent science in a new and previously quite unknown way. It became a quite separate science, without counterparts, detached from the natural sciences, more closely related to branches of art such as music and the visual arts, or to a large system of games as for example chess and draughts.

In 1854 Bernhard Riemann gave a lecture with the title *Über die Hypothesen, welche der Geometrie zu Grunde liegen* (*On the hypotheses which are fundamental to geometry*); it was published in 1868 after his death. In this lecture Riemann showed something that now could be seen to match perfectly with Beltrami's result, namely that if the other axioms of Euclid are relaxed somewhat, then Saccheri's and Lambert's hypothesis of the obtuse angle no longer leads to a contradiction. In this way one then gets a third type of plane geometry, in which there are no parallel lines and where it is a theorem that the sum of the angles in a triangle is more than 180 degrees. The sum must also be less than 900 degrees since the sum of the angles in the rest of the plane—which is a triangle with the same vertices and the same sides!—must also be more than 180 degrees. (If the angles of the original triangle are u, v, w, then the angles of this complementary triangle are $360° - u$, $360° - v$, $360° - w$). A suitable model in Euclidean space geometry for this Riemannian non-Euclidean geometry is simply the "intrinsic" geometry of the surface of a Euclidean sphere in which the great circles are taken as lines. Of course this geometry had been known since antiquity, but no one had seen it is this way before. In counterpoint to this, Beltrami's Euclidean surface, whose "intrinsic" geometry was a model for the Lobachevskian non-Euclidean Geometry, became known as a pseudo-sphere. Its Gaussian curvature is -1 at every point just as the Gaussian curvature of the usual sphere (with radius 1) is 1 at every point.

It should be mentioned that when Riemann gave his lecture Gauss was in the auditorium. More than that: the lecture was part of Riemann's doctoral work, and he had (as was customary) submitted three different subjects for his lecture; one of them was geometrical, and it was Gauss who had decided that this was the one he should speak about.

So now one had three plane geometries, all equally true: the Lobachevskian, the Euclidean and the Riemannian or, as they are also called, the hyperbolic, the parabolic and the elliptic geometry. This new insight—and also the understanding of its consequences for mathematics as a whole—spread rapidly in the mathematical world in the last quarter of the nineteenth century. It is significant that the English mathematician William Kingdom Clifford (184–1879) already in 1872 called Lobachevsky the Copernicus of mathematics, because he had opened a new world. Also it was very important that two of the leading mathematicians of the time, Felix Klein and Henri Poincaré, were both intensely engaged in the non-Euclidean hyperbolic geometry, and in the uses it could—in many surprising ways—be put to in other parts of mathematics. They each constructed a plane Euclidean model (as opposed to Beltrami's model on a curved Euclidean surface) of the Lobachevskian geometry, Klein in 1871, and Poincaré (even in two versions) in 1882. Klein's model and one of Poincaré's models use the inner points of a Euclidean circular disk D as their non-Euclidean points. The other of Poincaré's models uses the inner points of a Euclidean half-plane H. Klein's model uses the Euclidean chords of D as its non-Euclidean lines, while Poincaré's models use respectively the Euclidean circular arcs in D orthogonal to the boundary of D (including the diameters of D) and the Euclidean semicircles and half-lines in H orthogonal to the boundary of H. Of course distances are distorted in all three models; angles are also distorted in Klein's model but not in the two models of Poincaré.

A similar change took place in other parts of mathematics, which also contributed to this new understanding. In algebra the correspondence between complex numbers and the points of the plane had been established independently in 1797 by Caspar Wessel (1745–1818, Norwegian), in 1806 by Jean-Robert Argand (possibly 1768–

1822, Swiss), and at some point in between (but not used explicitly in any of his publications before 1848) by Gauss. Already in 1797 Wessel had tried—in vain—to generalize the complex numbers in such a way that a similar correspondence could be established between this generalization and the points in space, but his paper went unnoticed. In 1837 William Rowan Hamilton (1805–1865, Irish) expressed the same wish in the paper in which he had interpreted the complex numbers as pairs of real numbers. His work only bore fruit when in 1843 he discovered that one should not generalize to triplets of real numbers but to quadruples, and so invented his quaternions. With this invention, he did for algebra what Gauss, Bolyai and Lobachevsky had done for geometry.

If you ask somebody who is not a mathematician to give you an arithmetical fact you might get the answer that the order of the factors (in a product of two numbers) is arbitrary. This is precisely not the case with the quaternions; their multiplication is not commutative. It is interesting that the two Bolyais, father and son, also tried their hand in this area—from their home town in Transylvania (then in Hungary, now in Romania) where Farkas Bolyai had lived most of his life and János Bolyai in his youth and on and off since his retirement from the army in 1833. In 1837 they both entered a prize contest set up by a scientific society in Leipzig to give a rigorous geometric construction of imaginary numbers. Farkas Bolyai's solution was taken from his *Tentamen.* János Bolyai's solution resembled Hamilton's solution which, as mentioned above, was published in the same year, but which only became widely known in connection with his paper on quaternions 16 years later. Neither of the Bolyais won the prize; half of it was given to the third contestant, a Hungarian professor.

In our time mathematics really only deals with itself; to a mathematician it is in a way strange (but of course also pleasant) that it can be used to describe phenomena outside mathematics.

Non-Euclidean geometry and physics, philosophy and art

It appears from the previous sections that around 1830—and effectively from around 1870—a decisive change took place in mathematicians' understanding of their own science. Geometry was no longer a part of physics. Of course, one could still use it for description of physical situations and phenomena, but now one could choose among different geometries. This choice is not made by the mathematician but by the physicist, and he should choose the geometry that gives him the best description, the one that fits his experiments or his theory best. He cannot choose the "true" geometry, for mathematically they are equally true.

Over time mathematicians have given physicists many geometries to choose among. It may also happen that a new geometry is invented to fit the physicist's specifications; this was the case with Einstein's theory of relativity. Even so-called finite geometries have been invented. For instance, one can have a "plane" consisting of nine "points" distributed on twelve "lines" with three points on each. Even for such geometries there are eminently practical applications.

For a physicist or some other practitioner, geometry is therefore now a toolbox for descriptions, and maybe also a guideline for theories, while for a mathematician it is the study of all the many different geometries. They behave differently, and the behavior of a geometry depends on the choice of fundamental relations between points and lines and planes; here the mathematician chooses freely and independently. For the mathematician, geometry is something quite different from the intuitive ideas one can have of the organization of physical space. These ideas can be used for inspiration, but geometries can be built in many ways and can be studied without any regard to physical ideas.

Outside the worlds of mathematics and physics the new geometrical insight spread more slowly, and sometimes under protest. Many philosophers even raised very angry protests. One reason why Gauss never published anything on non-Euclidean geometry was probably that the German philosopher Immanuel Kant (1724–1804) in 1781 in his very famous and influential book *Kritik der reinen Vernunft* (*Critique of pure reason*) had placed Euclidean geometry as an *a priori* form of comprehension. When Gauss talked of the Boïotians, he can very well have meant the numerous students and followers of Kant.

Ever since the inception of non-Euclidean geometry there have been philosophers who could not come to terms with it. Let me mention just one curious example, Kristian Kroman (1846–1925, Danish) who was professor of philosophy at the University of Copenhagen for 38 years (1884–1922) and who lectured on pedagogy at my own institution until 1914. As late as 1920 he published a small book both in a Danish and in an English version in which he believed he had proved the parallel axiom and thereby refuted non-Euclidean geometry. What he did was just use the argument one can sometimes meet—as we have seen—in elementary school textbooks, of moving a small line segment along the sides and around the corners of a triangle.

Many artists however seized the new mathematical ideas with great interest. Much of modern art—from cubism to the very newest—would be unthinkable without the original inspiration from non-Euclidean geometry, and also from the geometries of more than three dimensions—both Euclidean and non-Euclidean—which mathematicians began to cultivate in the last years of the nineteenth century. In this connection one must paticularly mention the graphical artist M.C. Escher (1898–1972, Dutch); many of his striking pictures build directly on Poincaré's circle model of hyperbolic geometry.

Outside the circles of mathematicians, physicists, philosophers and artists, however, there are probably very few who are at all aware of the existence of non-Euclidean geometries. This is presumably also true for many mathematics teachers, especially at the primary and lower secondary level where most pupils get their one and lasting impression of geometry.

Non-Euclidean geometry and the mathematics teacher

Many mathematics teachers have not in their own education heard anything about non- Euclidean geometry. It may therefore come as something of a shock to them that geometry is really not a part of physics and fundamentally does not deal with our physical surroundings, and that geometrical theorems are not true in any straightforward physical sense. It may then be a consolation to know that this also surprised and annoyed many mathematicians when it all began to be known around 125 years ago, and that many of them became accustomed to these new ideas only with difficulty.

But does it matter if a teacher does not know about non-Euclidean geometry, one could ask. Nobody would after all expect him or her to teach it to children. The answer is that it does matter, profoundly. If a teacher knows that geometry is true on its own conditions and not as a sort of physics or a part of it, then this teacher will perhaps teach Euclidean geometry differently from how he or she would otherwise have done it. Moreover, knowledge of the existence of non-Euclidean geometry gives one a different view of what mathematics really is, and this may leave its mark everywhere on one's teaching of mathematics.

Also, in many of the subjects in school—physics, chemistry, biology, literature, history—one tries to be up to date. One would like to give the pupils an impression of or a feeling for what goes on in these subjects in our own time. It is very difficult to do the same in mathematics, one of the reasons being that the mathematics that one teaches is not new. Quadratic equations were solved and parabolas were drawn already in antiquity; coordinate systems, which created a connection between equations and curves, were invented around 1640; the first real textbooks of differential and integral calculus were written around 1700. It is of course not quite true—but nearly—that in school one can only teach mathematics which is at least a couple of hundreds of years old. It would be desirable if mathematics were taught in a spirit that could tell the pupils something about what mathematics actually is in our time. This can only happen if teachers of mathematics are themselves acquainted with this spirit, and therefore they should know about non-Euclidean geometry.

Another question is then: From where in their education or further education could teachers get such knowledge? Only seldomly will they have been offered a specialized course in non-Euclidean geometry, or will have taken one if offered. The natural place might be as part of a course in geometry. One can hope that this will happen more often in the future than has been the case up to now. Another obvious possibility is in a course in the history of mathematics. Such a course would be very incomplete if non-Euclidean geometry were not mentioned. In many countries history of mathematics is becoming more prominent in the education and further education of teachers, and many textbooks in the history of mathematics contain a whole chapter on non-Euclidean geometry, or sections which taken together amount to a chapter.

Finally, it could be asked which topics from non-Euclidean geometry one could include in school mathematics, assuming one wants to do more than just mention that Euclidean geometry is not the only possibility—with all the consequences this has for what mathematics really is. Such a topic should of course be rather concrete and intuitive; a possible choice could be the area of triangles. In Euclidean geometry the sum of the angles of a triangle is of course $180°$, whatever the area of the triangle might be. It is not so difficult to show that in elliptic geometry—modelled by the intrinsic geometry of the surface of an Euclidean sphere—the excess of a triangle, that is the amount by which the sum of its angles exceeds $180°$, is proportional to the area of the triangle (or equal to it if the area is measured in convenient units). It is a challenge to show that in hyperbolic geometry—e.g., modelled by the Poincaré circle—the defect of a triangle, that is the amount by which the sum of its angles is less than $180°$ (is defective in comparison with $180°$!), is proportional to the area of the triangle (or equal to it if the area is measured in convenient units). Indeed, the climax of such an investigation might be to find the area of a so-called limit triangle, that is a "triangle" with angle sum $0°$, with its vertices infinitely far away and its sides parallel! In the Poincaré model the vertices of such a "triangle" are on the Euclidean boundary circle and can

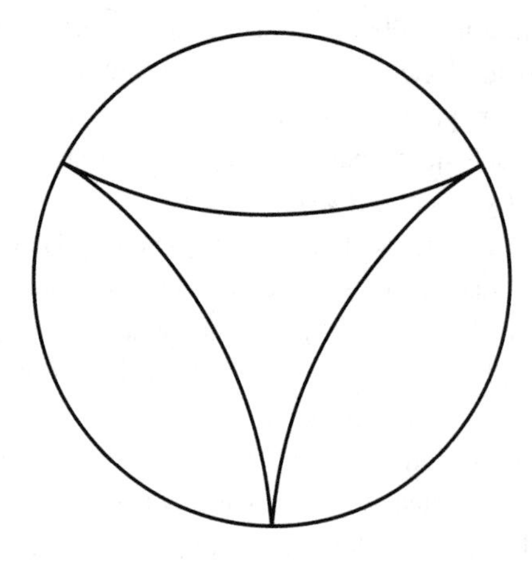

FIGURE 6

be chosen 120° from each other, and it is interesting to compare the non-Euclidean area of this "triangle" with its Euclidean area.

Another topic which might be touched upon in school mathematics is the tesselations with regular polygons of the elliptic, Euclidean and hyperbolic planes, the first and the last of these again modelled on the surface of a Euclidean sphere and in the Poincaré circle. In the elliptic case there are five such tesselations, corresponding to the five regular polyhedra (and also infinitely many tesselations with 2-gons!). In the Euclidean case there are three (triangular, square, and hexagonal). In the hyperbolic case there are (and this is the surprise) infinitely many, even infinitely many with triangles, infinitely many with 4-gons, infinitely many with 5-gons, infinitely many with 6-gons, infinitely many with 7-gons, etc. The variation in these beautiful patterns is endless and has also inspired artists; for example, see Escher's three "Angels and Devils", one elliptic (carved on a sphere), one Euclidean, and one hyperbolic (drawn in a Poincaré circle).

References

The following list is a selection of books and papers from the enormous literature on non-Euclidean geometry and related subjects. The list contains elementary and non-elementary expositions, short ones and long ones, more or less historical, more or less philosophical, more or less pedagogical, and with non-Euclidean geometry as the only subject or the main subject or a subject among other subjects. It is therefore my hope that the list can be useful for many different readers. Most of the books and papers in the list can be read by teachers and by college and university students, and some of them (especially [1], [10], [18], [21], [23], [34], [46], [48], [49], and [50]) also by secondary school pupils.

1. Bier, Martin: 1992, "A Transylvanian Lineage," *The Mathematical Intelligencer* 14 (2), pp. 52–54.
2. Bolyai, János: 1987, *Appendix: The Theory of Space,* Amsterdam. (New translation with commentary.)
3. Bolyai, John: 1955, *The Science of Absolute Space,* translated by George Bruce Halsted. (Supplement I in [4].)
4. Bonola, Roberto: 1955, *Non-Euclidean Geometry,* Dover, New York. (Reprint of the translation (1911) by H. S. Carslaw of the Italian original (1906).) Includes John Bolyai, *The Science of Absolute Space* and Nicholas Lobachevski, *The Theory of Parallels,* both translated by George Bruce Halsted.
5. Burton, David M.: 1985, *The History of Mathematics: An Introduction,* Allyn & Bacon, Newton, Massachusetts; 2nd ed., Wm. C. Brown, Dubuque, Indiana 1991; 3rd ed., McGraw-Hill, New York, 1997. (Contains a chapter on non-Euclidean Geometry.)
6. Calinger, Ronald (ed.): 1982, *Classics of Mathematics,* Moore, Oak Park, Illinois; 2nd ed. Prentice Hall, Englewood Cliffs, New Jersey, 1995. (Collection of sources with commentaries.)
7. Coxeter, H. S. M.: 1942, *Non-Euclidean Geometry* University of Toronto Press, Toronto; 3rd ed. 1957; 6th ed., Mathematical Association of America, 1998.
8. Dávid, L. v.: 1951, *Die beiden Bolyai,* Birkhäuser, Basel.
9. Engel, Friedrich & Paul Stäckel: 1895, *Die Theorie der Parallellinien von Euklid bis auf Gauss,* Teubner, Leipzig; repr. New York, London, 1968. (Collection of sources with commentaries; see also [33].)
10. Ernst, Bruno: 1976, *The Magic Mirror of M. C. Escher,* Ballantine, New York. (Translated from Dutch.) (Escher's non-Euclidean "Circle Limit" prints are treated pp. 108–109.)
11. Euclid: 1908, *The Thirteen Books of Euclid's Elements,* Translated and with commentaries by Thomas L. Heath, Cambridge University Press, Cambridge; 2nd ed. 1926; reprinted in three volumes, Dover, New York.
12. Eves, Howard: 1990, *Foundations and Fundamental Concepts of Mathematics,* Prindle, Weber & Schmidt, Boston; 3rd ed. of Howard Eves & Caroll V. Newsom, *An Introduction to the Foundations and the Fundamental Concepts of Mathematics,* Rinehart, New York, 1959. (Contains a chapter on non-Euclidean geometry.)
13. Faber, Richard L.: 1983, *Foundations of Euclidean and Non-Euclidean Geometry,* Marcel Dekker, New York.
14. Fauvel, John & Jeremy Gray (eds.): 1987, *The History of Mathematics: A Reader,* Macmillan, London, in association with Open University, Milton Keynes; repr. 1988. (Collection of sources with commentaries.)
15. Gillespie, Charles Coulston (ed.): 1970–80, *Dictionary of Scientific Biography,* Scribner, New York; 2nd ed. in 8 vols, 1981. (Contains, e.g., articles on Beltrami, F. Bolyai and J. Bolyai (by D. J. Struik), Euclid (by Ivor Bulmer-Thomas and John Murdoch), Gauss (by Kenneth O. May), Lambert (by Christoph J. Scriba), Lobachevsky (by B. A. Rosen-

feld), Proclus (by Glenn R. Morrow), Riemann (by Hans Freudenthal) and Saccheri (by D. J. Struik).)

16. Gray, Jeremy: 1987, "The Discovery of Non-Euclidean Geometry," In Esther R. Phillips (ed.), *Studies in the History of Mathematics,* Studies in Mathematics vol. 26, The Mathematical Association of America, Washington, DC.
17. Gray, Jeremy: 1979, *Ideas of Space: Euclidean, Non-Euclidean and Relativistic,* Clarendon, Oxford.
18. ——: 1987, *Non-Euclidean Geometry,* Topics in the History of Mathematics, unit 13. Open University, Milton Keynes. (Refers extensively to [14].)
19. Greenberg, Marvin J.: 1974, *Euclidean and Non-Euclidean Geometries: Development and History,* Freeman, San Francisco.
20. Halameisär, Alexander & Helmut Seibt: 1978, *Nikolai Iwanowitsch Lobatschewski,* B. G. Teubner, Leipzig.
21. Heiede, Torkil: 1992, "Why Teach History of Mathematics?," *The Mathematical Gazette* 76, pp. 151–157.
22. ——: 1996, "History of Mathematics and the Teacher," In Ronald Calinger (ed.), *Vita Mathematica: Historical Research and Integration with Teaching,* The Mathematical Association of America, Washington, DC.
23. Henderson, Linda Dalrymple: 1983, *The Fourth Dimension and Non-Euclidean Geometry in Modern Art,* Princeton University Press, Princeton, New Jersey.
24. Katz, Victor J.: 1993, *A History of Mathematics: An Introduction,* Harper Collins, New York. (Contains—in different chapters—long sections on the development of non-Euclidean geometry.)
25. Kelly, Paul & Gordon Matthews: 1981, *The Non-Euclidean, Hyperbolic Plane,* Springer, New York.
26. Kline, Morris: 1972, *Mathematical Thought from Ancient to Modern Times,* Oxford University Press, Oxford. (Non-Euclidean geometry is treated in Chapter 36.)
27. ——: 1980, *Mathematics: The Loss of Certainty,* Oxford University Press, Oxford.
28. Kroman, Kristian: 1920, *Mathematics and the Theory of Science,* Det Kongelige Danske Videnskabernes Selskab, Filosofiske Meddelelser I, 1. Høst, Copenhagen. (A philosopher's misguided attempt to refute non-Euclidean geometry.)
29. Lobachevski, Nicholas: 1955, *The Theory of Parallels,* translated by George Bruce Halsted. (Supplement II in [4].)
30. Mainzer, Klaus: 1980, *Geschichte der Geometrie,* BI-Wissenschaftsverlag, Mannheim.
31. Martin, George E.: 1975, *The Foundations of Geometry and the Non-Euclidean Plane,* Intext, New York; 2nd ed, Springer, New York, 1982; repr. 1986.
32. Meschkowski, Herbert: 1954, *Nichteuklidische Geometrie* Vieweg, Braunschweig; 3rd pr. 1965. Translation by Abe Shenitzer: *Noneuclidean Geometry,* Academic Press, New York, 1964.
33. Perron, Oskar: 1962, *Nichteuklidische Elementargeometrie der Ebene,* Teubner, Stuttgart.
34. Petit, Jean-Pierre: 1980, *Les Aventures d'Anselme Lanturlu: Le Geometricon,* Belin, Paris. Translation by A. Pierre and P. Weber, *Die Abenteuer des Anselm Wüsstegern: Das Geometrikon*; English translation expected. (A volume in a mathematically reliable cartoon series.)
35. Proclus: 1970, *A Commentary on the First Book of Euclid's Elements,* Translated and with commentaries by Glenn R. Morrow, Princeton University Press, Princeton, New Jersey. (See especially pp. 150–151 and pp. 284–295.)
36. Riemann, Bernhard: 1959, *Über die Hypothesen, welche der Geometrie zu Grunde liegen,* Wissenschaftliche Buchgesellschaft, Darmstadt. (Photo-mechanical reprint of the original from 1867.)
37. Rosenfeld, B. A., *A History of Non-Euclidean Geometry: Evolution of the Concept of a Geometric Space,* Springer, New York 1988. (Translated by Abe Shenitzer from the Russian original, Moscow 1976.)
38. Runion, Garth E. & James R. Lockwood: 1978, *Deductive Systems: Finite and Non-Euclidean Geometries,* National Council of Teachers of Mathematics, Reston, Virginia.
39. Saccheri, Girolamo: 1920, *Euclides Vindicatus (Euclides ab omni naevo vindicatus, Euclid Freed of all Blemish),* Latin text with English translation by George Bruce Halsted, Chicago; 2nd ed (with English translation of notes by Stäckel and Engel), Chelsea, New York, 1986.
40. Shirokov, P. A.: 1964, *A Sketch of the Fundamentals of Lobachevskian Geometry,* Noordhoff, Groningen. (Translated by Leo F. Boron from the Russian original.)
41. Sjöstedt, C. E. (ed.): 1968, *Le axiome de paralleles de Euclides a Hilbert,* Interlingue-Fundation, Uppsala, Natur och Kultur, Stockholm. (An extensive collection of sources in facsimile with parallel translations in Interlingua, with portraits and commentaries.)
42. Smith, David Eugene: 1959, *A Source Book in Mathematics 1–2,* Dover, New York, 1959 and later; original edition 1929. (A collection of sources, with portraits and commentaries.)
43. Stahl, Saul: 1993, *The Poincaré Half-Plane: A Gateway to Modern Geometry,* Jones & Bartlett, Boston.
44. Stillwell, John: 1989, *Mathematics and Its History,* Springer, New York. (Contains a chapter on non-Euclidean geometry.)
45. Sullivan, John Wiliam Navin: 1926, "Mathematics as an Art," In J. W. N. Sullivan (ed.), *Aspects of Science,* Second Series, Knopf, New York. Also in William L. Schaaf (ed.), *Mathematics: Our Great Heritage,* Harper, New York, 1948, and in James R. Newman, *The World of Mathematics 1–4,* Simon & Schuster, New York 1956 and later.
46. Sved, Marta: 1991, *Journey into Geometries,* with preface by H. S. M. Coxeter and illustrations by John Stillwell, The Mathematical Association of America, Washington, DC. (Alice, Lewis Carroll and others discovering a non-Euclidean wonderland.)
47. Trudeau, Richard J.: 1987, *The Non-Euclidean Revolution,* with introduction by H. S. M. Coxeter, Birkhäuser, Boston. (Treats also the role of Kant's philosophy.)
48. Wolff, Peter (ed.): 1963, *Breakthroughs in Mathematics,* Signet, New York. (Collection of sources with commentaries; contains a chapter with excerpts from Euclid and Lobachevsky.)
49. Zeitler, Herbert: 1970, *Hyperbolische Geometrie,* Bayerischer Schulbuch-Verlag, Munich. (A textbook for the Austrian gymnasium.)
50. Zheng, Yuxin: 1992, "Non-Euclidean Geometry and Revolutions in Mathematics," In Donald Gillies (ed.), *Revolutions in Mathematics,* Clarendon, Oxford.

Scientific Research and Teaching Problems in Beltrami's Letters to Hoüel

Livia Giacardi
University of Turin, Italy

> A process of mathematical reasoning is like a series of chords played on the lyre of the intellect, which is made up of the mathematical strings of human thought; and the discovery of a new branch of mathematics is comparable to the discovery of a new harmonic modulation.
>
> — Beltrami to Gustav Wolff

"In the science of mathematics the triumph of new concepts can never invalidate previously acquired truths: it can only alter their place or their logical basis, and increase or reduce their value and use. Nor can profound criticism ever damage the solidity of the scientific edifice, but rather lead to the discovery and clearer recognition of its true foundations."[1] These are the words with which, in 1868, Eugenio Beltrami opened his *Attempt at an interpretation of non-Euclidean geometry,* which offers an interpretation of Lobachevskian planimetry by means of surfaces of constant negative curvature or pseudospherical surfaces, thus providing a *real substratum* for hyperbolic geometry.

This was the period during which non-Euclidean geometries were just beginning to be known in Europe. One of the most active propagators of the new geometries in France was Jules Hoüel, as can be seen both from his tireless work in translating, reviewing and commenting on books and articles, and from his extensive correspondence. In Italy a similar role was played by Giuseppe Battaglini,[2] who had transformed the *Giornale di Matematiche,* of which he was editor, into an effective organ for the spread of Lobachevsky's hyperbolic geometry. The attitude in both France and Italy was, however, characterized by mistrust and, at times, flat rejection of non-Euclidean geometries: Joseph Bertrand called them *une débauche de logique,*[3] for Eugène Catalan the *non-euclidiens* were *inoffensifs et peut-être très utiles rêveurs*[4] and the Paris Académie des Sciences was inundated with supposed proofs of the postulate of the parallels. Beltrami outlined the situation in Italy in a letter to Placido Tardy: "I do not know if you have given any attention to the system of ideas which is being publicized under the name of non-Euclidean geometry. I know that Prof. Chelini is definitely against it, and that Bellavitis calls it madhouse geometry, while Cremona thinks it is debatable and Battaglini accepts it without reserve".[5]

1. Eugenio Beltrami's *Attempt*

The 65 letters[6] which Beltrami wrote to Jules Hoüel between 1868 and 1881 are valuable both for the clarification of the main doubts and misunderstandings which dominated the attitude of the scientific world to non-Euclidean geometries, and especially for the reconstruction of the genesis of Beltrami's research in this field. They also bring out lesser-known aspects of Beltrami, such as his interest in the problems of teaching mathematics and his material construction of the pseudospherical surface.

The main sources of inspiration lying behind the *Attempt* are to be found in Gauss's theory of surfaces as expounded in the *General Investigations of Curved Surfaces,*[7] in Lobachevsky's work on non-Euclidean geometry and in some of the results obtained by Ferdinand Minding on surfaces of constant negative curvature in 1839–1840. Some of Beltrami's own research work must be added to these: he had been prompted by reading a paper by Lagrange on geographical maps to try to discover whether there were surfaces which could be represented on a plane in such a way that their geodesic lines were represented by straight lines. In an 1865 paper,[8] he shows that these surfaces must necessarily have constant curvature.

The *Attempt* was not, however, in the least influenced by the innovative and fruitful ideas Riemann had expounded in his famous 1854 lecture *On the hypotheses which lie at the foundations of geometry,*[9] although Riemann had spent two years in Pisa at the very time when Beltrami was teaching geodesy there. As he wrote to Angelo Genocchi: "Last year, when no one knew about this fundamental work of Riemann's, I told our good friend Cremona of a paper of mine where I gave an interpretation of non-Euclidean plane geometry, which seemed satisfactory to me,"[10] and again, in a letter to Hoüel, he said: "What amazes me is that for all the times I talked with Riemann (during the two years he spent in Pisa, shortly before his sad end), he never mentioned these ideas to me, though they must have occupied him for quite a long time, for a fine draft cannot be the work of a single day, even for such a brilliant genius."[11]

Gauss's research on surfaces was basic for Beltrami: "The whole of my deductions," he wrote to Helmholtz, "rests on the representation of surfaces by Gauss's formula $ds^2 = E\,du^2 + 2F\,du\,dv + G\,dv^2$. Now, in this method, the relationship between the surface and the surrounding space is entirely overlooked: the surface is considered in itself, as it would be by a being who did not have any sense of the third dimension."[12] Indeed, as is well known, in his *Investigations* Gauss regards a surface "not as the boundary of a solid, but as a solid with one vanishing dimension, flexible but non-stretchable,"[13] and he formulates the general theory of the intrinsic geometry of a surface. He introduces the curvilinear coordinates of the points of the surface and expresses as a function of these the square of an element of length of an arc of a curve on the surface: $ds^2 = E\,du^2 + 2F\,du\,dv + G\,dv^2$ (the so-called first fundamental form).

The geometrical properties of a surface which are independent of deformations caused by bending, i.e., which can be expressed by means of the functions E, F and G alone which appear in the expression of the linear element and of their derivatives, constitute the intrinsic geometry of the surface. Gauss particularly proves that the curvature of a surface is a property that belongs to intrinsic geometry (*theorema egregium*). Two surfaces such that the expressions of their linear elements can be transformed so as to be identical have the same intrinsic geometry and can be applied to each other (locally only). This is one of the points which Hoüel did not grasp and which Beltrami explained to him again and again.[14] It is also important that Beltrami stressed that Gauss's theory of surfaces does not depend on the postulate of the parallels: "It seems to me," he wrote to Hoüel, "that this theory has not generally found complete *Würdigung* (appreciation), so much so that no one has yet noticed this crucial fact, namely that *it is wholly independent of the postulate of Euclid.*"[15]

Gauss's studies were continued by Minding, who was particularly interested in surfaces of constant negative curvature and, in an article in 1839,[16] found the three surfaces of revolution to which they can be applied, among them the surface generated by the revolution of the tractrix around its own asymptote, i.e., Beltrami's *pseudosphere*. In a later article (1840),[17] Minding arrived at another interesting result, though without perceiving its important implications. He observed that the trigonometric relations in geodesic triangles of a surface of constant negative curvature could be obtained from the corresponding formulas of spherical geometry on a sphere of radius R by multiplying R by $\sqrt{-1}$. While Minding failed to notice that these formulas agree with those for the hyperbolic plane, established by Lobachevsky in his *Imaginary Geometry* (1837), Beltrami was aware of this fact, which he developed in his *Attempt*.

He starts from the following specific expression of the square of the linear element of a surface of constant negative curvature equal to $-1/R^2$:

$$ds^2 = R^2 \frac{(a^2 - v^2)du^2 + 2uv\,du\,dv + (a^2 - u^2)dv^2}{(a^2 - u^2 - v^2)^2}, \tag{1}$$

where a^2 is an arbitrary constant.[18]

He chooses this particular expression because it has the advantage that every linear equation involving u and v represents a geodesic and vice versa. From the expressions which supply the sine and cosine of the angle of the two coordinate lines at the point (u, v) it can be seen that we have admissible values of u and v for $u^2 + v^2 \leq a^2$. Regarding the coordinates u and v as rectangular coordinates x and y of an auxiliary plane, Beltrami shows that the surface of constant negative curvature, or rather the totality of its real points, is represented biunivocally in the interior of the circle $u^2 + v^2 = a^2$ (*limit circle*). In this representation the geodesics of the surface are represented by the chords of the circle, and the limit circle corresponds to the line of the points at infinity of the surface. Furthermore, two points in the interior of the circle identify a unique chord, and hence any two real points of the surface identify a unique geodesic.

It must be noted that the surface here appears only as a two-dimensional manifold and the formula of ds^2 gives the law for measuring the distance between two infinitely close points, independently of the existence of an isometrical embedding of this manifold in Euclidean three-dimensional space.

Studying the relation between the angle of two geodesics and the angle of the chords representing them, Beltrami finds that:

1. Two chords, which intersect in the interior of the limit circle, correspond to two geodesics of the surface which intersect at a point at a finite distance, forming an angle different from $0°$ and from $180°$;
2. Two chords, which intersect on the circumference of the limit circle, correspond to two geodesics which intersect at a point at infinity, forming a zero angle;
3. Two chords, which intersect outside the limit circle, or are parallel, correspond to two geodesics which have no point in common on the whole real extension of the surface.

Beltrami calls point 2. geodesics *parallel* because they mark the passage from the ensemble of the secants to the non-secants. Thus, given a geodesic (represented by the chord AB, Fig. 1), from every real point on the surface, which does not belong to it, it is always possible to draw two geodesics parallel to the given one (represented by PA and PB). Hence the fifth Euclidean postulate is not valid.

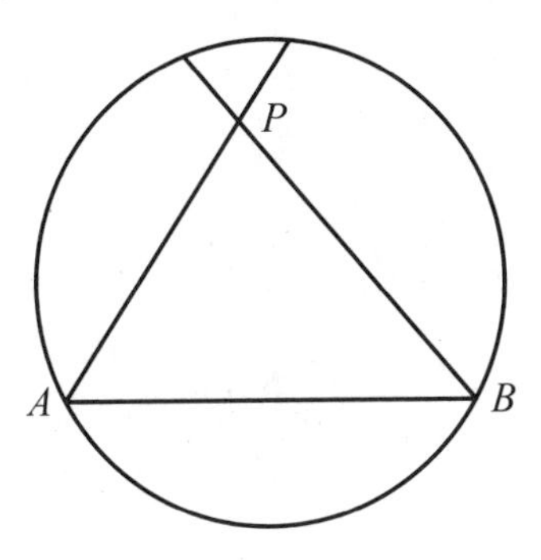

FIGURE 1
The chords PA and PB represent the two geodesics parallel to the geodesic represented by the chord AB.

Thus Beltrami shows that Lobachevsky's non-Euclidean plane geometry can be interpreted on surfaces of constant negative curvature, replacing the word "straight line" with "geodesic". The model described above provided the first proof of the consistency of Lobachevskian plane geometry, representing as it did the non-Euclidean plane in the Euclidean plane. This result also removed all doubt regarding the impossibility of proving the axiom of parallels by deducing it from the others relating to the straight line and the plane. In fact the latter are verified on a surface of constant negative curvature, while the fifth postulate is not. If the axiom of parallels could be deduced logically from the other axioms, then it would also have to hold true in the model, which is not the case.

Beltrami then goes on to develop certain problems regarding geometry on a surface of constant negative curvature, showing that the results obtained coincide with Lobachevsky's and Bolyai's. As far as trigonometry is concerned, he refers his readers to the already mentioned paper by Minding and the subsequent developments of Delfino Codazzi.[19] He himself simply obtains the theorem of the sum of the angles of a geodesic triangle and the relationship of this sum with the area of that triangle; he then finds the angle of parallelism which corresponds to a certain distance δ, arriving at the same formulae as Lobachevsky.

In the second part of the *Attempt,* Beltrami goes on to consider the geodesic circumferences with a real center, an ideal center and an infinite-distance center. Geodesic circles with a given center are orthogonal trajectories of the geodesics passing through a fixed real point. In the same way Beltrami is also led to consider orthogonal trajectories of a bundle of geodesics, which are represented in the limit circle by chords concurrent at a point outside the circle or on its circumference. If this point is outside, the orthogonal trajectories are called by Beltrami *geodesic circumferences with an ideal center*; if it is on the circumference they are called *geodesic circumferences with an infinite distance center.* The latter, being the orthogonal trajectories of a system of parallel geodesics, correspond to Lobachevsky's horocycles (Fig. 2).

For each of these three cases Beltrami considers a specific region of the pseudospherical surface which can be applied on a surface of revolution. In the third case, the surface of revolution is of the so-called parabolic type. If we take as coordinate lines $\sigma = k$ and $\rho = m$, a family of parallel geodesics and their orthogonal trajectories, the

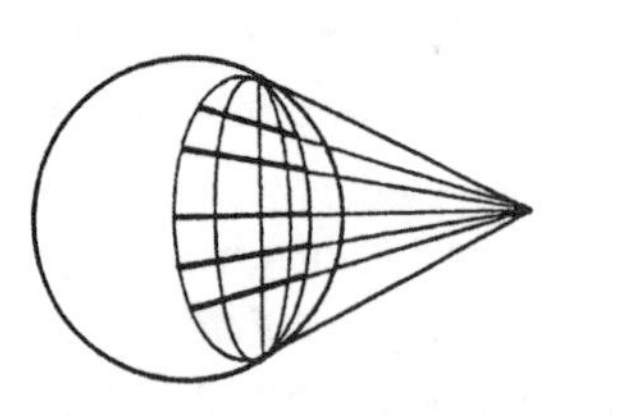

geodesic circumferences with an ideal center

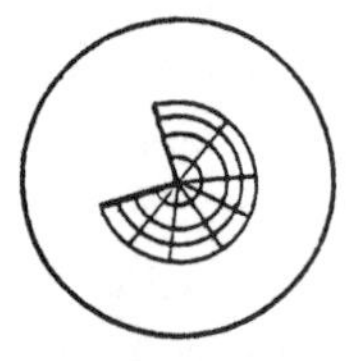

geodesic circumferences with a real center

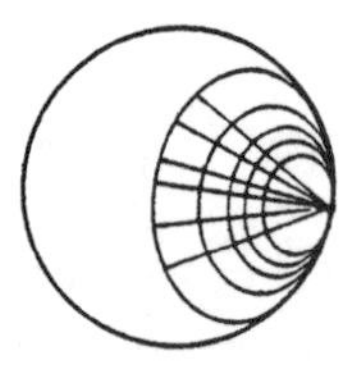
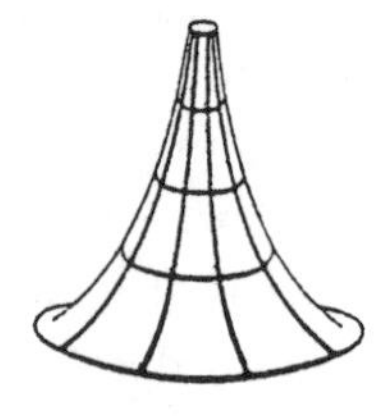

geodesic circumferences with an infinite distance center

FIGURE 2

linear element [1] takes the form

$$ds^2 = d\rho^2 + e^{-2\rho/R} d\sigma^2,$$

which is the linear element of the pseudosphere or tractroid, i.e., of that surface of revolution whose meridian curve is, as already stated, the tractrix.[20]

Since the geodesic circumferences with the center at an infinite distance correspond to Lobachevsky's horocycles, it may be said that a system of concentric horocycles is transformed, with a bending of the surface, into the system of parallels of the pseudosphere, and the parallel geodesics $\sigma = k$ then form the meridians. Beltrami says that the region of the pseudospherical surface situated on a specific side with respect to the line $\rho = 0$ (if the radius of the parallel is chosen equal to R) is, in his own words, "wrapped around" the tractroid, or pseudosphere, an infinite number of times.[21]

It is necessary to point out that the pseudosphere does not represent the whole hyperbolic plane, but only a restricted region, namely a horocyclic sector. This limitation renders the pseudosphere utterly useless as a means for drawing significant hyperbolic figures. Every geodesic that is not merely a meridian winds itself round the horn as it proceeds in one direction, whereas in the opposite direction it is abruptly cut by the cuspidal edge.

The first explicit criticisms of the interpretation of Lobachevskian plane geometry described in the *Attempt* came from Helmholtz and from Klein in the years 1870–1871, and were taken up again in more detail by Genocchi in 1877. They raised doubts as to the existence in Euclidean space of an infinitely extended pseudospherical surface: "But we cannot," wrote Helmholtz, "in our space, construct a pseudospherical surface which is indefinitely extended towards the axis of revolution: we always arrive either at a limit, as in the case of the champagne glass, or at two limits, as in the case of the ring."[22] (See Fig. 3.) Similarly, Klein held that the interpretation of hyperbolic

FIGURE 3
Plaster models of the pseudospherical surfaces of hyperbolic, elliptic and parabolic types, preserved at the Department of Mathematics, University of Turin.

geometry on surfaces of constant negative curvature cannot "provide understanding of the whole plane, since the surfaces of constant negative curvature are always limited by cuspidal edges."[23] Genocchi especially thought that "it must be proved that the partial differential equation which expresses the surfaces of constant negative curvature admits at least one integral which satisfies all the conditions required for a pseudosphere."[24]

The emphasis here is on a problem which Beltrami had left open, but of which, as his correspondence suggests, he seems to have been aware: whether there is or is not, in Euclidean space, a surface which represents the whole two-dimensional manifold of constant negative curvature and whose geometry coincides with that of Lobachevsky's whole plane. For example he wrote to Hoüel: "In my *Attempt* I said that the pseudospherical surface, in so far as it is represented by the variables u, v, is *indefinitely extended* in every direction and *simply connected* (einfach zusammenhängende, according to Gauss and Riemann), and this is perfectly true. However, since the *general* integral [solution] of this surface in ordinary coordinates x, y, z is not known (it is an integral which current calculus is a long way from being able to supply) and since, as a result, *the more general* form of this surface is not known, it cannot be proved, *a priori*, that it *can* exist in *ordinary* space in the double state of *indefinitely extended and simply connected*."[25]

This problem was solved in 1901 by Hilbert, who proved that there is no regular analytical surface in three-dimensional Euclidean space (i.e., no surface completely free of singularities) on which Lobachevsky's plane geometry is valid in its entirety.[26]

Shortly after the publication of Beltrami's *Attempt* Klein established the connection between the projective metric introduced by Arthur Cayley[27] and non-Euclidean geometry: he showed that if the Cayley *absolute* is a real non-degenerate conic then the part of the projective plane in its interior is isometric to the Lobachevskian plane.[28] Beltrami's model in a circle, described above, is a special case of Klein's model when the conic is a circle.

It is clear from his correspondence with Hoüel that, as early as the summer of 1869, Beltrami had had a suspicion that there was a link between Cayley's research and his own, for he wrote: "The second of these conjectures[29] would be more important, if I were able to give it a concrete form, because it does not exist thus far in my head except as a very vague concept, though it is undoubtedly based on truth. It is the conjecture of a strict analogy, perhaps identity, between pseudospherical geometry and Cayley's theory on *the analytic origin of metrical relations*, with the help of the conic (or quadric) *absolute*."[30] However, Beltrami did not follow up this conjecture, an omission which cost him some regret when, in the summer of 1872, learning of Klein's result, he realized he had let the distinguished German steal his thunder: "I deeply regret," he wrote to Hoüel, "having allowed Klein to anticipate me on this point which I had already gathered material, but to which I made the mistake of not giving enough importance. Besides, this way of looking at matters is not entirely new, and this is exactly why I did not hasten to publish my observation. It is closely connected to an observation already made by Chasles regarding the angle of two straight lines, considered as the logarithm of a cross ratio; it is also connected to a theorem of Laguerre Verlay."[31]

2. The cardboard models of the pseudospherical surface

In 1872 Beltrami devoted the brief paper, *On the surface of revolution which serves as a model for pseudospherical surfaces*, to the study of the pseudosphere. His aim here was, as he himself says, "to prepare the geometrical elements of a material construction, if possible simple and exact, of the surface itself."[32] Beltrami had actually been interested in the material construction of the pseudospherical surface since 1869, when he had written to Hoüel: "The meantime I have had a wild idea, which I shall tell you about . . . I wanted to try to construct materially the pseudospherical surface, on which the theorems of non-Euclidean geometry are realised . . . ."[33] In the same letter he gave a detailed description of two models of pseudospherical surfaces which he had produced by cutting out and then glueing together curvilinear paper trapezia: "if . . . you consider," he wrote, "the surface lying between two meridians, close enough together to allow it to be replaced, over a certain length, by a plane, you can, with little bits of paper cut into appropriate shapes reproduce the curved trapezia whose true surface can be supposed to be compounded."

We know that Beltrami made at least four cardboard models, one of which is still preserved in the Department of Mathematics of the University of Pavia. This is the one[34] he sent as a gift to his friend Luigi Cremona on 25 April 1869, with a covering letter which hints, among other things, at the possibility of an industrial production of the model, the idea which prompted Beltrami to write the already-mentioned paper of 1872.

The model (Fig. 4a) consists of curvilinear trapezia made of thick paper, cut out and glued together as required, each of them approximating a portion of a pseudospherical surface lying between two meridians and two parallels.

(a) Beltrami's cardboard model, preserved at the Department of Mathematics, University of Pavia.

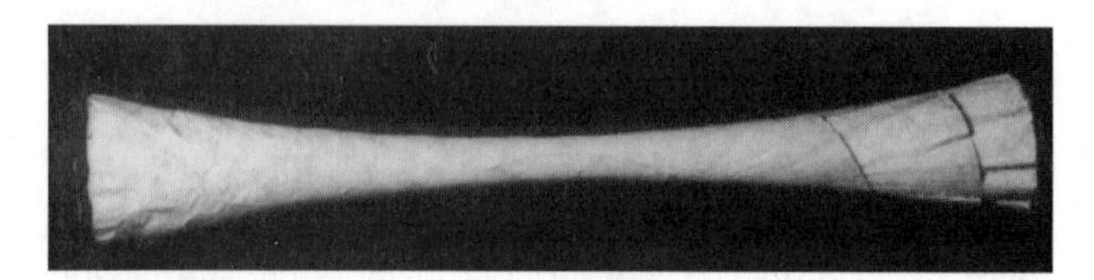

(b) The model folded according to the pseudospherical surface of the hyperbolic type.

Since overall the model approximates a geodesic circle, it can be described by means of the auxiliary plane, where it is represented by the circle of diameter AB, while the circle of diameter ab represents the limit circle (Fig. 5). In this figure the lines which correspond to those drawn by Beltrami on his model are identified by a heavier stroke. They are:

- the diameter AB, 1.029 metres long,
- the geodesic segment OC perpendicular to the diameter AB at its midpoint,
- the geodesic OM, symmetrical to ON with respect to OC (OM and ON are two geodesics parallel to AB),
- the horocycle arcs EF and $E'F'$, tangent to each other and having their center at infinity, at a and b respectively and the geodesic HK, perpendicular to AB and tangent to both the horocycles EF and $E'F'$.

The model can be folded (Fig. 4b) according to the pseudospherical surface of the hyperbolic type, referred to in the *Attempt* with the equation

$$ds^2 = d\xi^2 + \cosh^2 \frac{\xi}{R} d\eta^2$$

or according to the parabolic type (Fig. 4c) or simply the pseudosphere, whose linear element, as we have seen, is

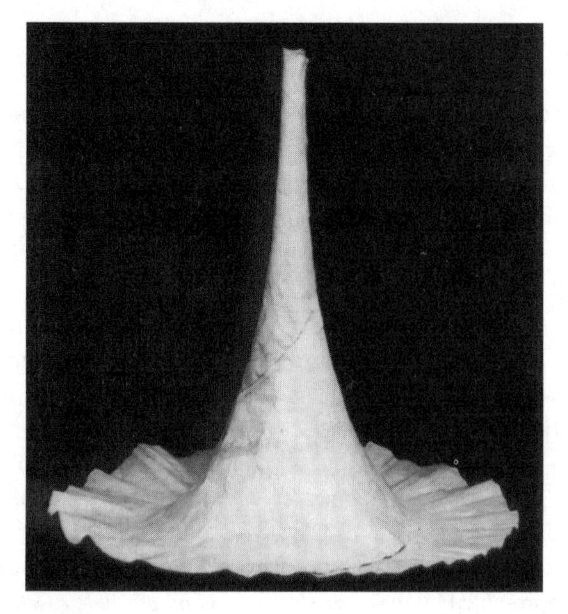

FIGURE 4
(c) The model folded according to the pseudospherical surface of the parabolic type.
These three photographs are reproduced here by kind permission of Professor Mario Ferrari.

given by the formula

$$ds^2 = d\rho^2 + e^{-2\rho/R} d\sigma^2,$$

whereas it is not possible to fold the model according to the pseudospherical surface of the elliptical type defined

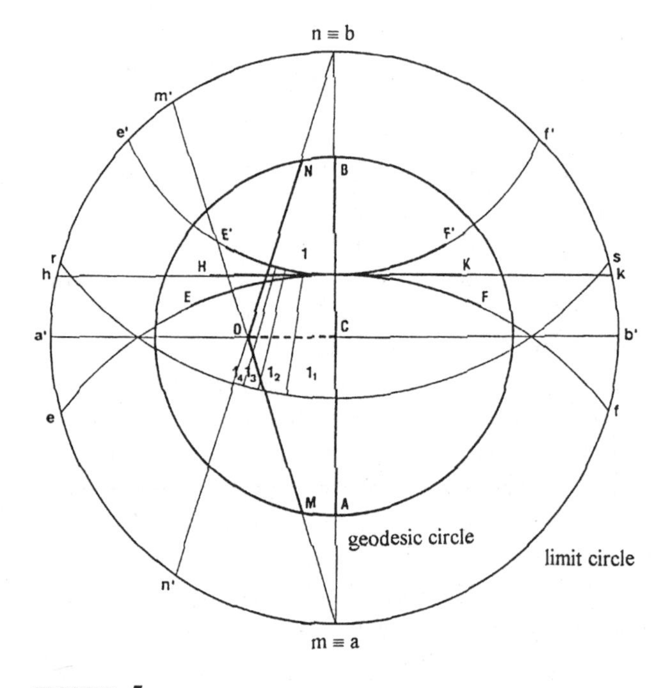

FIGURE 5
Representation of the model on the auxiliary plane.

by

$$ds^2 = d\rho^2 + \left(R \sinh \frac{\rho}{R}\right)^2 d\varphi^2$$

without making a cut.[35]

Beltrami was almost afraid that his interest in the material construction of the pseudosphere might be regarded as an eccentricity, and never missed an opportunity to insist, in his letters, on the importance of these material constructions, both as a tool for checking the results obtained and as a means of discovering new properties or theorems. One new and elegant result which he obtained is this: "Draw a straight line AB and at each of its points M draw the straight line MT which marks the direction of the parallel to AE relative to the distance AM, following Lobachevsky. *The envelope of these straight lines is the meridian of the pseudospherical surface*. It follows that the distance MN to the point of contact is constant."[36]

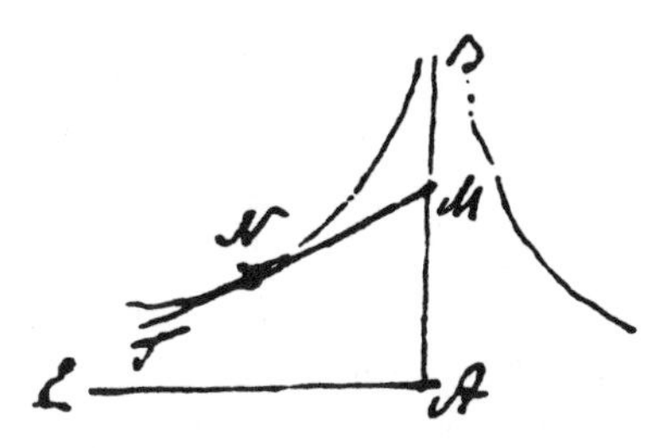

FIGURE 6
Drawing relating to the "Theorem of pseudospherical geometry" in Beltrami's letter to Hoüel, 13 March 1869.

3. Beltrami and the teaching of elementary geometry

A particularly interesting, but less well known, aspect of Beltrami which emerges from his correspondence with Hoüel is the attention he gave to the problems of teaching at both secondary school and university level. He deeply regretted the attitude of teachers in Italy at that time, reluctant as they were to open their minds: "The number of secondary school teachers who concern themselves with their own discipline is very small here: and even those who take an interest do not have that feeling of didactic and scientific solidarity which there is in Germany and which wins attentive readers for any fairly serious article."[37] This interest not only led him often to join examining boards or committees for school inspection, but also to collaborate with the Ministry of Education in 1884 in the modification of the secondary-school mathematics syllabus[38] and to become a member of the Higher Council for Public Education.

In addition, though indirectly, he took part in the heated debate over the problem of the teaching of elementary geometry provoked by the Act of Parliament issued by the Minister Michele Coppino on 10 October 1867. This Act, which introduced Euclid's *Elements* as a textbook in classical secondary schools, was really the brainchild of Luigi Cremona, who was at that time a member of a special committee whose task was to formulate new syllabi.

Cremona was convinced that the study of mathematics ought to be "a means of general culture, gymnastics for the mind designed to develop the faculty of reasoning and to assist that just and healthy criterion in the light of which we distinguish what is true from what only appears to be so,"[39] and he was also convinced that no text was better suited than the Elements to lead to the achievement of this aim. It was also at Cremona's prompting that in 1867, in Florence, the famous text was issued, known simply as *Betti-Brioschi* (the names of its authors), which offered a new translation of Euclid's *Elements* with notes and additions for secondary schools.[40]

Hoüel also took part in the debate caused by the publication in the *Giornale di Matematiche* of the Italian translation of J.M. Wilson's paper, *Euclid as a textbook of elementary geometry,* published in the *Educational Times* in 1868.[41] Wilson pointed out the deficiencies of the *Elements* in both scientific and didactic terms, concluding peremptorily that "Euclid is antiquated, artificial, unscientific and ill-adapted for a textbook."[42] Understandably, Cremona and Brioschi reacted rather violently to this article, writing a joint letter[43] to the editor, Giuseppe Battaglini, which appeared in the next issue of the *Giornale*. In their letter they tried to confute Wilson's criticisms, but they were not really convincing; in fact they concluded their article by admitting the defects of Euclid's *Elements* and by saying that they should be revised, but not distorted.

But Hoüel himself, who had been concerned with this subject for some time[44] and who was both a friend and a collaborator of Battaglini's, considered it his duty to intervene in defense of Euclid, and sent a letter to the *Giornale* in which he stated: "I could say a great deal about the proofs Mr. Wilson gives that Euclid is *antiquato, artifizioso, illogico(!!!) e inadatto come libro d'istituzione*. *Antiquato,* he may be: I have said so myself on occasion. *Artifizioso,* no more so than three-quarters of modern works. But *illogico,* I deny it, and I believe he seems so only to those who have not completely understood him.... As to being *inadatto come libro d'istituzione,* yes and no" and he goes on to say that the *Elements* is unsuitable as a textbook "if you follow the English system of making students learn Euclid by heart without explanations."[45]

In the course of his correspondence with Hoüel, Beltrami often mentions this debate, and his own opinion is clear from the following passage: "In mathematics, on the current question, that is to say the usefulness of the Euclidean method (recommended for some years), opinion is divided. Some teachers are comfortable with it, and think it good and useful; others prefer prior methods, which can be summed up definitively in the Geometry of Legendre. I thought, however, that I could note that these last belonged to the class of followers of routine, that is to say those who ask nothing better than to reduce teaching to a stereotype. Besides, there has been the usual phenomenon of *elementary papers* written by people who really need to study the classic papers, and whom I have often wanted to remind of our Giusti's extraordinary epigram:

"Write a book? don't trouble to do it
If your readers will learn nothing through it."[46]

Note

For a history of non-Euclidean geometry, see, for example, H.S.M. Coxeter, *Non-Euclidean geometry,* The Mathematical Association of America, 1998 (1st ed., University of Toronto Press, 1942) and B.A. Rosenfeld, *A history of non-Euclidean geometry: Evolution of the concept of a geometric space,* Springer-Verlag, Heidelberg 1988 (Original Russian ed. 1976).

Further details of the pseudosphere can be found in F. Schilling, *Die Pseudosphäre und die nicht-Euclidische Geometrie,* Teubner, Leipzig 1931.

Endnotes

[1] "Nella scienza matematica il trionfo di concetti nuovi non può mai infirmare le verità già acquisite: esso può soltanto mutarne il posto o la ragion logica, e crescerne o scemarne il pregio e l'uso. Né la critica profonda dei principi può mai nuocere alla solidità dell'edificio scientifico, quando pure non conduca a scoprirne e riconoscerne meglio le basi vere e proprie," E. Beltrami, "Saggio di interpretazione della geometria non euclidea," *Giornale di Matematiche* 6 (1868), 284–312, quotation on p. 284; *Opere Matematiche,* 4 vols., 1902–1920, Milano Hoepli, I, 374–405.

[2] Apropos of this cf. P. Calleri, L. Giacardi, "Le lettere di Giuseppe Battaglini a Jules Hoüel (1867–1878). La diffusione delle geometrie non euclidee in Italia," *Rivista di Storia della Scienza* (2) 3(1) (1995), 127–209.

[3] J. Bertrand, "Sur la somme des angles d'un triangle," *Comptes Rendus Hebd. des Séances de l'Académie des Sciences de Paris* 69 (1869) 1265–1269, quotation on p. 1266.

[4] P. Barbarin, "La correspondance entre Hoüel et de Tilly," *Bulletin des Sciences Mathématiques* (2) 50 (1926), 50–64 e 74–88, quotation on p. 78.

[5] "Non so se Ella abbia accordato alcuna attenzione a quel sistema di idee che ora si va divulgando col nome di geometria non-euclidea. So che il Prof. Chelini gli è decisamente avverso, e che il Bellavitis la chiama geometria da manicomio: mentre il Cremona lo crede discutibile ed il Battaglini lo abbraccia senza reticenze," letter from Beltrami to Tardy, 14 November 1867, *Cassetta Loria,* Genoa University Library.

[6] These letters are all preserved in the Archives of the Académie des Sciences of Paris, with the exception of one which is in the Bibliothèque of the Institut de France; the critical edition of the letters in L. Boi, L. Giacardi and R. Tazzioli, *La découverte de la géométrie non euclidienne sur la pseudosphère: Les lettre d'Eugenio Beltrami à Jules Hoüel (1868–1881),* Paris, Blanchard, 1998.

[7] K.F. Gauss, "Disquisitiones generales circa superficies curvas," *Commentationes Societatis Regiae Scientiarum Gottingensis Recentiores* 6 (1828), 99–147; *Werke,* 12 vols., 1863–1933, Göttingen Königlichen Gesellschaft der Wissenschaften, IV, 217–258.

[8] E. Beltrami, "Risoluzione del problema: riportare i punti di una superficie sopra un piano in modo che le linee geodetiche vengano rappresentate da linee rette" (Solution of the problem: represent the points of a surface on a plane in such a way that the geodesics are represented by straight lines), *Annali di Matematica Pura ed Applicata* 7 (1865), 185–204; *Opere* I, 262–280.

[9] B. Riemann, "Über die Hypothesen, welche der Geometrie zu Grunde liegen," Habilitationsvortrag (1854), Göttingen, published for the first time in *Abhandlungen der K. Gesellschaft der Wissenschaften zu Göttingen* 13 (1868), 133-152; cf. also *Bernhard Riemann gesammelte Werke, wissenschaftlicher Nachlass und Nachträge* (ed. R. Narasimhan) 1990 Leipzig Teubner, 304–319 (first ed.1876, second ed. 1892).

[10] "L'anno scorso, quando nessuno sapeva di questo lavoro fondamentale di Riemann, io aveva comunicato all'ottimo Cremona un mio scritto nel quale davo un'interpretazione della planimetria non-euclidea, che mi sembrava soddisfacente," letter from Beltrami to Genocchi, Bologna, 23 July 1868, *Fondo Genocchi, Busta M,* Municipal Library of Piacenza.

[11] "Ce qui m'étonne c'est qu'ayant eu bon nombre d'entretiens avec Riemann (dans les deux années qu'il a passé à Pise, peu avant sa fin regrettable) il ne m'ait jamais parlé de ces idées, qui ont dû cependant l'occuper assez longtemps, car une belle esquisse ne peut pas être l'oeuvre d'un seul jour, même pour un aussi beau genie," letter from Beltrami to Hoüel, Bologna, 4 December 1868. For all Beltrami's letters to Hoüel, see Boi, Giacardi, Tazzioli quoted in note 6.

[12] "L'ensemble de mes déductions repose sur la représentation des surfaces par la formule de Gauss $ds^2 = E\,du^2 + 2F\,du\,dv + G\,dv^2$. Or, dans cette méthode, les rapports de la surface et de l'espace environnant échappent entièrement: la surface est considérée en elle-même, telle qu'elle le serait par un être qui ne eût pas le sens de la troisième dimension," letter from Beltrami to Helmholtz, Bologna, 24 April 1869, cf. L. Koenigsberger, *Hermann von Helmholtz,* 3 vols., 1902–1903, Braunschweig Friedrich Vieweg und Son, II p. 154.

[13] "Non tamquam limes solidi, sed tamquam solidum, cuius dimensio una pro evanescente habetur, flexibile quidem, sed non extensibile," Gauss, *Werke* IV, quoted in note 7, Art. 13, p. 238.

[14] Cf. for example Beltrami's letters to Hoüel, Pordenone, 12 October 1869 and Bologna, 19 December 1869.

[15] "Il me semble que cette doctrine n'a pas trouvé généralement sa complète *Würdigung*, à tel point que personne n'a encore remarqué ce fait capital, savoir qu'*elle est entièrement indépendante du postulat d'Euclide*," letter from Beltrami to Hoüel, Bologna, 19 December 1869.

[16] F. Minding, "Wie sich entscheiden lässt ob zwei gegebene krumme Flächen auf einander abwickelbar sind oder nicht; nebst Bemerkungen über die Flächen von unveränderlichem Krümmungsmasse" (How to decide whether two curved surfaces are mutually applicable; including remarks on surfaces of constant measure of curvature), *Journal für die reine und angewandte Mathematik* 19 (1839), 370–387.

[17] F. Minding, "Beiträge zur Theorie der kürzesten Linien auf krummen Flächen" (Contributions to the theory of the shortest curves on curved surfaces), *Journal für die reine und angewandte Mathematik* 20 (1840), 323–327.

[18] Cf. Beltrami *Saggio*, quoted in note 1, p. 287ff.

[19] D. Codazzi, "Intorno alle superficie le quali hanno costante il prodotto dei due raggi di curvatura" (On those surfaces which have the product of the two radii of curvature constant), *Annali di Scienze Matematiche e Fisiche* 8 (1857), 346–355.

[20] The three different expressions of the linear element given above are obtained as follows. If on a surface of revolution we assume as coordinates u, v the arc of the meridian and the longitude, the linear element is given by

$$ds^2 = du^2 + r^2\,dv^2$$

where r is the radius of the parallel and is a function only of u. Then the curvature K of the surface will be

$$K = -\frac{1}{r}\frac{d^2r}{du^2};$$

supposing it to be constant and negative, i.e.,

$$K = -\frac{1}{R^2},$$

we have

$$\frac{d^2r}{du^2} = \frac{r}{R^2}.$$

By integrating we obtain

$$r = \gamma_1 e^{u/R} + \gamma_2 e^{-u/R},$$

the equation which represents the most general meridian curve of a pseudospherical surface of revolution. Depending on whether the arbitrary constants γ_1 and γ_2 are of the same or opposite signs, or whether one of them is zero, we have three types of meridian lines and therefore of pseudospherical surfaces of revolution, the hyperbolic, elliptic and parabolic types. Beltrami's symbolism has been preserved in the text.

[21] Cf. Beltrami *Saggio*, p. 303.

[22] "Mais nous, ne pouvons pas, dans notre espace, construire une surface pseudosphérique indéfiniment étendue dans la direction de l'axe de révolution: nous arrivons toujours, soit á une limite comme dans le cas du verre à champagne, soit à deux limites comme dans le cas de l'anneau," H. Helmholtz, "Les axiomes de la géométrie," *Revue des Cours Scientifiques de la France et de l'Étranger* 7 (1870), 498–501, quotation on p. 499.

[23] "Fournir l'intuition du plan tout entier, les surfaces de courbure negative constante étant toujours limitées par des arêtes de rebroussement," F. Klein, "Sur la géométrie dite non-euclidienne," *Bulletin des Sciences Mathématiques et Astronomiques* 2 (1871), 341–351, quotation on p. 345.

[24] "Il faudrait démontrer que l'équation aux dérivées partielles qui exprime les surfaces á courbure constante admet au moins une intégrale satisfaisant à toutes les conditions requises pour la pseudosphère," A. Genocchi, "Sur un mémoire de Daviet de Foncenex et sur les géométries non euclidiennes," *Memorie della R. Accademia delle Scienze di Torino* (2) 29 (1878), 365–404, cf. Appendice, p. 395.

[25] "J'ai dit, dans le *Saggio*, que la surface pseudosphérique, en tant qu'elle est représentée par les variables u, v est *indéfinie* en tous les sens, et *simplement connexe* (einfach zusammenhängende, suivant Gauss et Riemann), et cela est parfaitement vrai. Mais, comme l'on ne connaît pas l'intégrale *générale* de cette surface en coordonnées ordinaires x, y, z (intégrale que l'analyse actuelle est probablement bien éloignée de pouvoir donner), et que, par suite, on ne connaît pas la forme *la plus générale* de cette surface, on ne peut pas démontrer, *a priori*, qu'elle peut exister dans l'espace *ordinaire* au double état d'*indéfinité en tous les sens et de simple connexion*," letter from Beltrami to Hoüel, Bologna, 2 January 1870, cf. also the letter from Beltrami to Helmholtz, Bologna, 24 April 1869.

[26] D. Hilbert, "Über Flächen von konstanter Gausscher Krümmung" (On the surfaces of constant Gaussian curvature), *Transactions of the American Mathematical Society* 2 (1901), 87–99.

[27] A. Cayley, "A sixth memoir upon quantics," *Philosophical Transactions of the R. Society of London* 149 (1859), 61–90; *The Collected Papers*, 14 vols., 1889–1898, Cambridge University Press, II, 561–592.

[28] F. Klein, "Über die sogenannte nicht-Euklidische Geometrie" (On the so-called non-Euclidean geometry), *Nachrichten von K. Gesellschaft der Wissenschaften zu Göttingen* 17 (1871), 419–433; *Gesammelte mathematische Abhandlungen*, 3 vols, 1921–1923, Berlin Springer, I, 244–253.

[29] Beltrami is referring here to Hoüel's French translation of the *Attempt*, and suggests two possible additions.

[30] "La seconde serait plus importante, si je parvenais à lui donner une forme concrète, car elle n'existe jusqu'ici dans ma tête qu'à l'état de conception assez vague, quoique sans doute fondée dans le vrai. C'est la conjecture d'une étroite analogie, et peut-être identité, entre la géométrie pseudosphérique et la théorie de M. Cayley sur *l'origine analytique des rapports métriques*, à l'aide de la conique (ou de la quadrique) absolue," letter from Beltrami to Hoüel, Bologna, 29 July 1869.

[31] "Je regrette beaucoup de m'être laissé prévenir par M. Klein sur ce point, sur lequel j'avais déjá rassemblé des matériaux, mais auquel j'ai eu le tort de ne pas donner assez d'importance. Au reste cette manière de voir n'est pas absolument nouvelle, et c'est précisément à cause de cela que je n'ai pas eu hâte de publier ma remarque. Elle se rattache intimément à une observation déjà

ancienne de M. Chasles, concernant l'angle de deux droites considéré comme le logarithme d'un rapport anharmonique ... et à un théorème de M. Laguerre Verlay," letter from Beltrami to Hoüel, Bologna, 5 July 1872, cf. also the letter to Enrico D'Ovidio, 25 December 1872, published in E. D'Ovidio, "Eugenio Beltrami. Commemorazione," *Atti della R. Accademia delle Scienze di Torino* 35 (1899–1900), 3–8, on pp. 6–8.

[32] "Di preparare gli elementi geometrici di una costruzione materiale, possibilmente facile ed esatta della superficie stessa," E. Beltrami, "Sulla superficie di rotazione che serve di tipo alle superficie pseudosferiche" (On the revolution surface which serves as the type for pseudospherical surfaces), *Giornale di Matematiche* 10 (1872), 147–159, quotation on p. 147; Opere II, 394–409.

[33] "J'ai eu, dans cet intervalle, une idée bizarre, que je vous communique ... J'ai voulu tenter de construire matériellement la surface pseudosphérique, sur la quelle se réalisent les theorèmes de la géométrie non-euclidienne ... Si ... on considère la surface comprise entre deux méridiens, assez rapprochés pour qu'on puisse la remplacer, sur une certaine longueur, par un plan, on peut, par de morceaux de papier, convenablement decoupés, reproduire les trapèzes curvilignes dont la surface véritable peut être censée se composer," letter from Beltrami to Hoüel, Bologna, 13 March 1869.

[34] This model and how to make it are described in detail in Beltrami's letter to Hoüel, Bologna, 22 April 1869; cf. also A.C. Capelo, M. Ferrari, "La 'cuffia' di Beltrami, storia e descrizione," *Bollettino di Storia delle Scienze Matematiche,* 2 (2) (1982), 233–247.

[35] Cf. on p. 298 of the *Saggio.*

[36] "Que l'on trace une droite AB et qu'en chacun de ses points M, on tire la droite MT qui marque la direction de la parallèle à AE, suivant Lobatcheffsky, par rapport à la distance AM. *L'enveloppe de ces droites est le méridien de la surface pseudosphérique.* Par suite la distance MN au point de contact est constante," letter from Beltrami to Hoüel, Bologna, 13 March 1869; cf. also E. Beltrami, "Teorema di geometria pseudosferica" (Theorem of pseudospherical geometry), *Giornale di Matematiche* 10 (1872), 53; *Opere* II, 392–393.

[37] "Le nombre des professeurs des lycées qui s'occupent de leur science est très-petit chez nous: et même ceux qui s'en occupent n'ont pas le sentiment de cette solidarité didactique et scientifique qu'il y a en Allemagne et qui procure des lecteurs attentifs à tout article un peu sérieux," letter from Beltrami to Hoüel, Rome, 5 January 1875.

[38] Cf. *Gazzetta Ufficiale del Regno d'Italia,* 11 November 1884.

[39] "Un mezzo di coltura generale, una ginnastica del pensiero diretta a svolgere la facoltà del raziocinio e ad aiutare quel giusto e sano criterio che serve di lume per distinguere il vero da ciò che ne ha soltanto l'apparenza," cf. the Supplement to the *Gazzetta Ufficiale del Regno d'Italia,* 24 October 1867. Apropos the debate arising from the Coppino Act, cf. L. Giacardi, "Gli Elementi di Euclide come libro di testo. Il dibattito italiano di metà Ottocento," in E. Gallo, L. Giacardi and S. Roero (eds.), *Conferenze e Seminari 1994-1995,* Mathesis Subalpina e Seminario T.Viola, Turin 1995, pp. 175–188.

[40] E. Betti, F. Brioschi, *Gli Elementi d'Euclide con Note, Aggiunte ed Esercizi ad uso de' Ginnasi e de' Licei,* Florence, Successori Le Monnier, 1867.

[41] J.M. Wilson, "Euclide come testo di Geometria Elementare," *Giornale di Matematiche* 6 (1868), 361–368.

[42] From the specifically didactic point of view, Wilson's criticisms focus on the following points:

- "the *expression* is much more prominent than the *substance*: the geometrical facts are generally simple, but are disguised and overlaid by the formality of the diction,"
- "Euclid is *unsuggestive* ... The object of education is to give power; and if the result of teaching Euclid is in most cases only a certain degree of readiness in reproducing Euclid's proofs, without the power of solving problems, the education has been to a great extent a failure,"
- "the progress of geometry has been uninterrupted improvement; and it is simply incredible, *a priori,* that any textbook written so long ago could be a fit introduction to the science at the present time,"
- "the exclusion of the arithmetical applications from our geometrical exercises is a loss," J.M. Wilson, "Euclid as a textbook of elementary geometry," *Educational Times* (1868), 125–128, quotations on pp. 127, 128.

[43] "Presso di noi, l'introduzione dell'Euclide nelle scuole ha reso un altro grandissimo servigio: quello di sbandire innumerevoli libercoli, compilati per pura speculazione, che infestano appunto quelle scuole dove è maggiore pei libri di testo il bisogno del rigore scientifico e della bontà del metodo" (In our country the introduction of Euclid in schools has done us another great service: it has done away with innumerable worthless books, compiled by mere speculation, which in fact infest those schools where there is the greatest need of scientific rigour and good method in the textbooks), F. Brioschi, L. Cremona, "Al Direttore del Giornale di Matematiche ad uso degli studenti delle Università italiane. Napoli," *Giornale di Matematiche,* 7 (1869), 51–54. Part of this letter was translated by Hoüel into French, and published in the *Nouvelles Annales de Mathématique,* (2) 7 (1869), 278–283, under the title *L'enseignement de la géométrie élémentaire en Italie.*

[44] Cf. J. Hoüel, "Essai d'une exposition rationelle des principes fondamentaux de la géométrie élémentaire," *Archiv der Mathematik und Physik* 40 (1863), 171–211 e *Essai sur les principes fondamentaux de la géométrie élémentaire ou Commentaire sur les XXXII premières propositions des Eléments d'Euclide,* 1867, Paris Gauthier-Villars, second ed.1883.

[45] "J'aurais beaucoup de choses à relever dans les preuves que donne M. Wilson de ce qu'Euclide est *antiquato, artifizioso, illogico (!!!) e inadatto come libro d'istituzione. Antiquato,* soit: je l'ait dit moi-même à l'occasion. *Artifizioso,* pas plus que les trois quarts des ouvrages modernes. Mais illogico, je le nie, et je prétends qu'il ne le paraît qu'à ceux qui ne l'ont pas compris entièrement ... Quant à être *inadatto come libro d'istituzione,* oui et non" and he goes on to say that the *Elements* is unsuitable as a textbook "si l'on suit le système anglais consistant à faire apprendre Euclide par coeur sans l'expliquer," J. Hoüel, "Estratto di una lettera del Prof. Hoüel al redattore," *Giornale di Matematiche,* 7 (1869), 50; cf. also Hoüel's letter to Cremona, Bordeaux 3 February 1869, in L.Giacardi, "La corrispondenza fra Jules Hoüel e Luigi Cremona (1867–1878)," *Quaderni della Rivista di Storia*

della Scienza, n.1 (1992), 77–94, see pp. 81–84 and Battaglini's letter to Hoüel, Naples, 2 February 1869, in Calleri, Giacardi, "Le lettere di Giuseppe Battaglini...," quoted in note 2.

[46] "En fait de mathématiques, sur la question à l'ordre du jour, c'est-à-dire sur l'utilité de la méthode euclidienne (prescrite depuis quelques années), les avis sont partagés. Quelques professeurs s'en trouvent bien, et la croient bonne et utile; d'autres lui préfèrent les méthodes antérieures, qui se résument en définitive dans la Géométrie de Legendre. J'ai cru cependant pouvoir remarquer que ces derniers appartiennent à la classe des routiniers, c'est-à-dire de ceux qui ne demandent mieux que de stéréotyper l'enseignement. Il y a eu du reste le phénomène ordinaire des *traités élémentaires* compilés par des gens qui auraient bien besoin d'étudier les traités classiques, et auxquels j'ai eu souvent envie de rappeler le formidable epigramme de notre Giusti:

Il fare un libro è meno che niente
se il libro fatto non rifà la gente,"

letter from Beltrami to Hoüel, Bologna, 12 June 1870. I am grateful to R.A. Henderson for this translation of Giusti's rhyme.

NOTE TO PRINTER: BLANK PAGE

NOTE TO PRINTER: BLANK PAGE

A Window on the World of Mathematics, 1870

Reminiscences of Augustus De Morgan —a dramatic presentation

Gavin Hitchcock
University of Zimbabwe
Harare, Zimbabwe

Abstract. The figure of De Morgan as an old man in the last year of his life is evoked by means of a dramatic monologue constructed in the style of De Morgan's trenchant writings and colourful personality, incorporating extracts and paraphrases from his work in mathematics and education.

Our aim is to explore the possibilities of this dramatic form for presenting a vivid human perspective on mathematics as a living, growing subject. By sharing the vision of a great mathematician and educator, and his view of the achievements of his contemporaries, we attempt to capture the mood of that moment in mathematics in Britain—the excitement, the preoccupations, and the sense of intellectual community across national boundaries and personal rivalries.

Some themes of the presentation are: the influence of nationalism, elitism, sectarian prejudice, and the powerful institutions of the time, on the development of mathematics; the profound effects of the full acceptance of negative and complex numbers; the dawning vision of the nature and reach of abstract algebra; the late discovery of mathematical logic and its relation to mathematics; the role of the imagination in mathematical creativity; the joyful exuberance of the mathematical community living in the springtime of a liberated, truly 'pure' mathematics.

Introduction

If we describe an item of news as 'hot off the press', or an act of communication as issuing 'from the horse's mouth', we mean to imply immediacy, relevance, veracity and authenticity. Similarly, in the phrases 'first-hand information', 'eyewitness account', 'inside knowledge', a sense of value is conveyed; this thing is living truth—it is worth one's serious attention. In this article we exploit a time-honoured principle to serve the cause of mathematics education: we listen to an eyewitness account from a mathematical insider; we present a report from the man-on-the-spot. Augustus De Morgan was not only a great mathematician; he had a heart for communicating the excitement and fascination of mathematics, and he held strong views on education at all levels. It is unfortunately impossible to travel back in time and record an interview with him, but we can attempt to reconstruct what he might have said—and (much more) how he might have said it. For we are always being told by psychologists that tone of voice and body-language make up the greater part of any act of human communication, and maybe we have lost more than we realize in attempting to exempt mathematics from such subjectivity. While there may well be much to be gained from a sympathetic and imaginative reading of the text, this play is written primarily to be *played* before an audience, with appropriate passion, pathos and humour. Some suggestions for staging follow.

There is no need to commit the entire text to memory, as you can easily deceive an audience into believing you have memorized the lines by judicious use of a glass of water, a pipe or a cup of tea, and other objects apparently engaging your attention on the table in front of you. Some physical action will enhance the dramatic effect. For example, De Morgan may come in and sit down at the start, and make an exit during his last lines. He may walk across to bookshelves, or mantelpiece, at appropriate moments. It is not necessary to have great acting talent or experience. A few simple props can be very effective: a stick, a few photographs (of Trinity College, or people being discussed) on the wall or on the desk, and some books (his own, and those of George Boole).

We may distinguish three main motives in eavesdropping thus on the supposed reminiscences of Augustus De Morgan:

- to learn about De Morgan himself, and the community of contemporary mathematicians;
- to share De Morgan's perceptions about mathematics at that particular juncture (1870) in its development;
- to experience something of the communal adventure of mathematics-making, and so gain an insight into the nature of mathematics as an ongoing human enterprise.

All three of these represent realistic goals, in my opinion, even for a more general audience without knowledge of some of the technical terms used. If the piece is to be used to provide context and motivation for mathematics students, it is perhaps better presented when the students have already encountered some of the ideas and names. Areas of mathematics which are touched on here include: mathematical logic, Boolean algebra, negative and complex numbers, quaternions, abstract algebra, vector analysis, divergent series, and the theory of invariants. More general themes which occur include: the birth of pure mathematics (or the self-realization of mathematics), the relationship between mathematics and logic, the power of symbolism, and the nature of mathematical creativity.

This short play is offered here not only to be used in classes, but also to encourage others to explore the use of theatre in bringing new life to mathematics teaching. We close this introduction with a suggestion and some questions.

Consider assigning students projects of the following form:

Project. Write a monologue (it could take the form of a letter to a friend/student/parent/colleague) as from a certain mathematician in a certain year, expressing his/her viewpoint on where mathematics is going, what's been exciting in the recent past, what he/she is proud of, what he/she hopes to achieve, whom he/she is in touch with, what he/she has read. Alternatively, write (and enact) an imaginary interview with the mathematician.

Questions. How effective are such "windows on the world of mathematics" (either as student projects or as ready-made material for reading and enacting in class) in communicating the spirit of mathematics-making, and enlivening the teaching of mathematics and its history? How can these best be used at various levels of mathematics education? How can the window be rendered as clean and transparent as possible (for example, by up-dating archaic language) without compromising the authenticity of the view?

The Play

Scene: *Augustus de Morgan,*[1] *in 1870, during the last year of his life, aged 65, musing in his study.*

I think the most striking change in mathematics over my lifetime has been the joyous assertion of logical freedom! Our laws—whether of number, algebra or even geometry—are not absolute, not logically necessary after all. There are new geometries, new algebras, to explore—new entities, such as Sir William Hamilton's quaternions and Professor Arthur Cayley's matrices, obeying quite remarkable laws. And the way to all this was opened, I think, by the gradual acceptance of the negative numbers, closely followed by the imaginary numbers, as mathematicians began to realize the relative meaning of the terms "possible" and "impossible" or, indeed, the terms "real" and "imaginary"! It is human tradition, drawing upon the resources of human imagination, which sets the limits on the field of operation—which erects the fences and draws the horizons. We must, of course, ensure that any proposed law is logically *permissible*—that is, consistent with its fellows. Our structures are otherwise agreeably arbitrary, free creations of the human spirit, regulated by considerations of convenience and expediency (such as the principle of permanence), or by considerations of elegance or the desired applicability of the resulting theory.

This quality of freedom would have shocked the mathematicians of the last century. But they were nevertheless unconsciously preparing the way, as they were won over by the negative numbers and the imaginary numbers, and swept along by the exhilarating currents of symbolic algebra and analysis.

You know, I think the new vision of a pure, free mathematics really dawned on us in the year— 1847 it was—when I published my *Formal Logic,* and my friend George

Boole[2] published his wonderful little book on *The Mathematical Analysis of Logic*. I remember both books were published on the very same day! I recognized the prophetic voice at once. He was just a common boy from Lincoln, who was forced to leave school before he was sixteen, and taught himself Greek, Latin and mathematics. He became a school teacher to support his parents, brothers and sister, and eventually opened his own school. He first made a name for himself when his essay won the Royal Society's gold medal; he crowned his career by becoming a professor of mathematics at Queen's College in Cork, Ireland, and having a fine mathematical theory named after him! That's an honour few of us can hope for—I would be proud to have one law named for De Morgan![3]

It all started when he got himself heavily involved in a controversy some of us were having over the nature of logic. Sir William Hamilton (not Sir William Rowan Hamilton, the Irish one;—the *Scottish* one this time: he was a baronet and a philosopher)—he claimed that logic was the business of philosophers. [*chuckles*] Well, Boole and I showed that it was *our* business—not only can logic be used to increase the power of mathematical language in striking ways, but it can be treated as a branch of mathematics. We call it Mathematical Logic now, and the Algebra that Boole invented to be a kind of Calculus of Logic we now call Boolean Algebra! Some laws in this Algebra look very surprising at first; maybe that's why it took so long in coming! For example: [*writes*] $x + x = x$ and $x \cdot x = x$ for any x; what's more, it has no negatives. Boole thoroughly developed his Algebra in his next book, *An Investigation of the Laws of Thought*,[4] published seven years later, and no one will ever again be able to define mathematics as the science of number and magnitude![5] Benjamin Pierce, the American mathematician and philosopher, recently defined mathematics as *the science which draws necessary conclusions*. And Hermann Grassmann, the German mathematician, calls it *the science of forms*. It's the science of formal systems of rules operating with symbols ... it can be about anything, or nothing, like music—and it's just as beautiful!

Mathematics and logic—these two have had a curious relationship! In spite of being bound together forever by Boole's revolution, hardly anyone besides Jevons and myself can claim to have worked in both disciplines; the mathematicians and logicians live in two camps aloof from each other. The mathematicians care no more for logic than logicians for mathematics. Here's the irony of it: the two eyes of exact science are mathematics and logic. Boole's genius taught us to see with both of them; but the mathematical sect puts out the logical eye, and the logical sect puts out the mathematical eye, each believing that it sees better with one eye than two! No one knows better than I how great is their loss; not only do I have one foot in each camp, but I am blind in one eye![6]

Boole died a few years ago; he would have been delighted with some recent events. It seems the Germans have finally discovered our Peacock, a decade after his death, and the Irish Hamilton too. What a transformation has taken place since our Cambridge Analytical Society set out bravely to rescue British mathematics from the doldrums! Here we are, leading the world in algebra half a century later! I suppose old Peacock should be given the credit for getting the thing started—and Herschel and Babbage too; and then there was Robert Murphy, and Gregory, and of course Boole and myself.

And, I must say, the new generation promises to outshine us; the line of great British algebraists continues with Cayley, the brilliant young Clifford, Jevons, and of course Sylvester (my students, the last two). I can't resist mentioning the part my old College at Cambridge has played: the illustrious line of Trinity men includes Peacock, Cayley and Clifford. Sylvester was a St. John's man—but do you know that the University would not give Sylvester a degree, on the grounds that he was Jewish? Nor could he teach at Cambridge ... that sort of thing makes my blood boil. I have refused all these years to allow my name to be posted for fellowship in the Royal Society, for similar reasons. Fellows are supposed to be nominated on merit, but the process is too much open to social influences! At least they recognized Sylvester's genius—when he was only twenty-five they elected him a fellow.

I had Sylvester join me at University College for a few years; then we lost him briefly to America, where (they say) a man is accepted for himself. But—poor Sylvester, [chuckles] he seems to be ever at war with the world (those were his own words!)—he was not very happy over there for some reason, and felt obliged to come back to England, in, let me see—1843, I think it was. He then spent many years (like Cayley) in the wilderness of actuarial work and law. No one would have guessed that these two were really great mathematicians in exile, strolling through the Courts of Lincoln's Inn deep in mathematical conversation—together creating the beautiful theory of invariants! At last he became a professor of mathematics again, in a military academy—a post quite unworthy of him—and he was forcibly retired last year.

[*sighs deeply*] Ah—when will the great Universities of England honour such a man for his mathematical gifts, disregarding his birth and creed, and age?[7] My own convictions have seriously affected my mathematical profession twice—at the beginning and at the end of my career. After thirty years' service as the first Professor of Mathematics at University College, I felt constrained to resign. The council refused to appoint a good man to the chair of logic and philosophy on sectarian grounds. As I wrote in my letter to the chairman of council: "It is not necessary for me to settle when I shall leave the college; for the college has left me." And long ago, I found myself, like Sylvester, ineligible for a fellowship at my own University of Cambridge, because I could not in all conscience sign certain theological articles considered necessary to proceed to the Master's Degree. I described myself as "Christian unattached." God knows I have striven to be true to the light I have been given! At the end of my life, I can look back and affirm that I would do the same again. . . . Lest anyone misunderstand my motives and I cause such a one to stumble, I have caused to be written in my will this article of faith: "I commend my future with hope and confidence to Almighty God; to God the Father of our Lord Jesus Christ, whom I believe in my heart to be Son of God but whom I have not confessed with my lips, because in my time such confession has always been the way up in the world." [*chuckles*] That will set a gaggle of tongues wagging![8]

Well, now, where was I? Ah, yes—those Germans! We may still be in front, but they have been running hard to catch up, and they've done some remarkable work. It seems that this Hermann Grassmann[9] has proposed an abstract science of directed quantities in many dimensions. And Hankel has vindicated my suspicion that the "double algebra" of imaginary numbers constitutes the ultimate algebra in which the laws of arithmetic are preserved intact. No more general "arithmetical" algebra is possible! Peacock would like that! And my self-esteem is somewhat restored; it was not so very long ago that I refused to believe in triple or quadruple algebras of *any* kind. That was before Boole and Hamilton—and now there are Grassmann's algebras with exotic laws similar to Hamilton's quaternions and Cayley's matrices. Multiplication depends on order; $x \cdot y$ may not be equal to $y \cdot x$. Ah! The mathematical menagerie has an inexhaustible store of surprises!

It is truly astounding, when one comes to reflect upon it, how great a degree of unanimity we mathematicians have achieved over previously contentious issues. Not that we don't still have our petty differences, but I believe it would have been generally admitted, by about the middle of this century, that the only subject yet remaining (of an elementary character), on which a serious schism existed among mathematicians, as to the absolute correctness or incorrectness of results, was the question of divergent series.[10] And even those monstrous creatures are rapidly becoming domesticated and their somewhat embarrassing uses regarded as legitimate.

What I said back in the forties about the way we should react to anomalies and embarrassments has been proved true in striking ways. The history of algebra shows us that nothing is more unsound than the rejection of any method which naturally arises, merely because of one or more apparently valid cases in which such a method leads to erroneous results. Such cases should indeed teach *caution*, but not *rejection*. For if the latter had been preferred to the former, negative quantities, and still more, their square roots, would have been an effectual bar to the progress of algebra. And think of those immense fields over which even the rejecters of divergent series now roam without fear! Those fields would not even have been discovered, much less cultivated and settled.[11]

How singular, in retrospect, that the burning issue of the *reality* of negative numbers should have appeared a *logical* one, and turned out in the end as a victory, not for logic, but for the imagination! Babbage[12] saw it earlier than most of us in England, I think; and poor Peacock fought for the *logical* status of his Principle of Permanence, only to see it become a handmaid to the imagination! The realization has dawned slowly, but is now clear to all of us: the moving power of mathematical invention is not reasoning but imagination![13]

The great difficulty of the opponents of algebra—the so-called "pure arithmeticians"—lay in a lack of ability or will to accept *extension* of terms. They refused to admit any use of symbols which outstrips the limits of absolute number. They would forbid all extension of language,[14] and so cut themselves off from one of the great creative forces of the imagination, which is operative in all poetry and great literature: to allow the words, the symbols, to carry one beyond oneself!

Perhaps this sect is extinct now. During the last century, its chief writers were Robert Simson, Francis Masères and William Frend. So far as these opponents [of negative numbers and symbolic algebra] set out their objections, it can be seen that there is much force in them against the mode of elementary writing then in vogue. I was casually brought into contact with Mr Frend[15]—he later became my father-in-law, of course—but this first contact was in early life at Cambridge, at a time when I was engaged in examination of the first principles of everything mathematical. Having had many discussions with him, and been led thereby to an attentive examination of Masères, Simson, and others, I long ago came to the conclusion that there is a very strong bias in the minds of such people, who are thus irresistibly led to a sweeping condemnation of almost everyone else, on matters of subjective nature. The bias is a craving for simplicity—a craving that will, in the end, find a way of rejecting whatever cannot be immediately reduced to earliest axioms. A very unfortunate state of mind ... But I suspect that even those opponents played a useful part in the strange story of algebra—by goading others to defend and analyse their own principles.[16]

A strange story indeed!—where will algebra go in the future, I wonder? How will students of the twentieth century be taught these ideas that have been forged in such creative fires of the human imagination? Will they simply take them for granted, unquestioning, unmoved by the triumphs of previous generations of mathematicians?

[*EXIT with aid of walking stick*]

Endnotes

[1] De Morgan was active until the autumn of 1870, although in poor health, and he died on 18 March, 1871. More about the man and his writings can be found in his biography, written by his widow: Sophia Elizabeth De Morgan, *Memoir of Augustus De Morgan,* (London: Longmans, Green, 1882). See also H. Pycior, "Augustus De Morgan's algebraic work: the three stages," *Isis,* 74 (1983), pp. 211–226; Ivor Grattan-Guinness, "An eye for method: Augustus De Morgan and mathematical education," *Paradigm,* no. 9 (December 1992), pp. 1–7; there are articles on him in the *Dictionary of National Biography,* the *Dictionary of Scientific Biography* and the *Encyclopaedia Britannica.* See also Joan Richards, *Isis,* 78 (1987), pp. 7–30.

[2] For an insight into the relationship between Boole and De Morgan, as well as a chronological list of De Morgan's papers and books, see: G. Smith (ed.), *The Boole-De Morgan correspondence 1842–1864,* (Oxford: Clarendon Press, 1982). For an excellent biography of George Boole (1815–1864) see: Desmond MacHale, *George Boole: his life and work,* (Dublin: Boole Press, 1984). See also Eric Temple Bell: *Men of Mathematics* (New York: Simon & Schuster, 1937), and William Kneale: "Boole and the revival of logic," *Mind,* Vol. 57, April 1948, pp. 149–175.

[3] The name De Morgan is best known today for "De Morgan's Laws". He originally gave them in the form: "the contrary of an aggregate is the compound of the contraries of the components; the contrary of a compound is the aggregate of the contraries of the components." Augustus De Morgan, *Trans. Camb. Phil. Soc.,* 10 (1858), pp. 173–230. In the notation of symbolic logic:

$1-(x+y)=(1-x)(1-y)$ and $1-xy=(1-x)+(1-y)$

In the language of sets:

$$S - \cup_i A_i = \cap_i (S - A_i)$$

$$S - \cap_i A_i = \cup_i (S - A_i)$$

[4] George Boole, *An Investigation into the Laws of Thought, on Which are Founded the Mathematical Theories of Logic and Probabilities,* 1854. (New York: Dover Publications, 1953); also as Vol. 2 of *Logical Works* (La Salle, Ill.: Open Court Pub. Co., n.d.)

Bertrand Russell, the great twentieth century mathematician and philosopher (not normally given to exaggeration!), has ascribed to Boole the greatest discovery of the nineteenth century—the discovery of "pure mathematics". Albert Einstein and Russell himself have each described memorably what this means:

Einstein: Insofar as mathematics speaks about reality it is not necessary [i.e., a logically necessary consequence of axioms], and insofar as it is necessary, it does not speak about reality.

Russell: Mathematics is the subject in which we do not know what we are talking about, nor whether what we are saying is true.

[5] However, common usage, as represented by the Oxford English Dictionary, has not reflected this transformation, as some twentieth century definitions of "mathematics" illustrate:

- *Science of space and number in the abstract* (OED Pocket ed., 1924);
- *Originally the collective name for geometry, arithmetic, algebra, etc. (pure mathematics), and in a wider sense those branches of research which consist in the application of this abstract science to concrete data (applied or mixed mathematics)* (OED, The Shorter, 1936).
- *Abstract science of space and quantity* (OED Concise, 1964).

[6] Augustus De Morgan, "Review of a book on Geometry," *The Athenaeum* (1868), Vol.2, pp. 71–73. Boole and De Morgan derived mutual stimulus from each other, mutually acknowledged. De Morgan ascribed to "Dr. Boole's genius" the "most striking results ... in increasing the power of mathematical language," and the binding together of the "two great branches of exact science, Mathematics and Logic." Sophia De Morgan, *Memoir,* as in ref. 1, p.167. As to whether the mathematicians and logicians remain aloof from each other in the late twentieth century, here is an extract from the Preface to a popular text written, not only for intending logicians, but for mathematicians in general: "Every mathematician must know the conversation-stopping nature of the reply he gives to an enquiry by a non-mathematician about the nature of his business. For a logician in the company of mathematicians to admit his calling is to invite similarly blank looks, admissions of ignorance, and a change in the topic of conversation. The rift between mathematicians and the public is a difficulty which will always exist (though no opportunity should be missed of narrowing it), but the rift between logicians and other mathematicians is, in my view, unnecessary." A. G. Hamilton, *Logic for Mathematicians* (Cambridge: Cambridge University Press, 1978; revised 1988).

[7] James Joseph Sylvester (1814–1897) was given his degrees at last, *honoris causa,* when the offending prescription was revoked in 1871. As if to recompense this colourful mathematician for his protracted struggle, he came into his own in later life. He returned to America, where he took up a post at Johns Hopkins University (1876–1884), played a major role in initiating pure mathematical research in the United States, and founded the *American Journal of Mathematics.* See Karen H. Parshall & David E. Rowe, *The Emergence of the American Mathematical Research Community 1876–1900,* (Providence, RI: American Mathematical Society, London: London Mathematical Society, 1994). He finally became Savilian Professor of Mathematics at Oxford University in the mid 1880's, holding the post until his death. For a fascinating account of the lives and work of Cayley and Sylvester,

see Bell, *Men of Mathematics,* as in ref. 2. Sylvester, in a wry moment, once remarked "that they both lived as bachelors in London, but that Cayley had married and settled down to a quiet and peaceful life at Cambridge; whereas he had never married, and had been fighting the world all his days:" Alexander Macfarlane, *Lectures on Ten British Mathematicians* (New York: 1916), p.66. (This includes George Peacock among the ten.) Clifford, in contrast, attended King's College, London, 1860–3 and Trinity College, Cambridge, until 1867; he was soon appointed Professor of Applied Mathematics and Mechanics at De Morgan's long-time institution University College, London, in 1871—the year De Morgan died.

[8] Some of the remarks of De Morgan on matters of conscience and principle are taken from James Roy Newman, "Commentary on Augustus De Morgan," in James Roy Newman, ed. *The World of Mathematics,* 4 vols. (New York: Simon & Schuster, 1956; paperback, 1962), pp. 2366–2368 (vol. 4); this draws from articles on De Morgan in the *Dictionary of National Biography* and the *Encyclopaedia Britannica.*

[9] Hermann Gunther Grassmann (1809–1877) was a high school teacher at Stettin in Germany. He published his *Die Lineale Ausdehnungslehre* (The Calculus of Extension) in 1844. There appears to be a mythology surrounding this work to the effect that it was difficult to read, and that this accounts for the fact that Grassmann's novel ideas did not become widely known until some time after he published them in revised and simplified form in 1862. However, some historians believe that he was simply ahead of his time.

[10] The paragraph up to this point is based upon Augustus de Morgan, *Trans. Camb. Phil. Soc.* 8, Part II (1844), pp. 182–203; pub. 1849.

[11] This paragraph is taken largely from Augustus De Morgan, *The Differential and Integral Calculus* (London: Society for the Diffusion of Useful Knowledge, 1842), p. 566.

[12] Charles Babbage, "On the influence of signs in Mathematical reasoning," (*Proc. Camb Phil. Soc.,* c. 1827). Babbage's espousal of the importance of convenience, over logical necessity, in the framing of mathematical laws, is not surprising. He was an essentially practical man, devoting most of his life to the design and construction of a series of mechanical calculating machines, with the idea of aiding the production of mathematical tables. He resigned his chair at Cambridge, after 11 years as Lucasian Professor of Mathematics, in order to devote all his energies to his great project. Although Babbage's work on his difference machine and his analytic engine did not reach satisfactory conclusion in his lifetime, due to severe financial and technical constraints, his prophetic vision and practical laying of the groundwork give us good reason to call him "Father of the Computer". Throughout his life, he demonstrated the importance and power of the application of pure science to the work-a-day world. He became heavily involved in the economic functioning of the Post Office, as well as the pin-making industry and the printing trade.

[13] The last assertion is quoted by Morris Kline in *Mathematics in Western Culture* (New York: Oxford University Press, 1953). The quote appears on page 170 of the Pelican edition.

[14] Up to this point the paragraph is drawn from Augustus De Morgan, "On Infinity and On the Sign of Equality," *Trans. Camb. Phil. Soc.* XI, Part 1 (1864), footnote on p.38.

[15] William Frend (1757–1841) was a Fellow of Jesus College, Cambridge, and was mathematics Tutor to the University until he was dismissed in 1788, for propagating heretical theological views. For similar offences in 1793 he was put on trial in the University Vice-Chancellor's Court, and banished from University and College residence when he refused to retract. He subsequently practised as an Actuary for the Rock Life Assurance Company. With his henchman Francis Masères (1731–1824), lawyer and constitutionalist, Frend fought a bitter, rearguard action against the evils of symbolical algebra, fictitious and imaginary numbers, priding himself on being a "pure arithmetician and a noted oppugner of all that distinguishes Algebra from Arithmetic". For a good biography, see Freda Knight, *University Rebel: The Life of William Frend, 1757–1841* (1971).

[16] This paragraph is based upon De Morgan, "On Infinity."

Some Notes on the History of Mathematics in Portugal

António Leal Duarte, Jaime Carvalho e Silva, João Filipe Queiró
Universidade de Coimbra, Portugal

In the last eight to ten years, following the enthusiasm around the commemorations of the bicentenary of the death of José Anastácio da Cunha, there has been a growing interest in the history of mathematics in Portugal. This has led to the organization of several meetings, to the publication of a few papers, and above all to the creation of a National Seminar on the History of Mathematics, which has been meeting once a year.

We present here a brief summary of the three main periods in which the history of mathematics in Portugal may be divided, with emphasis on some points which have deserved, and should go on deserving, the attention of mathematicians. We conclude with some remarks on the historiography of mathematics in Portugal.

1. Mathematics in Portugal before 1772

In this period, and in fact in the whole History of Mathematics in Portugal, the most important name is Pedro Nunes (1502–1578), the first Professor of Mathematics in the University after its definitive transfer to Coimbra in 1537. Nunes' writings have been analyzed and commented on by Portuguese and foreign authors and, although further studies and researches are needed, it is possible today to have a clear picture of the significance of the work of the great Portuguese mathematician.

We describe the main works of Pedro Nunes:

1) In a sequence of studies, which culminated in *De arte atque ratione navigandi* (Opera, Basel, 1566), Pedro Nunes, following a query by sea-captain Martim Afonso de Sousa returning from an expedition to Brazil, made clear that rhumb lines—that is, the courses followed when maintaining a constant angle with the compass needle—are not geodesics (arcs of great circles). He understood their true nature; apart from trivial cases (meridians and parallels) in which they are circular, rhumb lines are spiral curves approaching the poles and winding an infinite number of times around them.

2) In one of his works dealing with rhumb lines, the *Tratado em defensam da carta de marear* (Lisbon, 1537), Pedro Nunes stated two desirable properties for maps: angle preservation and the representation of rhumb lines by straight lines. These requirements are precisely what made Mercator's 1569 great world map so useful in navigation. A possible influence of Pedro Nunes on Mercator has been in the past a matter of controversy, which deserves further study.

3) The book, *De erratis Orontii Finoei* (Coimbra, 1546; Basel, 1592), contains a list of sharp and detailed corrections to two works by the French mathematician Oronce

Fine (1494–1555). In those works Fine presented "solutions" to several classical problems, including the duplication of the cube, the quadrature of the circle, the construction of regular polygons (all completely solved only in the nineteenth century) and even the determination of longitude. Nunes severely criticizes these "solutions".

4) One of Pedro Nunes' major works is the *Libro de Algebra en Arithmetica y Geometria* (Antwerp, 1567). An expert on sixteenth century algebra, H. Bosmans, studied this book in detail some decades ago. The central subject is the solution of equations, mainly of first, second and third degree. A distinctive feature is the abstraction and generality with which theories and problems are presented. Pedro Nunes uses literal notation in his rigorous algebraic proofs, and reasonings with letters are independent of geometrical considerations. Operations with polynomials are studied. An important work of transition before Viète's notational advances (Bosmans calls Pedro Nunes "one of the most eminent algebraists in the sixteenth century"), the *Libro de Algebra* was well known and often quoted in Europe (by Wallis among others). It was translated into French and Latin, but these versions remained unpublished.

5) In *De Crepusculis* (Lisbon, 1542; Coimbra, 1571; Basel, 1573) Pedro Nunes studied—in answer to a question by Prince Henrique—"the extent of twilight in different climates". Among other results, he determined the date and duration of the shortest twilight for each place on the globe. This matter became a classic among problems of extremes and was tackled again a century and a half later by the Bernoulli brothers. Gomes Teixeira has made some interesting comparative remarks on the methods used by the Portuguese and the Swiss brothers: while the latter applied the newly invented calculus, Nunes resorted to the results and methods available to him, i.e., classical geometry and trigonometry. The *De Crepusculis* has been considered Pedro Nunes' master work by several commentators, and it deserves a modern reappraisal. Upon mentioning the book's impact in Europe, including Tycho Brahe's applause and Clavius' quotations, the Portuguese historian Joaquim de Carvalho says that "it reached the recognition due to scientific explanations, entering and flowing, often anonymously, in the stream of exact knowledge which constitutes mankind's heritage" (*Notes on De Crepusculis,* 'Works of Pedro Nunes', vol. II, Lisbon, 1943). This view is appropriate to the *Libro de Algebra* as well, and in fact to all the mathematical work of Pedro Nunes.

Pedro Nunes is Portugal's first example of a "pure" scientist, for whom precision and rigor were an imperative. (In this respect, it is interesting to mention the quarrels he had with sea pilots, in which he proudly defended the superiority of scientific knowledge.) The Portuguese mathematician was a unique case in the Iberian Peninsula. "Spiritually he lived outside both nations [Portugal and Spain]", says J. Rey Pastor (*Los matemáticos españoles del siglo XVI,* Madrid, 1926).

The Academy of Sciences in Lisbon started in the 1940s a remarkable project of publication of Pedro Nunes' Works. Of the six volumes planned, four were published. The series was suspended almost 40 years ago, at the death of Coimbra professor Joaquim de Carvalho, who was the main force behind it.

There were a few other mathematical activities in Portugal in the sixteenth to eighteenth centuries, some of which we mention briefly.

In 1547 Pedro Nunes was appointed to the newly created position of royal cosmographer, which carried the requirement of daily instruction on mathematics (applied to navigation). The Royal Academy of the Navy replaced the office of royal cosmographer in 1779.

In the Jesuit college of Santo Antão, in Lisbon, a public 'Aula de Esfera' (Sphere school) was in operation from the end of the sixteenth century until the eighteenth century. Besides mathematics applied to navigation, also astronomy, geometry and arithmetic were studied there.

Apart from Santo Antão and their University in Évora, the Jesuits organized courses in mathematics throughout the country, in particular in the Coimbra colleges. To teach these courses, many foreign professors, especially Italians and Germans, came to Portugal. We note the following names, some with works on astronomy, navigation and cartography: C. Grienberger (1564–1636) (later Clavius' successor in the Roman college), C. Borri (1583–1632) (who in the early seventeenth century made Galileo and sunspots known in Portugal), I. Stafford (1599– 1642), D. Capassi (1694–1736) and G. Carbone (1694–1750). The latter two collaborated in the establishment of an astronomical observatory at Santo Antão.

Capassi left in 1729 for Brazil with another Jesuit professor, Diogo Soares, to carry out King João V's instructions to draw the map of the great transatlantic state. Map drawing is one of the main themes of Portuguese mathematical activity in this period. Another example was the Swiss Jesuit João König, who, appointed in 1682 to the chair of mathematics at the University of Coimbra, left four years later, by government order, to draw a map of Portugal.

Around the middle of the seventeenth century, with the war of independence after 60 years of Spanish rule, mathematical studies applied to military activities received some impetus. A School of Fortification and Military Architec-

ACADEMIA DAS CIÊNCIAS DE LISBOA

PEDRO NUNES

OBRAS

NOVA EDIÇÃO REVISTA E ANOTADA
POR UMA COMISSÃO DE SÓCIOS
DA ACADEMIA DAS CIÊNCIAS

VOL. VI

LIBRO DE ALGEBRA
EN ARITHMETICA Y GEOMETRIA

IMPRENSA NACIONAL DE LISBOA
MCML

Regla de la ygualacion en los quebrados y raizes.

No tratamos al presente de la ygualacion de los quebrados de primera intencion, como son $.\frac{1}{2}$ $.\frac{1}{3}.$ y $.\frac{1}{4}.$ y qualquier otro quebrado cuyo numerador y denominador sean conoscidos, porque estos son auidos por enteros, porque por la misma arte de los enteros tienen su ygualacion. Mas queremos tratar de la ygualacion de los quebrados de segunda intencion, a que llaman esimos de cosas o censos o cubos, y tambien de las raizes. Y la arte que ternemos sera esta, que assi como quando la ygualacion viene a dignidades mayores, las abatemos ygualmente, *reduziendolas a otras menores*, para las quales son dadas Reglas, o por ygual diminuició de las denominaciones desas dignidades mayores, o partiendolas por vn commun partidor, o haziendo la ygualacion en las sus raizes, assi tambien en los quebrados y raizes, q̃ a nuestro entendimento se representan como quantidades defectuosas, porque son referidas a otras mayores, obrando por el cõtrario, haremos semejantemente nuestra ygualacion, haziendolas crescer proporcionalmẽte, porque por este modo si las primeras quantidades eran yguales, tambien las segundas necessariamente seran yguales. Por lo qual si las quantidades q̃ queremos ygualar fueren quebrados, o enteros con quebrados, por qualquier modo que el quebrado entre, multiplicaremos en ✠ el numerador de vno por el denominador del otro, y sera hecha la ygualacion q̃ buscauamos. Y por la misma arte si vuiere raizes,* siendo reduzidas a otras de vna misma denominacion, multiplicaremos cada vna dellas en si, y sera hecha la ygualacion. Y porque esta doctrina se comprehende mejor por exemplos, traeremos algunos, por los quales los otros se podran entender. *P. 138 r.

Si queremos ygualar $\frac{.20.}{.1.co.}$ con $\frac{.30.}{.1.ce.}$ multiplicaremos en ✠, diziendo assi: .20. por .1.ce. hazen .20.ce. y .30. por .1.co. hazen .30 co. y ternemos por tanto .20.ce. yguales a .30.co. los quales abreuiados, seran .20.co. yguales a .30. numero, y queda hecha la ygualacion.

Iten, si queremos ygualar $\frac{.20.co.}{.1.ce.}$ con .12. nu. pornemos la vnidad debaxo de los .12., porque es denominador de qualquier entero, y multiplicaremos en ✠, y ternemos .20.co. yguales a .12.ce. q̃ *abreuiados* vernan .20., yguales a .12.co.

Iten, si queremos ygualar $\frac{.20.co.\ \tilde{m}\ .4.}{.3.ce.}$ con $\frac{.5.co.}{.1.cu.}$ multiplicaremos en ✠, y haran .20.ce.ce. m̃ .4.cu. yguales a .15.cu. restau[ra]remos lo diminuto, y ternemos .20.ce.ce. yguales a .19.cu. que abreuiados vernan .20.co. yguales a .19.nu.

Iten, si queremos ygualar $\frac{.20.\ \tilde{m}\ .1.co.}{.1.ce.\ \tilde{p}\ .1.co.}$ con $.3\frac{.1.co.}{.1.ce.}$ primeramente conuerteremos la segunda quantidad, la qual es intero con quebrado,

— 165 —

FIGURE 1
The cover of the 1946 edition of the 'Libro de Algebra' by Pedro Nunes, and one of its pages (where the polynomial fractions are discussed

ture was created, and also in Santo Antão some attention was given to these topics. Later, in the eighteenth century, there are mathematical studies applied to artillery in military units and academies.

What stands out from this very brief survey is the practical or applied nature it suggests about mathematical activity in Portugal in this period. Under the patronage of the state or in the Jesuit schools, people studied subjects seen as corresponding to immediate concrete needs of the nation.

An inspection of the list of mathematical works written in this period in Portugal, or by Portuguese authors, reveals the same overall picture. We note a clear predominance of works on subjects in what may be called applied mathematics: navigation, atlases and maps, astronomical calendars, geometry applied to fortification, arithmetic applied to financial activities. Worthy of note is the frequency of records of astronomical observations (lunar eclipses, comets). Many of these texts exist only in manuscript form.

It seems clear that, in the century after Pedro Nunes, mathematics in Portugal did not take part in the great advances of the time. The persistent need to enlist the help of foreign professors seems related to the fact that "pure" mathematics was not studied *per se* in the country.

This is not the place to recall the Portuguese intellectual and cultural frame of mind in the period under study. Its reflections on the mathematical life (or lack of it) stem primarily from the fact that the great scientific advances were generally associated to philosophical propositions against which national political and religious authorities were on permanent watch. This led to an explicit attitude of rejection of novelty in teaching.

In the first half of the eighteenth century, there is evidence of a change in the scientific atmosphere. Inside and outside the country, there appear several Portuguese with an interest in modern scientific trends. We note, among others, the names of Jacob de Castro Sarmento, with a Newtonian theory of tides (*Theorica verdadeira das marés,* London, 1737); José Soares de Barros e Vasconcelos, astronomer for many years in Paris; the engineer Manuel de Azevedo Fortes, author of a *Logica racional, geometrica e analytica* (Lisbon, 1744); Teodoro de Almeida, of the Oratorian Congregation, with his *Recreação Philosophica* and *Cartas Phisico-Mathematicas*; and the Jesuits Eusébio da Veiga, astronomer in Lisbon, Manuel de Campos and Inácio Monteiro (whose books reveal he had a wide knowledge of the scientific advances of the time). All these authors deserve modern studies.

2. Mathematics in Portugal from 1772 to 1910

The Portuguese University was transferred to Coimbra in 1537 by King João III. Its reform in 1772 represents the biggest qualitative and quantitative change of the mathematics panorama ever carried out in a short time in Portugal. In December 1770 a 'Junta da Providencia Literaria' (Committee of Literary Providence) was created. It presented a report in August 1771 under the title "Historical Compendium of the State of the University". Here we can see that Science in general and Mathematics in particular were in bad shape in the University. For example, in the 60 years preceding the reform there was only one chair of Mathematics and there was no professor to fill the position. In that document mathematics is considered "an important science to the good being of the Kingdom, and navigation, and ornament of the University."

In the new University Statutes, approved in October 1772 with great pomp and solemnity in the presence of the Marquis of Pombal, Prime Minister of King José I and the force behind these transformations, two new faculties were created: Mathematics and (Natural) Philosophy.

Mathematics is placed in a very high position in these statutes: "Mathematics has such an indisputable perfection amongst all natural knowledge, as well as in the luminous exactitude of its Method, as in the admirable speculation of its doctrines, that it not only but with rigor or with property merits the name of Science, but also is the one which has credited singularly the force, the ingenuity, and sagacity of Man." And penalties are specified to the ones who may diminish the importance of the mathematical studies: "All those, who directly or indirectly discourage or dissuade somebody from the mathematical studies; . . . will loose all the oppositions to the chairs of their respective Faculties."

Other aspects of these statutes should be mentioned. The mathematics course was made up of four disciplines: geometry, algebra, mathematical physics and astronomy (plus a supplementary discipline of drawing and architecture and some disciplines from the Faculty of Philosophy). The discipline of geometry was compulsory in the first year for all University students (including those in law and theology). This brought some problems. For example, in 1787, a letter from the King directed that separate books of geometry should be arranged for students of law and theology. In another letter, in 1790, it was ordered that law students should not be allowed to attend the first year without passing the geometry exam.

For the new Faculty of Mathematics, two Italian and two Portuguese professors were hired. The latter two, José Monteiro da Rocha and José Anastácio da Cunha, who had both learned Mathematics essentially by themselves, deserve a special mention because of the original work they produced.

José Monteiro da Rocha (1734–1819) studied in the Jesuit college of Bahia, Brazil, and was the main author of the statutes of the new Faculty of Mathematics. He organized the Astronomical Observatory of the University and translated books by Bezout, Bossut, and Marie into Portuguese. His scientific work was concentrated in the areas of numerical methods and astronomy. In the work, *Addition to the Rule of Mr Fontaine to solve by approximation the problems that reduce to quadratures,* published in the *Memoirs of the Academy of Sciences of Lisbon,* he studied methods to accelerate the convergence of Fontaine's formula of approximate integration, which, as remarked by Tiago de Oliveira (in 'Jozé Anastásio, o geómetra exilado no interior,' an article published in the volume *Em homenagem a José Anastácio da Cunha,* Coimbra, 1987), is Richardson's extrapolation formula. This work presents upper estimates of the approximation error and deals with the case of improper integrals. Another work on numerical analysis concerns the approximate calculation of the volume of a barrel, a problem proposed by Kepler. Monteiro da Rocha wrote several works on astronomy, most of which were published in Paris under the title *Mémoires d'Astronomie Pratique* (Paris, 1808), with translation of Manuel Pedro de Mello. One of the most important of these memoirs was the practical calculation of orbits of comets, discovered before Olbers, to whom it is credited. Another

FIGURE 2
The plan of the Astronomical Observatory of the University of Coimbra made by José Monteiro da Rocha

work, praised by Delambre, concerns the prediction of the sun's eclipses.

José Anastácio da Cunha (1744–1787) wrote a treatise, *Principios Mathematicos,* in which he tried to give rigorous bases to all of mathematics. Here we find for the first time, with surprising rigor, the definition of a convergent series (using what later became known as the Bolzano-Cauchy condition), the definition of exponential functions from their power series for real and complex variables, and the definition of the differential of a function in a way similar to what is done nowadays. These definitions have been studied by several authors, including A.P. Youschkevitch (in *Revue d'Histoire des Sciences,* vol. XXVI and XXXI), E. Giusti, J. Mawhin, I. Grattan-Guinness (their texts appear in the volume *Anastácio da Cunha—1744/1787—o matemático e o poeta,* Lisboa, 1990) and the present authors (in *The Mathematical Intelligencer,* vol. 10, no.1, 1988 and also in the volume just mentioned). Unfortunately, his book, although it had two editions in French, was read by few people and apparently did not much influence the development of mathematics. Cunha also wrote an *Essay about the Principles of Mechanics,* where he proposed an axiomatic view of mechanics, and other works presumably lost.

A major task of the Faculty of Mathematics was the training of specialists in mathematics. One of the first to get a doctoral degree after the 1772 reform was Frei Alexandre de Gouveia, who would later become bishop of Beijing at a time when scientific knowledge was important in order to be accepted in China. Details about his action are not known, but he was a member of one of the most important scientific committees in China, the Mathematics Tribunal. In order to have a better idea of the impact of the training of specialists, we present a table of the number of doctoral degrees awarded in the Faculty of Mathematics until the end of the nineteenth century (Figure 3).

Many of the new holders of the doctoral degree stayed as professors of the Faculty of Mathematics, but others became professors at the Military and Polytechnic Schools

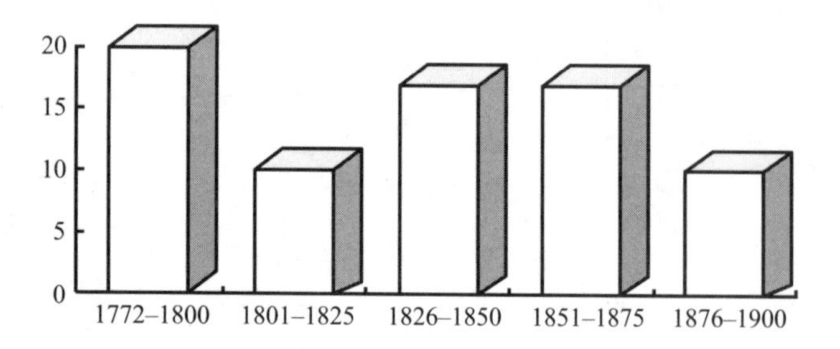

FIGURE 3

of Lisbon and Oporto, as did many of the mathematics graduates.

In the university, the quality of the professors was a concern. In a meeting between the Faculties of Mathematics and Philosophy it was agreed that professors would be hired, not according to seniority, but only with "the annual literary works in which they should exercise and qualify in front of the Congregations with written and verbal memoirs, according with the plan that the same Congregations . . . will propose to His Majesty." This indicates some concern with research.

Unfortunately, the full impact was not the one intended by Pombal or by José Monteiro da Rocha, for a long time Chairman of the new Faculty of Mathematics and Vice-Rector of the University. There were very few students; the publication of mathematics texts was not easy; and the political instability of the country did great harm to university life. From 1807, the year of the first Napoleonic invasions, until the end of the Civil War in 1834, all scientific activity was considerably reduced. Many professors were forced to abandon the university; others went into exile; some were killed; and the university was closed in the school years of 1810–1811, 1828–1829, 1831–1832, 1832–1833, 1833–1834. (Afterwards, other periods of political instability perturbed the work in the University, which was closed in 1846–1847 and other short periods.)

The majority of the mathematical works published in Portugal in the end of the eighteenth and beginning of the nineteenth century emanated from the Academy of Sciences, founded in 1799 by the Duke of Lafões, uncle of Queen Maria I. The Academy began the publication of its Memoirs in 1787, but only in 1797 were the first mathematical papers published. (The first was by José Monteiro da Rocha on Kepler's barrel problem.) The Academy was very active until the French invasions and the Civil War.

As pointed out by Portuguese mathematician Luís Woodhouse in 1925, "the period in the history of mathematical studies in Portugal that goes from the rebirth after Pombal's reform of the University of Coimbra in 1772, until the end of the eighteenth century and afterwards in the first years of the nineteenth century until the deep and constant political perturbations eventually separated and dispersed the vital elements of the Portuguese mathematical science, is not at all without interest; it is not vain and sterile, but it is short."

Only after the end of the Civil War was it possible to make some new real reforms of the educational system. The University and the Academy of Sciences resumed their activities, not without some difficulty. Despite the opposition of the University (of Coimbra), Polytechnic Academies were created in 1837 in Lisbon and Oporto, and in 1911 these originated the Faculties of Sciences of the newly created Universities of Lisbon and Oporto. In Coimbra, that same year, the Faculties of Mathematics and Philosophy of Pombal's reform were merged in a Faculty of Sciences. Conditions were thus created, with three competing schools, for a qualitative jump in the mathematical development of Portugal.

Direct contacts with foreign mathematicians were scarce during this period. Only after 1908 did the Faculty of Mathematics have its own budget. For the first time money was made regularly available for "missions ... to study a subject of its chairs or other related with the Faculty of Mathematics", with a three month duration. The first professors to go abroad were Luciano Pereira da Silva, Henrique de Figueiredo and Sidónio Pais. This last professor (later to become Portugal's President for a short period: he was murdered in 1918), was twice in Paris (in 1909 and 1910), where he visited several scientific institutions and attended talks, for example, by Mme. Curie, Langevin, Goursat, Appell and Picard.

Among Portuguese mathematicians of the nineteenth century we should mention:

Manuel Pedro de Mello (1765–1833), a student of José Anastácio da Cunha, was the first hydraulics professor in Coimbra, a post for which he prepared himself during a study visit in Europe; in his lessons he followed, among others, Poisson's treatise on fluid mechanics. Mello's most important work, now lost, was a "Memoir about the program for the proof of the force parallelogram", written for a contest of the Royal Academy of Sciences of Copenhagen in 1806, in which he won the first prize. (About this contest we only know there were a lot of entries.)

Daniel da Silva (1814–1878) studied in Coimbra and was a teacher at the Naval School. In his work "Memoir about the rotation of forces around a point" (1851) he developed a theory on the subject without knowing Möbius' work, published in 1837; but this work of Möbius had some flaws, corrected by Darboux in 1876 and 1877. Silva's theory did not have these flaws, but Darboux never read the work of the Portuguese mathematician. He published other original works in number theory, geometry, and insurance, of which only a small part became known outside Portugal in 1903, through a work of the Italian mathematician Alasia de Quesada published in a Pavia journal. In a letter to Gomes Teixeira in 1877, Daniel da Silva wrote: "My Memoir, having many more things beyond what Möbius did, including a correction of an error he made, of whose correction Darboux is very proud, lies ignored, for almost twenty-six years, in the libraries of almost all Academies of the world. What profit do I get from writing in Portuguese!"

Francisco Gomes Teixeira (1851–1933) was a professor in the Universities of Coimbra and Oporto. He was by far the most prolific Portuguese mathematician of this period, having published more than a hundred papers in many international scientific journals. He corresponded with and met personally the most distinguished European mathematicians of the time. Gomes Teixeira published essentially in analysis (series expansions, interpolation, functions of a complex variable, differential equations, special functions, etc) and in geometry (properties of remarkable curves). He started the first Portuguese mathematics journal, the *Jornal de Sciencias Matematicas e Astronomicas* (*Journal of Mathematical and Astronomical Sciences*) in 1877, replaced by a broader publication in the Sciences in 1905. He also published several works about the history of mathematics in Portugal, namely studies on Pedro Nunes, Monteiro da Rocha, Anastácio da Cunha and Daniel da Silva, and a book on the history of mathematics in Portugal, *História das Matemáticas em Portugal* (1934).

Other mathematicians of this period had activities that should be mentioned. Dantas Pereira is the author of the first textbook containing references to elementary probability, "Course of Studies for the Commerce and Exchequer" (1798) and also of a method for the approximate calculation of the roots of algebraic equations (1799), not very different from the method later published by Horner (1819), according to the study of Gomes Teixeira (in his *História das Matemáticas em Portugal,* 1934). Garção Stockler published in 1819 the first history of mathematics in Portugal (and the first history of mathematics related to a single country) and also several works on analysis. Simões Margiochi published in 1821 a work where he obtained certain formulas for the roots of the equations of degree not greater than 4, later rediscovered by Luiz Olivier (*Crelle's Journal,* 1826). Adrião Pereira Forjaz de Sampaio is the author of the first Portuguese scientific text on statistics, *First Elements of Statistical Science* (1841). Henrique de Figueiredo published in 1887 a work about Riemann surfaces. Sidónio Pais was the author of the first scientific work on probability theory published in Portugal, with his thesis *Introduction to the theory of the observation errors* (1898).

Let us sketch briefly the panorama of mathematical publishing in Portugal.

Until the middle of the nineteenth century, the main mathematical publications were the *Memórias da Academia das Ciências*. Only after 1857 were doctoral dissertations published by the University, and among these we can find many interesting works, showing that Portuguese mathematicians of the time were aware of what was being done in the rest of the world. We can point out as an example the already mentioned dissertations of Henrique de Figueiredo and Sidónio Pais.

Scientific research was a part of the 1772 Statutes, which established the creation of a "General Congregation of the sciences for the advancement, progress and perfection of the natural sciences". But this Congregation was never actually in operation, and contacts with mathematicians from other countries were limited. It is true that the best foreign publications of the time were received in Portugal, but Portuguese mathematicians often discovered results or methods that were later rediscovered by others, as was the case with Monteiro da Rocha, Anastácio da Cunha, Garção Stockler, Dantas Pereira or Daniel da Silva. It was only with the launching of the *Jornal de Sciencias Matematicas e Astronomicas* in 1877 that it was really possible to develop relations between Portuguese and foreign mathematicians, mainly from Europe. Portuguese mathematicians published articles in this journal and these were read outside Portugal; also many foreign mathematicians published papers there, thus strengthening Portuguese awareness of mathematical research done outside Portugal.

The 1772 Statutes established that there should be books in Portuguese for the students of every course. Some translations were made, but few originals were published. Mathematics professor Luís da Costa e Almeida even proposed in 1886, without success, at the 'Conselho Superior de Instrução Pública' (High Council for Public Instruction), that money should be given to professors in charge of writing textbooks and that this money should be equivalent to teaching a full year course. The first Portuguese original textbook that had a real impact, nationally and internationally, was the *Curso de Analyse Infinitesimal,* in three volumes (1887–1892), by Francisco Gomes Teixeira, which was the object of favorable reviews in the *Bulletin des Sciences Mathématiques* and the *Bulletin of the American Mathematical Society.* This book was adopted for a long time in the University and in the Polytechnic Academies of Oporto and Lisbon. It made widely available, for the first time in Portugal, many topics of advanced analysis (some based on research papers of Gomes Teixeira) such as partial differential equations, complex functions and functions defined by series.

One of the best known works of this period is an encyclopedia of curves by Gomes Teixeira. The Madrid Academy of Sciences had proposed a contest in 1896 with the theme "Ordered catalogue of all the curves". Gino Loria and Gomes Teixeira both took the prize, although there was supposed to be only one prize and Gomes Teixeira wrote in a language not accepted by the rules. Later Gomes Teixeira expanded his work and published it in French, with the title *Traité des courbes spéciales remarquables planes et gauches,* which received a prize from the Paris Academy of Sciences (the referee was Paul Appell). This contains a wealth of information about all the classical curves, including their history. It was reissued recently by two publishers (Chelsea, New York, 1971; Jacques Gabay, Paris, 1995).

3. Mathematics in Portugal since 1910

One of the main features of mathematical activity in Portugal up to the end of the nineteenth century is the isolation of the Portuguese mathematical community. In fact (with the exception of Pedro Nunes) this community and its activity were mostly unknown abroad. In contrast, one of the main features of mathematical activity in Portugal in the twentieth century is the breakdown of this isolation. The breakdown started, in fact, at the end of the nineteenth century, with Gomes Teixeira. During the first quarter of the twentieth century Teixeira remained the only Portuguese mathematician of international stature. By the end of the twenties other Portuguese mathematicians began to be known abroad and to publish their results in some of the best foreign mathematical journals.

This was the case of Aureliano de Mira Fernandes (1884–1958). He studied at Coimbra University, where he received his PhD in 1910 with a thesis (mainly expository) on Galois Theory. In 1911 he was appointed professor in the recently established 'Instituto Superior Técnico' of Lisbon (Lisbon School of Engineering), where he remained until his retirement in 1954. In the twenties he became an active researcher in differential geometry and tensor calculus. From 1924 until his death he published a series of papers on those subjects (several of them in Italian journals) and had many contacts with Italian mathematicians (namely T. Levi Civita).

Another active mathematician in the same period was José Vicente Gonçalves (1896–1985). He studied in Coimbra, earning his PhD in 1921 with a thesis on complex function theory. He was a professor in the Universities of Coimbra (until 1942) and Lisbon (1942–1966), and worked on classical analysis (inequalities, zeros of functions, infinite series and products, continued fractions, orthogonal polynomials). Although his papers were published in Portuguese journals and some of them were written in Portuguese, they became known outside Portugal (some of them gave rise to further research by foreign mathematicians, like A. Ostrowski). He was also the author of some very influential university manuals and some high school textbooks on mathematics.

As to the institutions related to mathematical activity, these were, at the beginning of the century, the Faculty of Mathematics of Coimbra University, the Escola Politécnica in Lisbon and the Academia Politécnica of Oporto. The last two were mainly engineering schools (although since 1902 the three institutions gave degrees in mathematics for teachers of high schools). We should mention that the number of students who graduated was usually very small (varying from zero to six).

In 1910 Portugal's political regime changed from a monarchy to a republic. The new leaders were greatly concerned with educational matters. The Universities of Lisbon and Oporto were created. As mentioned before, the Escola Politécnica of Lisbon and the Academia Politécnica of Oporto were renamed as Faculties of Sciences and became part of the respective Universities (the Faculty of Mathematics and the Faculty of Natural Philosophy of Coimbra were also merged in a new Faculty of Science). Also some other schools were created or changed into schools of university level (this was the case of Lisbon's Instituto Superior Técnico, where Mira Fernandes was appointed professor). But more important than this increasing of the number of schools was the attempt in 1923 by the Minister of Instruction, António Sérgio, to organize a government agency for scientific research. This agency, which was established only in 1929 (the Junta de Educação Nacional), had as one of its functions to give grants for scientific research and to send students to work in foreign centers of scientific research.

During the thirties and forties a modernization of mathematical teaching in the Universities occurred with the publication of new textbooks by Vicente Gonçalves, on analysis, and by António Almeida Costa (1903–1978), on algebra.

Almeida Costa graduated in 1924 from Oporto University, where he taught from 1924 until 1952. In that year he moved to Lisbon University, where he remained until his retirement in 1973. His first interests were in mathematical physics and astronomy. In 1937 he received a scholarship to study these subjects in Berlin, where he was advised to study group representation theory as part of his training. With that study his interests shifted to abstract algebra, and this turned out to be his main field of activity. After his return to Oporto in 1939, he became very active in promoting abstract algebra, and it was mainly through him that the subject was introduced in Portugal. In addition to the publication of several research papers and textbooks in that area, he organized seminars, trying to attract young mathematicians to work on algebra. Almeida Costa supervised some PhD students and was one of the very few Portuguese mathematicians who founded a "School".

The end of the thirties and beginning of the forties were a period of intensive and exciting mathematical activity, thanks to a new generation of mathematicians such as Ruy Luís Gomes (1905–1984), António Aniceto Monteiro (1907–1980) and Hugo Ribeiro (1910–1989).

Ruy Luís Gomes graduated from Coimbra, where he was a student of Vicente Gonçalves, and there received his PhD in 1928, with a thesis on mathematical physics written under Mira Fernandes. In the following years he published several papers on this subject and also on real analysis (integration theory). He was professor in Oporto from 1928 until 1947.

António Monteiro graduated from Lisbon in 1930 and obtained his PhD in Paris in 1936 under M. Fréchet with a thesis on topology. Besides this subject he also worked on algebra and algebraic logic. After his returning to Lisbon, in 1936, he began to organize courses and talks for young students trying to initiate them in research activities. One of these students, Hugo Ribeiro, said much later: "Since then I have never met anyone that, at the level we had, was so effective in promoting young people". As a result of this activity a mathematics research center was established in Lisbon in 1940, and, inspired by this one, another was founded in Oporto in 1941 by Ruy Luís Gomes, with the collaboration of Almeida Costa. Monteiro played also a major role in developing this center (he lived in Oporto in the years 1943–45). Several monographs were published by these centers. In 1943 a Council for Mathematical Research ('Junta de Investigação Matemática') was created under the direction of Mira Fernandes, Ruy Luís Gomes and Monteiro.

Another mathematician engaged in these activities was Hugo Ribeiro. He finished his undergraduate studies in 1939 and was initiated in research activities by António Monteiro. He received his PhD in Zürich in 1946. His main works were in mathematical logic (in Zürich he was a student of P. Bernays, F. Gonseth and H. Hopf).

It was this generation (mainly António Monteiro) who promoted, in 1937, the creation of *Portugaliae Mathematica* (still in publication), a research journal of international quality, and in 1940 of *Gazeta de Matemática,* an expository and informative journal for the mathematical community (including students and high school teachers). These initiatives received support from senior Portuguese mathematicians. For instance Mira Fernandes and Vicente Gonçalves published several papers in *Portugaliae Math-*

PORTUGALIAE
MATHEMATICA

VOLUME 1
1937-1940

FACULDADE DE CIÊNCIAS
LISBOA — PORTUGAL

Publicação subsidiada pelo Instituto para a Alta Cultura

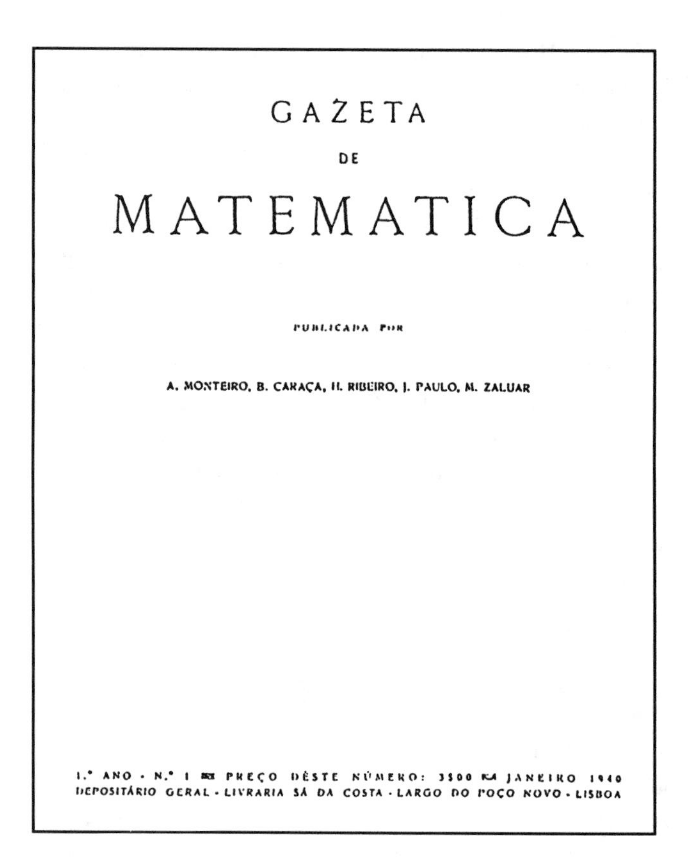

GAZETA
DE
MATEMATICA

PUBLICADA POR

A. MONTEIRO, B. CARAÇA, H. RIBEIRO, J. PAULO, M. ZALUAR

1.° ANO · N.° 1 · PREÇO DÊSTE NÚMERO: 3$00 · JANEIRO 1940
DEPOSITÁRIO GERAL · LIVRARIA SÁ DA COSTA · LARGO DO POÇO NOVO · LISBOA

FIGURE 4
The first issue of the journals *Portugaliae Mathematica* and *Gazeta de Matemática*

ematica. It was also this generation who founded in 1940 the Portuguese Mathematical Society.

We should add that each Faculty of Science (and also the Engineering and Economics Schools) published journals containing research and expository mathematical papers, mainly by Portuguese mathematicians. Gomes Teixeira founded the journal of the Oporto Faculty in the nineteenth century; the others were founded in the twenties and thirties of the twentieth century. This shows the increasing mathematical activity during the twentieth century.

Unfortunately, a great part of the above described activity was interrupted, because most of its originators (and some of its participants) were, for political reasons, expelled from their posts (including Ruy Luís Gomes, who was also imprisoned more than once because of his political activities) or could never find a position in the Universities (including Monteiro and Hugo Ribeiro; the PhD obtained by Ribeiro in Zürich was never recognized by Portuguese authorities) and were forced into exile. (Portugal was under a right-wing authoritarian regime from 1926 until 1974.) Hugo Ribeiro went to the United States. Among other positions he held a visiting professorship at the University of California, Berkeley, and a professorship at Pennsylvania State University. Monteiro and Ruy Luís Gomes (and some other young mathematicians) went to South America (Brazil and Argentina). We should add that Monteiro and Ruy Luís Gomes played a major role in developing Mathematics in those countries. Monteiro, for example, had a strong influence on L. Nachbin. Ribeiro supervised several PhD students in the U.S. (M. Keedy, K. Supruonowicz, D. Smith, Y. Wu, J. Griffin, E. Goldwasser). With the end of the regime in 1974 Ruy Luís Gomes returned to Oporto University and Ribeiro was appointed Professor there.

In the fifties and sixties we should mention the activity of J. Sebastião e Silva (1914–1972), one of the most distinguished Portuguese mathematicians of the twentieth century. He graduated from Lisbon in 1937, was initiated in research by A. Monteiro in the Lisbon mathematics center and studied in Rome (1943–1945) with L. Fantappié. He got his PhD in Lisbon in 1948. He was professor at the Lisbon Agronomy School (1950–1960) and the Lisbon Faculty of Sciences. He published research papers on logic, numerical analysis and functional analysis (locally convex spaces and distribution theory). He was also active as a pedagogue, namely in the rewriting of the secondary school mathematics curriculum, with the introduction of some set theory, abstract structures, numerical anlysis, applications of mathematics and statistics. He also authored textbooks for the new curriculum and gave special training courses for high-school teachers.

There was also some activity in the history of mathematics (mainly about Portuguese mathematicians). Besides Gomes Teixeira in the beginning of the century, there was Vicente Gonçalves and his penetrating studies on Anastácio da Cunha (he also published several small notes on the history of mathematics). Luciano Pereira da Silva (1864–1926) and Luís de Albuquerque (1907–1992) were two mathematicians who worked mainly in the history of science (on the fifteenth and sixteenth century European explorations and navigations). We should also mention the activity of Bento de Jesus Caraça (1901–1948), a Mathematics Professor at the Lisbon School of Economics (the Instituto Superior de Ciências Económicas e Financeiras). Although he didn't publish research papers, he was an inspiring lecturer and author of some influential university manuals. He published in 1942–43 a very interesting book, *Conceitos Fundamentais da Matemática* (*Fundamental Concepts of Mathematics*), on the history and philosophy of mathematics (aimed for the layman) which became a bestseller with several reprints since its first publication. Bento Caraça also published some papers on those subjects.

In the sixties, university teaching in Portugal underwent a big change at the level of the undergraduate mathematics curriculum, with the emergence of new topics (like topology). There were also changes in the structure of the university career, namely the creation of post-graduate courses, which put young graduates very early in contact with research.

In spite of the increasing mathematical activity in this century, we may note that there was no tradition of training disciples and PhD supervising. In fact the majority of the Portuguese mathematicians of this century were either self-made men (the cases of Mira Fernandes and Vicente Gonçalves) or they got their PhD abroad. There were some exceptions. As mentioned above, Ruy Luís Gomes was a disciple of Mira Fernandes and Vicente Gonçalves; the latter advised João Farinha (1910–1957) on his work on continued fractions; the work of J. Santos Guerreiro (1910–1987) on distributions was done under the influence of Sebastião e Silva. But the main exceptions occurred with Almeida Costa and António Monteiro, who did supervise PhD students and tried to create graduate schools of mathematics.

In the late sixties there began an exponential growth of the educational system which eventually reached the universities. In the ensuing decades there was a huge increase in the number of undergraduate students and in the number of universities. This led to an increase in the numbers of doctorates and research centers. An increasing number of students began to get their PhD in Portugal under the supervision of Portuguese mathematicians. New areas were developed and new research groups of international level were founded, namely in applied mathematics: numerical analysis, dynamical systems, control theory, optimization, statistics (to the development of this area we must attach the name of J. Tiago de Oliveira (1928–1992)) and theoretical computer science. Besides these there exist in Portugal other research groups of an international level: we mention the school of functional analysis and differential equations (founded by J. Sebastião e Silva) and two algebra schools (one founded by A. Almeida e Costa and the other by Luís de Albuquerque).

4. The historiography of mathematics in Portugal

Up to now, only four texts have been published which deserve to be called "Histories of Mathematics in Portugal":

Memorias históricas sobre alguns matemáticos Portugueses, e estrangeiros domiciliados em Portugal, ou nas conquistas—António Ribeiro dos Santos, 1812;

Ensaio histórico sobre a origem e progressos das Matemáticas em Portugal—Francisco de Borja Garção Stockler, 1819;

Les Mathématiques en Portugal—Rodolfo Guimarães, 1900–1909;

História das Matemáticas em Portugal— Francisco Gomes Teixeira, 1934.

Note the publication years. In the last 60 years there has not been a single attempt of a synthesis in this field. This is significant because the existing "Histories" in some way establish a paradigm, a point of view, which makes its influence felt on any reader trying to obtain information on the subject.

It should be clear that all four works are very valuable, and each in its own way was an important contribution to the study of mathematics history in Portugal.

Rodolfo Guimarães, for instance, after a historical introduction, aims to list all mathematical texts of Portuguese authors, or published in Portugal, up to the end of the nineteenth century. Gomes Teixeira carries out a deep analysis of the works of four important authors: Pedro Nunes, Anastácio da Cunha, Monteiro da Rocha and Daniel da Silva. As to Stockler, he was a pioneer, and his *Ensaio* had great influence. In the following paragraphs we do not wish to lessen these personalities and their works.

The truth is that the above mentioned works share an "old fashioned" vision of Portuguese history, and this vision tends to obscure certain periods and to distort the approach to the past, in our case the mathematical past.

This vision is the dominant one in general Portuguese history in the nineteenth century. It is the vision of the Enlightenment, of French Revolution influence, of liberalism. Very understandably, this vision has almost a "fighting" outlook on the recent past, and in this outlook it is practically an axiom that the two centuries before the Marquis of Pombal were a time of darkness, intolerance and ignorance. The blackness is even greater when one thinks of sixteenth century humanism, which in Portugal was associated with the flowering of art, culture and science brought on by the Discoveries.

This vision is faithfully reflected in Stockler's *Ensaio,* which dates from 1819: Portugal has a period of splendor in mathematics in the sixteenth century, in which Pedro Nunes stands out, then a 200-year period of decadence until Pombal's reform of the University of Coimbra, when important names like José Anastácio da Cunha make their appearance.

Almost 100 years after the publication of Stockler's *Ensaio,* Rodolfo Guimarães repeats the same views almost verbatim in the historical introduction to his book. As to Gomes Teixeira, he does not transcribe nor imitate, but the general outlook is still the same.

Now, the study of Portuguese general history has evolved a long way since the last century. In particular, the distance to the era of the Inquisition and of Jesuit influence in Portuguese society has increased, and that period is studied and analyzed nowadays like any other in Portuguese history.

But in the history of Mathematics we stayed in the same place. The general vision is still the same. Since studies about particular mathematicians and points of detail are rare (among the few relevant authors, besides Gomes Teixeira, we should mention Vicente Gonçalves and Luís de Albuquerque), we find ourselves reduced to the endless repetition of the same views, without new information, making permanent appeal to the same sources, which are basically the ones indicated.

At the present time, it seems appropriate to have clear ideas and propose what might pompously be called a change of paradigm. The above described vision, the importance of authors like Stockler, Guimarães and Teixeira notwithstanding, seems today exhausted, truncated, hopelessly dated and, what is worse, anaesthetizing. If we go on like this, long periods and many authors will remain in obscurity in the history of mathematics in Portugal.

For this change we propose the following lines of approach:

1) Priority (of course non-exclusive) to the study of the periods 1578 (death of Pedro Nunes) to 1772 (Pombal's reform of the University of Coimbra), and the nineteenth century. These are the periods that have received less attention up to now.

2) Priority to what may be called "positive history": listing of authors and works (extending and completing Guimarães' extraordinary catalogue), locating and microfilming printed and manuscript mathematical texts, publishing or reissuing the most significant of them. Study of these works. Study of the reception in Portugal of the great mathematical advances of the seventeenth and eighteenth centuries. As to the nineteenth century, study of the Coimbra dissertations and the Academy of Sciences memoirs. All this should be done, and is beginning to be done, with the consciousness that the history of mathematics in Portugal must be one of primary sources, with complete abandonment of acritical repetition of secondary sources.

3) Study, with basis on the collected material, of some hypotheses, which might serve as guidelines and orientation in this effort. Examples (among others possible):

- Was there an almost exclusive emphasis on applied mathematics in Portugal in the sixteenth to eighteenth centuries?
- If so, was this related in a fundamental way to the effort of the Discoveries, the routine of long transoceanic voyages, the concrete needs resulting from contact with the landing territories and their populations?
- Did that emphasis have an impact on the noticeable lack of professors of mathematics (even at the elementary level) in Portugal in the eighteenth century?
- If a picture of mathematical poverty in Portugal in the period 1578–1772 is confirmed, is it appropriate to use the word "decadence", which suggests a fall of some (even if small) height? In other words, was Pedro Nunes the leader of a group of sixteenth century Portuguese mathematicians or was he a singularity?
- Did Portuguese mathematicians keep abreast, in a systematic way, of developments in European mathematics in the nineteenth century?

We end with the obvious. Nothing of this will be done if there is no one to do it. We believe that in a few years it will be possible to write a new History of Mathematics in Portugal.

5. Short bibliography about the history of Mathematics in Portugal

Albuquerque, L.: 1974–1978, *Estudos de História* (six volumes), Coimbra: Acta Universitatis Conimbrigensis.

Bosmans, H.: 1907–1908, "Sur le 'Libro de Algebra' de Pedro Nuñez," *Bibliotheca Mathematica*, 3rd series, t. 8, Leipzig, 154–169.

——: 1908, "L'algèbre de Pedro Nuñez," *Anais da Academia Politécnica do Porto*, 3, 222–271.

Cunha, José Anastácio da: 1790, *Principios Mathematicos*, Lisboa. Facsimile edition 1987, Coimbra: Dep. Matemática Univ. Coimbra.

——: 1811, *Principes Mathématiques*, Paris. Facsimile edition 1987, Coimbra: Dep. Matemática Univ. Coimbra.

Carvalho, J. F., Oliveira, M. P., & Queiró, J. F. (ed.): 1987, *Em homenagem a José Anastácio da Cunha*, Coimbra: Dep. Matemática Univ. Coimbra.

Ferraz, M. L., Rodrigues, J. F., & Saraiva, L. (ed.): 1990, *Anastácio da Cunha—1744/1787—o matemático e o poeta*, Lisboa: INCM.

Freire, F. C.: 1872, *Memória Histórica da Faculdade de Matemática*, Coimbra: Imprensa da Universidade.

Gomes, Alfredo Pereira: 1997, *Portugaliae Mathematica. Um marco na história da investigação matemática portuguesa*, Coimbra: Publicações de História e Metodologia da Matemática (CMUC), 5.

Gomes, Ruy Luís: 1983, "Tentativas feitas nos anos 40 para criar no Porto uma Escola de Matemática," *Boletim da Sociedade Portuguesa de Matemática* 6, 29–48.

Gonçalves, J. V.: 1940, "Análise do Livro VIIII dos 'Principios Mathematicos' de José Anastácio da Cunha," In *Actas do Congresso do Mundo Português*, 123–140.

——: 1971, "Aureliano de Mira Fernandes, investigador e ensaísta," In *Obras Completas de Aureliano de Mira Fernandes* (vol. 1), v–xxi. Lisboa: ISCEF.

Guimarães, R.: 1909, *Les Mathématiques en Portugal*, Coimbra: Imprensa da Universidade.

Köthe, G.: 1973, "J. Sebastião e Silva et l'Analyse Fonctionnelle," *Anais da Faculdade de Ciências do Porto*, XVI-4, 339–349.

Morgado, J.: 1995, *Para a história da Sociedade Portuguesa de Matemática*, Coimbra: Publicações de História e Metodologia da Matemática (CMUC), 4.

Nunes, Pedro: 1567, *Libro de Algebra en Arithmetica y Geometria*, Antwerp. New edition 1946 Lisboa: Imprensa Nacional.

Oliveira, A. J. F.: 1988, "Anastácio da Cunha and the concept of convergent series," *Archive for Hist. of Exact Sciences*, 39, 1–12.

Oliveira, J. T.: 1989, *O essencial sobre a História das Matemáticas em Portugal*, Lisboa: INCM.

——: 1995, *Obras Completas* (vol. 2), Évora: Pendor.

Portugaliae Mathematica vol. 39: 1980, Volume honoring A. A. Monteiro.

Queiró, J. F.: 1988, "José Anastácio da Cunha: a forgotten forerunner," *The Mathematical Intelligencer*, 10, 38–43.

Rodrigues, M. A.: 1982, *Actas das Congregações da Faculdade de Matemática (1772–1820)*. Coimbra: Arquivo da Universidade.

Santos, A. R.: 1812, "Memorias historicas sobre alguns Mathematicos Portuguezes, e Estrangeiros Domiciliarios em Portugal, ou nas Conquistas," *Memórias de Literatura Portuguesa publicadas pela Academia Real das Ciências de Lisboa*, VIII, 148–229.

Saraiva, L.: 1993, "On the first history of Portuguese mathematics," *Historia Mathematica*, 20, 415–427.

Saraiva, L., & Oliveira, A. J. F.: 1992, "Mathematics and Mathematicians in Portugal (1800–1950)," To appear in the Proceedings of the meeting *European Mathematics 1848–1939*.

Silva, Clóvis Pereira da: "Sobre a Matemática no Brasil após o período colonial," to appear in *Revista da SBHC*, no. 16.

Stockler, F. B. G.: 1819, *Ensaio historico sobre a origem e progressos das Mathematicas em Portugal*, Paris.

Teixeira, F. G.: 1934, *História das Matemáticas em Portugal*, Lisboa: Academia das Ciências.

Youschkevitch, A. P.: 1973, "J. A. da Cunha et les fondements de l'Analyse Infinitésimale," *Revue d'Histoire des Sciences*, 26, 3–22.

Mathematics in South and Central America: An Overview

Ubiratan D'Ambrosio*
State University of Campinas, Brazil

1. Introductory remarks

It is impossible to give in a short paper an account of the developments in mathematics in Central and South America. A number of individuals, institutions and events relevant in these developments will not be mentioned. Most of those mentioned will appear with only brief notice. Others, not mentioned, do not imply a lesser academic standing. I elected a few events and names which I deem to be a good starting point for further research. The history of mathematics in South and Central America is a field open for research. Several notes and a list of references in the end are starting points for research projects.

The origin of mathematical knowledge is an unresolved issue. It is difficult to disagree that the search for explanations (religions, arts and sciences), systems of values and behavior styles (communal and societal life), the psycho-emotional and the imaginary, and models of production and of property are related to mathematical thinking. There is growing scholarship in the search for different styles and modes of building up knowledge in different natural and cultural environments, where the development of mathematical ideas is recognized. Obviously, in the Americas these developments followed different paths from Europe, Asia and Africa. There is no indication at all of mutual influences between the so-called Old World and the New World.

The conquest and colonization of the lands now known as the Americas had as a consequence an enormous redirectioning of the course of development of the civilizations of the continent. This is a common fact of cultural dynamics and easily seen when we observe cultures which had no previous contacts with the colonizer. The incorporation of modes of thought of the colonizer is frequently noticed. Reading the chroniclers of the conquest we easily recognize different ways of explaining the cosmos and the creation and of dealing with the surrounding environment. Religious systems, political structures, architecture and urban arrangements, sciences and values were, in a few decades, suppressed and replaced by those of the conqueror. A few remnants of the original behavior of these cultures were and still are outlawed or treated as folklore. But they surely integrate the cultural memory of the peoples descending from the conquered. Many of these behaviors are easily recognized in everyday life. And mathematical ideas are present in everyday life. Mathematics is both a human endeavour and a cultural form, hence it is subjected to cultural dynamics.[1]

The relationshiop of mathematics to cultural dynamics is one focal point of the research program known as Ethnomathematics, which deals with the the generation, the intellectual and social organization and the diffusion of different ways, styles, modes (tics) of explanation, understanding, learning, coping with and probing beyond (mathema) the immediate natural and socio-cultural environment (ethno). The dynamics of this process is a major problem we face in doing the history of ideas in every region of this world.

The conquest paved the way to colonization. The early colonizers of the Americas, the Spanish and the Portuguese, paved the way for the French, the English and the Dutch colonizers and later on for African, European, and Asian immigrants. With them came new forms of coping with the environment and of dealing with daily life, and new ways of explanation and learning. The result was the emergence of a synthesis of different forms of knowing and explaining which were generated by and available to the different communities, to workers and to the people. We recognize the emergence very soon of new religions and new languages—the creoles—in the Americas, of new cuisines, new music and new arts. All of these are absolutely interrelated as a synthesis of the cultural forms of the ancestors.

As a cultural form, mathematics and mathematical behavior becomes part of societal development. Modes of production, labor and social organization are intimately connected with mathematical ideas. Particularly in the Americas, the variety and peculiarities of the expositions of cultures and the specificities of the populational migrations reveal an effort of the colonizer to transfer, with minor adaptions, the forms of social, economical and political organization and administration prevailing in the metropolises, including schooling and scholarship (academies, universities, monasteries). The new institutions in the Americas were based on the styles prevailing in the metropolises, mostly under the influence, and even control, of religious orders.

All this, which took place during most of the sixteenth, seventeenth, and eighteenth centuries, occurred while new philosophical ideas, new sciences, new ways of production and new political arrangements were flourishing in Europe. Cultural facts produced in Europe were assimilated in the Americas under specific, mostly precarious, conditions. Indeed the Americas were the consumers of some of these new cultural facts. There is a clear co-existence of cultural goods, particularly knowledge, produced in the Americas and produced abroad. The former were consumed mostly by the lower strata of society, the people and workers, and the latter by the dominant classes. These boundaries are not clearly defined and the mutual influence of the resulting intellectual productions are evident.

This poses the following question, which permeates my entire research program on the colonial era: what are the relations between the producers and consumers of cultural goods? The so-called civilizatory mission, as the colonial expansion used to be justified, is intrinsically associated with this question.

This is a question affecting the relations between academia and society in general, hence between the ruling elites and the population as a whole, and it is particularly important for understanding the role of intellectuality in the colonial era. An immediate consequence of this broader historical view is a critique of epistemology, which generates profound historico-epistemological conflicts.[2] The basic question above guides my proposal for a historiography of mathematics and for what I have called "the basin metaphor", for which ethnomathematics comes as a fundamental instrument of historical analysis.[3]

Curiously enough, the factors influencing the consumption of what we might call Academic Mathematics produced in an alien cultural environment, and what "outsiders" of the profession—that is, nonmathematicians—have to say about mathematics, have not been given attention in the prevailing historiographies. My proposal incorporates into the History of Mathematics, in an essential way, the views about mathematics of aliens, in both senses, migrants and non-mathematicians.[4]

2. Specificities of the Americas

This paper deals with Central and South America. For this region we need a specific chronology. The chronology adopted in current historiographies of science, particularly of mathematics, does not apply to this region. If we consider mathematics as a cultural endeavor, we have to accept what we might call a "situated" chronology.

My proposal is a chronology based on five major periods:

1. Pre-columbian;
2. Conquest and early colonial times (roughly sixteenth and seventeenth centuries);
3. The established colonies (eighteenth century);
4. Independent countries (nineteenth century);
5. The twentieth century.

This division is justified when we look into the most relevant turning points in the development of the region. Of course, mathematical development is subordinated to the overall scenario of society.

Also, geographic divisions are very important. For the pre-Columbian period, sources are available mainly for the Aztec, Maya and Inca civilizations. An enlarged concept of sources, mainly drawn from anthropologists, is needed to look into other civilizations, such as, for example, those of the prairies and of the Amazon basin. Much finer divisions, taking into account both political and cultural specificities, are needed for a special study of pre-Columbian mathematics. A similar situation occurs in studying traditional African cultures.

For the period after the conquest of the Americas, the most appropriate method is to follow the administrative organization in Viceroyalties: Nueva España (roughly what is today Mexico and upper Central America), Nueva Granada (southern Central America, approximately Costa Rica, Colombia, Venezuela, Ecuador), Peru (roughly Peru and Bolivia), La Plata (roughly what is now Chile, Paraguay, Argentina and Uruguay) and the Viceroyalty of Brazil, which was a Portuguese conquest. Since independence, we have roughly the current political division.

In what follows, historical periods are defined according to the general chronology associated with the conquest and colonization of the Americas. Beginning with the independence movements in the late eighteenth and early nineteenth century, until present times, the cultural map is roughly the same.

In this paper I will not cover the developments in the pre-Columbian period. But a few remarks are necessary.[5] The imposition of the culture of the conquerer obviously depended on the culture of the conquered. But our knowledge of the pre-Columbian period is still very incomplete. There was a clear effort made by the colonial regimes to ignore or obliterate any sense of the history or historic achievement of the native cultures. Today we are faced with the difficult task of reconstructing the histories of these cultures, both looking into the chronology of the events and understanding the important migratory currents that shaped their developments. Of course, this leads us to look into the history of mathematics in pre-Columbian times.

The emergence of this scholarship relies heavily on a new reading of the chroniclers who described the Maya stellae, reported on the Peruvian quipus, described Aztec daily life, and indeed reported on every aspect of the conquered people. But these views are biased and, understandably, they failed to identify and barely recognized any form of mathematical knowledge in these cultures. There are many references from the period.[6]

3. Conquest and early colonial times

Although Mexico is not covered in the delimitation of this paper, it is imperative to mention the developments in the Viceroyalty of Nueva España in this period. Most of the developments in Central and South America are dependent on the important and strategic position of Mexico in the New World.

In the early colonial times, the Spanish and the Portuguese tried to establish schools, mostly run by Catholic religious orders. The demands for mathematics in these schools were essentially for economic purposes related to trade, but there was also an interest in mathematics related to astronomical observations. Reliance on indigenous knowledge was limited, but there was some interest in the nature of native knowledge.

An important source justifying this assertion is the first non-religious book published in the Americas, an arithmetic book related to mining, the *Sumario compendioso de las quentas de plata y oro que en los reinos del Pirú son necessarias a los mercaderes y todo genero de tratantes. Con algunas reglas tocantes al arithmética,* by Juan Diez freyle, printed in New Spain in 1556. It is a book on arithmetic as practiced by the natives, to which the author adds some questions on the resolution of quadratics.

Another important source is the *Historia del Nuevo Mundo* [1653], by Bernabé Cobo.[7] The archives of the Jesuit missionaries, as well as of other religious orders, are rich in historic material, but they are as yet to be explored.

Already in the first century after the conquest, we have practical books published in Mexico, such as the *Arte menor de arithmetica,* by Pedro de Paz, in 1623, and *Arte menor de arithmética y modo de formar campos,* by Atanasio Reaton, in 1649. One should also notice the book *Nuevas proposiciones geométricas,* written by Juan de Porres Osorio, in Mexico. It might be interesting to compare this literature with books used in the English colonies in North America.[8]

Astronomy was a major area of interest in Latin America in the seventeenth century. There are important discussions on the meaning of comets. Many of the interpretations relate to their purpose of conveying divine messages and messages to mankind. But there were also searches for scientific explanations. The figure of Don Carlos de Sigüenza y Góngora, of Mexico, towers. His works focus on astronomical observations and calculations. His *Libra astronómica y filosófica,* written in 1690, is considered one of the most important works of Latin American science. In it Sigüenza y Góngora refutes prevailing astrological arguments about comets.

In Brazil, research on comets was of major importance. We see the same tone of the reflections of Sigüenza y Góngora in the work of Valentin Stancel (1621–1705), a Jesuit mathematician from Prague who lived in Brazil from 1663 until his death. His astronomical measurements are mentioned in Newton's *Principia.*[9] Several polemical exchanges of letters and papers from these times reveal interesting epistemological arguments. Particularly revealing of the internal tensions in such established schools of thought as the Jesuit order are the reports on Stancel by his superior, Antonio Vieira (1608–1697), commenting on his views about the nature of comets.[10]

In the Viceroyalty of Peru we also have the same concerns. The first to be recognized as a mathematician in Peru is Francisco Ruiz Lozano (1607–1677), who wrote *Tratado de los Cometas,* essentially a treatise of medieval mathematics explaining the phenomenon.

4. The established colonies

In late colonial times, since the middle of the eighteenth century, a good number of expatriates and *criollos* played an important role in creating a scientific atmosphere in the colonies. This happened under the influence of the *Ilustración* [Enlightenment], the important intellectual revival that began in Spain under Charles III and in Portugal under José I and his strong minister, the Marquis of Pombal.[11]

A number of intellectuals well versed in a variety of areas of knowledge were responsible for introducing mathematics to the colonies. These include Juan Alsina and Pedro Cerviño in Buenos Aires, who lectured on Infinitesimal Calculus, Mechanics and Trigonometry. In Peru, Cosme Bueno (1711–1798), Gabriel Moreno (1735–1809) and Joaquín Gregorio Paredes (1778–1839) are best known. I do not have access to details of the life and work of these individuals.

In Brazil, José Fernandes Pinto Alpoim (1695–1765) wrote two books, *Exame de Artilheiros* (1744) and *Exame de Bombeiros* (1748), both focused on what we might call military mathematics, and both written in the form of questions and answers. Not much is known of his life and mathematical work. Where and with whom did he study? These are questions open for research.

In South America, Colombia had a privileged situation in the pre-independence days, which was reflected in the developments in the nineteenth century. A rather distinguished figure was José Celestino Mutis (1732–1808), who was the author of an unpublished translation of Newton, but who was also responsible for bringing to Colombia new ideas of mathematics in Colombia, mainly relying on the books by Christian Wolff. He was the founder of the *Observatorio de Bogotá,* in 1803. His most distinguished disciple, Francisco José de Caldas (1771–1816), became the director of the Observatory. Caldas was deeply involved in the Independence War and was shot by the Spaniards.

In Venezuela, a *Real y Pontificia Universidad de Caracas* was founded in 1721, with no mathematics contemplated in its plan of studies. On the other hand, in Chile, the *Universidad Real de San Felipe,* which was inaugurated in 1747 in Santiago, was provided with a 'catedra' of Mathematics. Fray Ignacio León de Garavito, a self-instructed *criollo* mathematician, was responsible for this chair. But the most significant developments of mathematics in Latin America in those days took place in Mexico. The preponderance of New Spain in the Americas in that period was responsible for effects of these developments in the rest of the Spanish colonies.

It is important to recognize that much of the development of mathematics in Europe in the Renaissance was due to the incipient industrialization and to the emergence of new metropolises. The same is true in the colonies. Practical problems related to building up the economy of the colonies and establishing the centers of administrative power resulted in the development of special kinds of applied mathematics. In all the colonies in the Americas urbanization was a major challenge. The transfer of the idea of a European city to the new world posed quite interesting problems, which influenced mathematical development. A sort of applied mathematics attempted to answer the most immediate questions posed by the economy of the region, such as mines and urban development.[12]

The influence of Mexico is particularly felt in Guatemala, which included Costa Rica. The most renowned scholar in the period is José Antonio Liendo y Goicoechea (1735–1814), who had a Mexican background. He taught at the *Universidad de San Carlos de Guatemala,* which had already become an important academic center after a new plan of studies was published in 1785. This plan was written in the form of 25 theses, under the title *Temas de Filosofia Racional y de Filosofia Mecánica de los sentidos, de acuerdo con los usos de la Física; y de otros tópicos físico-teológicos según el pensamiento de los modernos para ser defendidos en esta Real y Pontificia Academia Guatemalteca de San Carlos....*"[13] This was essentially a medieval proposition. Goicoechea was responsible for modernizing this plan of studies and incorporating experimental physics into the project. He introduced mathematics incorporating Newtonian ideas, based on the texts of Christian Wolff.

5. Independent countries

The independence of the Viceroyalties of Nueva España, Nueva Granada, Peru, La Plata and Brazil was achieved in the first quarter of the nineteenth century. The process of modernization of the newly independent countries, which is proclaimed in the declarations of independence, did not change the prevailing attitude towards mathematics.

The political division in countries following independence is practically the same as today. The independence of 1821 lessened the influence of Mexico in Central and South America. The establishment of new universities and the renewal of the old ones, immediately preceding and after independence, generated open attitudes with respect to sources of knowledge on which to build up the newly established countries of Latin America, formerly restricted almost exclusively to influences coming from Spain and Portugal. The new countries attracted considerable attention from the rest of Europe, and a number of scientific expeditions were sent to South America. They had a great influence in creating new intellectual climates throughout the region. This new source of intellectual interest is seen very strongly in the building up of large and diversified libraries, both public and private, and the acquisition of modern literature and instruments.

In Costa Rica the colonial authorities established the *Casa de Enseñanza de Santo Tomas* in 1814, in which Rafael Francisco Osejo, born in 1780, taught. He wrote in 1830 *Lecciones de aritmética,* written in the form of questions and answers, a common feature in that period. In 1843 the *Casa de Enseñanza* was transformed into the *Universidad de Santo Tomas,* where chairs in Engineering were established. But no chair in mathematics was created.

Colombia soon attracted foreign mathematicans. The Italian Agustín Codazzi (1793–1859) contributed to the creation of *Colejio Militar,* founded in 1846, which attracted a Frenchman, Aimé Bergeron, a graduate from the École Polytechnique. He seems to have been responsible for introducing Descriptive Geometry and creating an interest in Colombia in a proof of the Fifth Postulate of Euclid. Particularly noticeable are the attempts of Indalecio Liévano (1834–1913), possibly a student of Bergeron, and of Hermógenes Wilson. Very interesting is a later position taken by Julio Garavito Armero (1865–1920) in discussing non-Euclidean geometry and criticizing studies about the Fifth Postulate and ruler and compass constructions.[14] Indeed, Geometry was a major subject not only in Colombia, but in all of South America. An indication of this is the fact that the *Élements de Geométrie* of André Marie Legendre, published in 1794, was translated in Brazil by Manoel Ferreira de Araújo Guimarães in 1809, in Colombia by Luis M. Lleras in 1866, and in Venezuela by Jesús Muñoz Tebar, in 1879.

In Venezuela, the *Pontificia Universidad* was transformed into the *Universidad Central de Venezuela* in 1826. There Jose Rafael Acevedo taught the first course in Mathematics in 1827. Lino Pombo (1797–1862) had a role in founding the *Academia de Matematicas de Venezuela.* He was concerned with the preparation of teachers of mathematics and in 1821 proposed the creation of a *Escuela Normal de Matemáticas.* He wrote a complete course of mathematics.

In Brazil, the evolution of mathematics followed a different path. While as a colony Brazil was practically deprived of cultural facilities (no universities, no libraries, not a single printing press), the transfer of the royal family of Portugal to escape the Napoleonic invasion, in 1808, was decisive and changed cultural life in the colony. The Portuguese court settled in Rio de Janeiro, where they had to create an infrastructure to run, from a former colonial town, the Kingdom of Portugal. They founded a major library, a botanical garden, an astronomical observatory and a military academy in 1810, where higher mathematics (calculus, analytic geometry, algebra and trigonometry) was taught. The military academy was given a higher status in 1839 under the name *Escola Militar da Corte.* There doctorates in mathematics could be granted. This institution had a very strong influence in the development of mathematics in Brazil.[15] The translation of the textbooks of Legendre, already mentioned above, of Lacroix and of other French authors, was an important factor in generating what we might call a mathematical style in Brazil.

Particularly interesting is the case of Joaquim Gomes de Souza (1829–1863), known as "Souzinha", the first Brazilian mathematician who visited and submitted papers to European academies. He presented his results in the *Academie des Sciences de Paris* and in the *Royal Society.* Only short notice of the papers were given,[16] and they were posthumously published as *Mélanges du Calcul Intégral,* in an independent printing by Brockhaus, of Leipzig, 1889. This work, dealing mainly with partial differential equations, is permeated by very interesting historical and philosophical remarks, revealing familiarity of the author with the most important literature then available. This was possible due to the existence of private collections in Maranhão, his home state in the Northeast. The contents of these libraries, as well as the details of their acquisition, is as yet an open field of research.

Argentina, independent since 1816, had a remarkable intellectual development. In 1822 there was founded the ephemeral *Sociedad de Ciencias Físicas y Naturales.*[17] We soon see the emergence of private libraries in Buenos Aires. Particularly important is the private library of Bernardino Speluzzi (1835–1898), which included the main works of Newton, D'Alembert, Euler, Laplace, Carnot and several other modern classics. Valentin Balbin (1851–1901), while Rector of the *Colegio Nacional de Buenos Aires* proposed in 1896 a new study plan which included history of mathematics as a distinct discipline. This is probably the first formal interest in the history of mathematics in South America, which eventually led to an important school of History of Science in Argentina.

In Peru, developments in statistics should be mentioned, beginning with the the book *Ensayo de estadística completa de los ramos económico-políticos de la provincia de Azángaro...* by José Domingos Choquechuanca (1789–1858), published in 1833.

In Chile, the Universidad de Chile was created in 1842, with a *Facultad de Ciencias Físicas y Matemáticas* [Faculty of Physical and Mathematical Sciences]. A most distinguished member of the faculty is Ramón Picarte, a lawyer, who had his paper *La división reducida a una adición,* accepted and published by the *Academie des Sciences de Paris* in 1859.[18] Much emphasis was given, in the new university, to teacher training. A convention with the government of Germany provided the pedagogical support to reform education in the country. Fifteen German mathematicians, most with a doctorate, emigrated to Chile in 1889. Again, this is an as yet unexplored field of research.

In the course of the nineteenth century, we notice a growing interest in the philosophical ideas of Auguste Comte (1798–1857). His philosophical doctrine, known as Positivism, placed mathematics in a central position. The positivistic movement was impeded by the demands of building up the ideological framework of the new countries. This was a major concern of the emerging political elites. But it equally had important consequences in the development of mathematics and the sciences in general.[19]

6. The twentieth century

The developments of early twentieth century mathematics in South and Central America cannot be appreciated without an overall perception of the economic and political forces at the turn of the century. There is an enormous need for identifying documents and for the preservation of extant sources in the countries, as well as in smaller administrative units, the states and provinces.

When we look into the scenario at the turn of the century, we see significant efforts of Germany to establish areas of influence in the southern part of South America. What Lewis Pyenson called the 'German Cultural Imperialism' is clearly illustrated by looking into the development of the exact sciences in Argentina and in Chile. A major step to consolidate this influence was the efforts for the development of the *Observatorio Astronómico de La Plata* [Astronomical Observatory of La Plata]. Richard Gans (1890–1954), a physicist who emigrated to Argentina in 1912, had an important role in the development of Argentinian science, in particular of mathematics.[20]

Unrelated to these efforts, the Spanish mathematician Julio Rey Pastor (1888–1962) visited Argentina in 1917 and decided to stay there. He was responsible for introducing modern approaches to the university curriculum. His books were widely used in several universities in Latin America.[21] In addition to making important contributions to mathematics, mainly to projective geometry, Rey Pastor is particularly noteworthy for his contributions to the history of mathematics, especially of Iberian mathematics in the sixteenth century. Rey Pastor also marked new directions in historiography by drawing attention to the mathematical achievements that made possible the great age of navigation.[22] Although Rey Pastor remained in Argentina for several years, he frequently returned to Spain, where he was responsible for considerable developments in mathematics.[23]

A disciple of Rey Pastor in Argentina, José Babini (1897–1983), became one of the most distinguished historians of science and mathematics in Latin America. His career as a driving force of mathematics in Argentina is significant. He was a founder of the *Unión Matemática Argentina,* and in 1920 he became Professor at the *Universidad Nacional del Litoral.* Besides having written many books and articles in non-specialized periodicals, Babini increased the scholarship on the Jewish medieval contributions to mathematics. He wrote, in collaboration with Julio Rey Pastor, a major work on the history of mathematics.[24] Unfortunately, this interesting book has not as yet been translated into other languages, and it is even more regrettable that it is not easily available.

In the 1930s, some European mathematicians emigrated to Argentina. Among them was the distinguished Italian mathematician Beppo Levi (1875–1961), who established an important research center in Rosario and founded a well known journal, *Mathematica Notae.* Recognized for his seminal theorem on the theory of integral, Beppo Levi devoted much of his research to the history of mathematics.[25]

A disciple of Rey Pastor who had students from several countries of the region is Luis Alberto Santaló. Born in Spain in 1911, Santaló had studied with W. Blaschke in Germany and already had an international reputation when he emigrated to Argentina during the Spanish Civil War. Santaló became world renowned as one of the founders of modern integral geometry and became a most influential scholar in mathematics, mathematics education and the history of mathematics in all of Latin America. Besides his relevant contributions to integral geometry, Santaló investigated the history of geometric probability and published relevant studies on Buffon.

In neighboring Uruguay, an important tradition of mathematical research was established early in the twentieth century. A representative of this movement, particularly devoted to the history of mathematics, was Eduardo García de Zuñiga (1867–1951). Garcia de Zuñiga succeeded in creating a most important library concentrating in the history of mathematics at the *Facultad de Ingenería de la Universidad de la Republica,* in Montevideo. His research was mainly in Greek Mathematics and his collected works have been recently published.[26] In mid-century, Rafael Laguardia and José Luiz Massera were responsible for the creation of a most distinguished research group in the stability theory of differential equations in the *Instituto de Matemática y Estadistica de la Facultad de Ingeneria de la Universidad de la Republica,* in Montevideo. This research group, of world reknown, attracted young mathematicians from all of Latin America and abroad. The military dictatorship established in Uruguay in 1971 saw in the declared political position of José Luiz Massera and Rafael Laguardia a reason simply to close the excellent mathematical library of the university and to interrupt all mathematical research in the country. Of course, Uruguayan mathematicians went to several countries where they were responsible for stimulating the creation of research groups. Massera spent all the period of the military dictatorship in jail and afterwards abandoned mathematical research to pursue a political career. Rafael Laguardia died in Montevideo during the political repression. More than in any other country under military dictatorship in South America in the sixties, Uruguay is an example of how a flourishing scientific community can be immobilized by a governmental decision.[27]

In Brazil, positivistic ideas were building up during the Empire and culminated with the proclamation of the Republic in 1889. The *Escola Militar* was later transformed into the *Escola Politécnica* and was the only institution granting degrees in mathematics. Most of the theses submitted reveal, in the choice of the themes, the bibliography and the style, a strong influence of Comptean ideas.[28] But in the beginning of the century, a number of young Brazilian mathematicians were absorbing the most recent progress of mathematics in Europe. Among them were Otto de Alencar, Manuel Amoroso Costa, Teodoro Augusto Ramos and Lelio I. Gama. In 1916, the *Academia Brasileira de Ciências* was founded. With the inauguration of the *Universidade de São Paulo* in 1934, new possibilities opened for mathematics in Brazil. We might say this was the beginning of systematic research in mathematics in the country. Luigi Fantapiè and Giàcomo Albanese, distinguished Italian mathematicians contracted by the University of São Paulo, respectively in the fields of functional analysis and algebraic geometry, were responsible for initiating an important research school in São Paulo.

7. Contemporary developments: after the end of the Second World War

After World War II, a number of European mathematicians emigrated to Latin America. Particularly important is the presence of Antonio Aniceto Monteiro, from Portugal, in Rio de Janeiro and in Bahia Blanca, Argentina.[29] Also at this time, there began an unprecedented cultural and economic interest of the United States in South America. This resulted in an increasing influence of the United States in the development of mathematics in the region. Before the war, Europe was the main source of visitors and the place chosen by Latin Americans to go abroad for studies. After the war we see a considerable number of European mathematicians going to Latin America in different circumstances. Some were looking for employment, scarce in post-war Europe. This reason, as well as political reasons, brought to South America mathematicians like André Weil, Jean Dieudonné, Alexander Grothendieck and several other Bourbakists to the *Universidade de São Paulo,* Wilhelm Damköhler to Argentina, John M. Horváth to Colombia, Kuo-Tsai Chen of China to Brazil. The Portuguese mathematicians Antonio A. Monteiro, Ruy L. Gomes, Manoel Zaluar Nunes, Antonio Pereira Gomes, and José Morgado went to Brazil and to Argentina. Manuel Balanzat from Spain went to Argentina and Venezuela, among many others. Others employed by the former colonial empires, specifically France and England, which were interested in preserving their presence in what became known as the Third World, were employed by cooperation agencies. Instrumental in these efforts were national organizations such as the British Council, ORSTOM and the *Coopération française,* and, on an international basis, UNESCO and the Organization of American States.

The growth of American influence is evident. The Organization of American States was instrumental in favoring United States influence and exchanges. Fellowships for doctoral studies became relatively easy to obtain. The United States became the main destination of a generation of young students pursuing their doctorates abroad. The creation of the National Science Foundation set up the model to be soon followed by practically every Latin American country through the CONICYTs, CONACYTs and the like. An effort of the AAAS to cooperate with homologous organizations in Latin America is also noted. This is a period in which we see the emergence of a scientific policy in the Latin American countries, with important consequences for the development of mathematics.[30]

In the late forties, the need of an overview of mathematical research going on in Latin America became clear. In 1951, the *Centro de Cooperación Científica de la UNESCO para América Latina,* located in Montevideo, Uruguay, organized a meeting in Punta del Este, to report on mathematical research going on in the region. The proceedings of this meeting tell us about research conducted in Latin America by national mathematicians and by foreigners, both immigrants and visitors.[31]

One of these mathematicians was Leopoldo Nachbin, of the *Universidade do Brasil,* who was doing research on the Theorem of Stone-Weierstrass and launching the study of holomorphy and approximation theory in Brazil. Luis Santaló, then at the *Facultad de Ciencias de la Plata,* Argentina, reported on his research in integral geometry. Francis D. Murnangham, from Johns Hopkins University, who was in Brazil with the mission of building up a research group in applied mathematics in the *Instituto Tecnológico de Aeronáutica* in São José dos Campos, a model instituion of advanced technology sponsored by the Brazilian Armed Forces and academically modelled upon MIT, reported on his research on matrix theory. Mischa Cotlar, of the *Facultad de Ciencias de Buenos Aires* reported on his research on ergodic theory in cooperation with R. Ricabarra. Mario O. González, of the *Universidad de la Habana,* reported on his research in differential equations. Alberto González Dominguez, of the *Facultad de Ciencias de Buenos Aires,* was working on distributions and analytic functions. Carlos Graeff Fernández, of the *Universidad Nacional Autonoma de México,* was working on Birkhoff's gravitational theory. Godofredo Garcia, of the *Facultad de Matemáticas de Lima,* reported on general relativity; and Rafael Laguardia, of the *Instituto de Matematica y Estadistica de la Facultad de Ingeneria de Montevideo,* reported on his research on Laplace transforms.

Foreigners included Wilhelm Damköhler, a German specialist in the calculus of variations, who emigrated to the *Universidad Nacional de Tucuman,* Argentina, and later went to the *Universidad de Potosí,* Bolívia; Peter Thullen, employed by the International Office of Labor in Paraguay, who was working on the theory of several complex variables; and Kurt Fraenz, of the *Facultad de Ciencias de Buenos Aires,* who reported on the mathematical theory of electric circuits. Paul Halmos, who at that time was visiting the *Instituto de Matemática y Estadístico de la Facultad de Ingenieria,* in Montevideo, gave an opening lecture on "Operators in Hilbert Spaces". The introductory lectures given by him in Montevideo circulated widely in Latin America. Augustin Durañona y Vedia, of the *Facultad de Ciencias de La Plata*; Roberto Frucht, of the *Facultad de Matematicas y Fisica de Santa Maria,* Chile; Pedro Pi Calleja, of the *Facultad de Ciencias de La Plata*; and Cesario Villegas Mañe, of the *Facultad de Ingeneria de Montevideo,* were invited discussants.

This "dropping of names" should not be regarded as an account of what was going on in South America in 1950. Many more individuals were active in mathematics. Each one of the mathematicians attending the Symposium deserves a study of his life and work and of his influence in his own country. This should be a priority theme for research in the history of mathematics in Latin America.

Three years later a second symposium was convened, in which Julius Rey Pastor was invited to give an account of research in mathematics going on in Latin America. Although incomplete, this was an attempt to cover the main areas which were being developed.[32]

In the fifties the *Conselho Nacional de Pesquisas/ CNPq* [National Research Council] was created in Brazil. Among its institutes was the *Instituto de Matemática Pura e Aplicada/IMPA* [Institute of Pure and Applied Mathematics]. IMPA organized in 1957 the *Primeiro Colóquio Brasileiro de Matemática* [First Brazilian Mathematics Colloquium] in the tourist resort Poços de Caldas. Approximately 50 mathematicians from Brazil, including researchers and graduate students, were invited. A few foreign lecturers were also invited. The Brazilian Mathematics Colloquium has since then met regularly, every two years for two weeks in July in Poços de Caldas. These research schools were followed by the *Escuelas Latinoamericanas de Matemática/ELAM* [Latin American Schools of Mathematics], held in different places each time. An overview of the current development of mathematics in Latin America is given by the study of the *Colóquios Brasileiros de Matemática* and of the *Escuelas Latinoamericanas de Matemática,* looking for participants, their geographical distribution, themes treated, and so on. I believe this is a research project needed in Latin America.

To analyze contemporary history is a difficult task, since we have to refer to processes still going on and we risk stumbling into personal and political sensibilities. Several academicians were active in the period when military regimes took control of the governments of the countries which were showing the greatest vitality in mathematical research in South America. The military coups, which occurred sequentially in the four countries which were most active in mathematical research: Brazil 1964, Argentina 1966, Uruguay 1971, Chile 1973, precipitated an important migratory flux of mathematicians, indeed scientists in all areas, among these countries. These movements soon were directed to the few Latin American countries which were able to keep democratic regimes, particularly Mexico and Venezuela. After the redemocratization of Argentina (1983), Brazil (1984), Uruguay (1984) and Chile (1989), some scientists returned and reclaimed their positions. Others were able to maintain their positions during the military regimes and kept these positions after democratization. The dividing line between opponents and sympathizers or even collaborators with the military regimes is very difficult to draw. Obviously, personal conflicts are still latent. Political issues played a role in the development of mathematics in Latin America and continue to do so. The intense migration of mathematicians, the same as in other areas, in the countries of Latin America, due to economic and political reasons, is a theme that deserves research.

Even the participation in the UNESCO symposium reveals a power play in the academic scenario. A number of mathematicians active in several countries of the region were not invited to the meeting. Looking into *Mathematical Reviews* we might be able to see that the invitations were selective. Although it is very difficult to identify the reasons behind these invitations, this is also a theme which deserves research.

An international symposium on "*La Migración de Científicos en los Países del Cono Sur: determinaciones económicas y políticas* [The Migration of Scientists in the Countries of the Southern Cone: Economical and Political Determinants]", was convened by the FEPAI: *Fundación para el Estudio del Pensamiento Argentino,* in July 1986. The interventions and debates revealed open wounds which remain from the period of military dictatorship. Although unpleasant and somewhat painful, it is important to look into this period and its consequences while some of the protagonists are still alive. I deem this as an important and needed research project.

8. Mathematics Education

An area of research which is growing very fast in Latin America is mathematics education. Until the end of World War II there was practically no coordination and not even any interchange about progress and difficulties in the teaching of mathematics at the various levels of education. A link between all the educational systems was the result of the influence of colonial times and the use of the colonial languages. Thus, the entire bloc of Spanish speaking countries would show similarities, and Brazil would show a slight difference. These links should be a positive factor for mathematical education in the Americas. But sometimes they brought charges of interference in the educational systems of the individual countries.

In the fifties we see the begining of efforts to improve the educational systems of the countries. Examples of these efforts are the several waves of the modern mathematics movement. A decisive move was the creation of the Inter-American Committee of Mathematics Education (IACME/CIAEM) by initiative of Marshall H. Stone (1903–1989). The committees, in addition to studies and researches, promotes the Inter-American Conferences of Mathematics Education, which take place every four years.[33] The international contacts of Latin American mathematics educators and colleagues from different parts of the world were intensified during the so-called modern mathematics movement.[34]

9. Increasing interest in the History of Mathematics

Although we recognize some interest in the history of mathematics since colonial times, in the last decades it has become a growing area of academic interest throughout South America. The founding of the *Sociedad Latinoamericana de Historia de las Ciencias y la Tecnologia,* in 1983, stimulated the organization of national societies devoted to the history of science, which include sections of the history of mathematics. Young mathematicians have recently obtained doctorates in history of mathematics in both Europe and in North America, which is a hopeful sign of maturity and continuing professionalization of the subject throughout Latin America. Among the research areas we see both European mathematics and Latin American progress in the mathematical sciences. A growing interest in contemporary mathematics in Latin America can also be noticed.

10. Additional References

I have tried to give an overall, although very incomplete, account of a vast subject. Mexico has advanced far in the very attractive area of research in the History of Latin American mathematics. In Central and South America the field is just beginning. Practically all the names mentioned in this paper, and several others not mentioned, are open to investigation. A few results of research are very partial and disperse. The project of an *Enciclopedia de las Ciencias y las Técnicas Iberoamericanas,* proposed by Mariano

Hormigón, will certainly put together more information on Central and South America.

In addition to references given as notes, I suggest:

Arboleda, Luis Carlos: 1985, "Dificultades estructurales de la profesionalización de las matemáticas en Colombia," in Peset, José Luiz (ed.), *La Ciencia Moderna y el Nuevo Mundo*, CSIC/SLAHCT, Madrid, pp. 27–38.

Azevedo, Fernando de (org.): 1994, *As Ciências no Brasil*, Editora UFRJ, Rio de Janeiro, (ed. original 1955).

Babini, José: 1992, *Páginas para una Autobiografia*, Prólogo y notas de Nicolás Babini, Asociación Biblioteca José Babini/Ediciones Letra Buena, Buenos Aires.

D'Ambrosio, Ubiratan: 1994, "O Seminário Matemático e Físico da Universidade de São Paulo. Uma Tentativa de Institucionalização na Década de Trinta," in *Temas e Debates*, ano VII, no. 4, pp. 20–27.

González Orellana, Carlos: 1985, *Historia de la Educación en Guatemala*, Editorial Universitaria, Guatemala.

Orellana C., Maurício: 1991, *Resumen de las Clases del Curso de Historia de la Matemática en América Latina y Venezuela*, Universidad Pedagógica Experimental Libertador, Caracas, (mimeographed).

Ramos, Gerardo: *El desarrollo de la Matemática en el Perú in Algunos aportes para el estudio de la historia de la ciencia en el Peru*, editor Ernesto Yepes, CONCYTEC, Lima, s/d, pp.15–19.

Santaló, Luis A.: 1970, "La Matemática en la Facultad de Ciencias Exactas y Naturales de la Universidad de Buenos Aires en el período 1865–1930," in *Bol. de la Acad. Nacional de Ciencias, Cordoba*, Tomo 48, pp. 255–273.

——: 1972, "Evolución de las Ciencias en la Republica Argentina 1923–1972," in *Tomo I: Matemática , Sociedad Científica Argentina*, Buenos Aires.

Silva, Clóvis Pereira da: 1999, *A Matemática no Brasil. Uma História do seu Desenvolvimento*, 2ª edição, Editora Unisinos, São Leopoldo, RS.

Trabulse, Elías: 1994, *Ciencia y Tecnologia en el Nuevo Mundo*, El Colegio de México/Fondo de Cultura Economica, Mexico.

Zuñiga, Angel Ruiz (ed.): 1995, *Historia de las Matemáticas en Costa Rica*, Editorial de la Universidad de Costa Rica, San José.

Endnotes

* An elaboration of the talk of the same title given at the ICME-8 Satellite Meeting of the HPM, in Braga, Portugal, 24–30 July 1996.

[1] This is well documented in several cultural forms, mainly religion, the arts and language. This is no less true in mathematical ideas. Among much research going on, those being conducted by Chateaubriand Nunes Amancio among the Kaingang and Pedro Paulo Scandiuzzi in the Xingu region, in Brazil, and by Samuel Lopez Bello, in Peru, point to this.

[2] This broader look, suggested by new historical scholarship, came under severe attack, in what became known as the Science Wars. See the special issue devoted to the theme in *Social Text*, 46–47, Spring/Summer 1996. Very revealing of the current status of these "wars" are the paper of Alan Sokal, and the attacks on Afrocentrism, the warnings against a "new dark age of irrationalism" and other controversial disputes going on in the academic world. All come as a consequence of challenging current epistemological order, what might be seen as an intellectual fundamentalism.

[3] See my papers entitled "Ethnomathematics, History of Mathematics and the Basin Metaphor," in *Histoire et Epistemologie dans l'Education Mathématique/History and Epistemology in Mathematics Education*, (Actes de la Première Université d'Eté Europeenne, Montpellier, 19–23 juillet 1993), eds. F. Lalande, F. Jaboeuf, Y. Nouaze, IREM, Montpellier, 1995, pp. 571–580; and "Ethnomathematics: An Explanation," in *Vita Mathematica. Historical Research and Integration with Teaching*, ed. Ronald Calinger, The Mathematical Association of America, Washington, DC, 1996, pp. 245–250.

[4] See my paper "Mathematics and Literature" in *Essays in Humanistic Mathematics*, ed. Alvin M. White, The Mathematical Association of America, Washington, DC, 1993; pp.35–47.

[5] The reader may consult, for a brief introduction, my paper "Science and Technology in Latin America During the Discovery" in *Impact of Science on Society*, vol. 27, no. 3, 1977, pp. 267–274.

[6] A good survey of pre-Columbian Mathematics is the book *Native American Mathematics*, ed. Michael Closs, University of Texas Press, Austin, 1986.

[7] Atlas, Madrid, 1964.

[8] A basic reference for these books is Louis C. Karpinsky: 1940, *Bibliography of Mathematical Works Printed in America Through 1850*, The University of Michigan Press, Ann Arbor.

[9] See the paper by Juan Casanovas S.J. and Philip C. Keenan, "The Observation of Comets by Valentine Stansel, a seventeenth century missionary in Brasil," *Archivum Historicum Societaatis Iesu*, LXII, 1993, pp. 319–330.

[10] For details see the paper by Carlos Ziller Camenietzki, "O Cometa, o Pregador e o Cientista. Antônio Vieira e Valentin Stancel observam o céu da Bahia no século XVII," *1° Seminário Nacional de História da Ciência e da Tecnologia*, Ouro Preto, 1995.

[11] A recent book on the Marquis of Pombal brings new elements to understand science and mathematics in this period. See Kenneth Maxwell, *Pombal, paradox of the enlightment*, Cambridge University Press, Cambridge, 1995.

[12] See the book by José Sala Catalá, *Ciencia y Técnica en la Metropolización de América*, Theatrum Machinae, Madrid, 1994. The author studies the problems posed by mining and the urban development of three cities: Mexico, Lima and Recife.

[13] "Themes of Rational Philosophy and of Mechanical Philosophy of the senses, according to the uses of Physics; and of other physical-philosophical topics following the thought of moderns to be defended in this Royal and Pontifical Guatemalan Academy of San Carlos. . . "

[14] An analysis of this period is given in Victor S. Albis G., "Vicisitudes del Postulado Euclideo en Colombia," *Revista de la Academia Colombiana de Ciencias, Exactas, Físicas y Naturales*, vol.XXI, no. 80, Julio de 1997; pp. 281–293.

[15] Clovis Pereira da Silva analyses these theses in his book on the history of Mathematics in Brazil, *A Matemática no Brasil. Uma história de seu desenvolvimento*, 2ª edição, Editora Unisinos, São Leopoldo, RS, 1999.

[16] *Comptes-Rendus de l'Académie des Sciences de Paris*, tomes XL, p. 1310, and XLI, p. 100 and *Proceedings of the Royal Society*, 1856, pp.146–149. It is quite interesting to read the referee's reports and the reaction of Gomes de Souza to the fact that Liouville did not give an appraisal of the paper, according to Gomes de Souza, because of "la petite jalousie". A thorough study of the scientific works of J. Gomes de Souza is still due.

[17] See the paper by Juan Carlos Nicolau, "La Sociedad de Ciencias Fisicas y Matematicas de Buenos Aires (1822–1824)," *Saber y Tiempo*, 2(1996), vol. 2, pp. 149–160.

[18] I did not have personal access to these papers and to the records of Picarte's presence in Paris.

[19] For details, see any book on the History of Latin America. Specifically dealing with mathematics, see the doctoral dissertation of Circe Mary Silva da Silva, *Positivismus und Mathematikunterrichte: Portugiesche und franzsische Einflüsse in Brasilien im 19. Jahrundert*, IDM, Bielefeld, 1991.

[20] See the book of Lewis Pyenson, *Cultural Imperialism and Exact Sciences. German Expansion Overseas 1900–1930*, Peter Lang, New York, 1985.

[21] Probably, the book in which we can see the curricular innovations is Julio Rey Pastor, *Elementos de Analisis Algebraico, 3ª Edicion Argentina Corregida y Ampliada*, Buenos Aires, 1948. Curiously, Rey Pastor did not send some of his textbooks to a publishing house. He would take care himself of the production and distribution of the books.

[22] A representative of his contribution in this area is the book *La Ciencia y la Técnica en el Descubrimiento de América*, Espasa-Calpe Argentina S.A., Buenos Aires, 1942.

[23] For several studies on the life and work of Julio Rey Pastor, see *Estudios sobre Julio Rey Pastor (1888–1962)*, Luis Español Gonzalez, ed., Instituto de Estudios Riojanos, Logroño, 1990.

[24] José Babini and Julio Rey Pastor, *Historia de la Matemática:* vol. 1, De la Antiguedad a la Baja Edad Media. Prefacio de J. Vernet, Gedisa, Barcelona, 1984; vol. 2, Del Renascimiento a la actualidad, Gedisa, Barcelona, 1985.

[25] His book, *Leyendo a Euclides*, Editorial Rosario S.A., Rosario, 1947, is a critical analysis of the general organization of the *Elements*.

[26] García de Zuñiga, E., *Lecciones de Historia de las Matemáticas* (ed. Mario H. Otero), Facultad de Humanidades y Ciencias de la Educación, Montevideo, 1992.

[27] An account of Mathematics in Uruguay is given in the article by Mario H. Otero, "Las Matematicas Uruguayas y Rey Pastor" in *Estudios sobre Rey Pastor (1888–1962)* cited in note 23.

[28] Clovis Pereira da Silva analyses, in the book cited in note 7, 24 of these theses.

[29] The influence of Antonio Aniceto Monteiro in both Brazil and Argentina was a central theme in a Luso-Brazilian meeting which took place in March 23–26, 1997 in Águas de São Pedro, Brazil. See *Anais-Actas do 2° Encontro Luso-Brasileiro de História da Matemática e 2° Seminário Nacional de História da Matemática*, ed. Sergio R. Nobre, UNESP, Rio Claro, 1997.

[30] Those interested in this theme may see the book *Historia social de las ciencias en America Latina*, coordinador Juan José Saldaña, Coordinación de la Investigación Científica/UNAM, Mexico, 1996, particularly the papers by Regis Cabral, "El Desarrollo de las Ciencias Exactas en América Latina y la Política International" (pp. 493–510), and of Francisco Sagasti, "Evolución y Perspectivas de la Política Científica y Tecnológica en América Latina" (pp. 511–533).

[31] The proceedings were published as *Symposium sobre Algunos problemas matemáticos que se están estudiando en Latino America*, Centro de Cooperación Cientifica de la UNESCO para America Latina, Montevideo, 1951.

[32] Julio Rey Pastor, "La matemática moderna en Latino América," *Segundo Symposium sobre Algunos problemas matemáticos que se están estudiando en Latino América, Villavicenzio-Mendoza, 21–25 Julio 1954*, UNESCO, Montevideo, pp. 9–20.

[33] A history of the commission is now available, in a double language (Spanish and English) edition: Hugo Barrantes & Angel Ruíz, Project Coordinator Eduardo Luna: *La História del Comité Interamericano de Educación Matemática/The History of the Inter-American Committee on Mathematics Education*, Academia Colombiana de Ciencias Exactas, Físicas y Naturales, Santa Fe de Gogotà, Colombia, 1998.

[34] An account of the influences of the movement can be seen in the doctoral thesis of Beatriz Silva D'Ambrosio, *The Dynamics and Consequences of the Modern Mathematics Reform Movement for Brazilian Mathematics Education*, School of Education, Indiana University, Bloomington.

Notes on Contributors

Abraham Arcavi received his PhD in mathematics education from the Weizmann Institute of Science in Israel. He presently holds a position as Senior Scientist in the Science Teaching Department of the Weizmann Institute. He is involved in curriculum development projects in mathematics, in in-service teacher education, and in research on mathematical cognition. His main research interests are in the areas of mathematics learning and teaching at the secondary and college levels, and history and philosophy of mathematics.

Evelyne Barbin is Maître de Conférences in Epistemology and History of Sciences of the IUFM (Institut Universitaire de Formation des Maîtres) of the Academy of Creteil. Her historical researches concern mainly mathematical proof in history, and history of mathematics in the seventeenth century. She is editor of many books concerning the history of mathematics, including *History of Mathematics, Histories of Problems* (1997). She is president of the inter-IREM National Commission on Epistemology and History of Mathematics, a body which has worked for twenty years, in the IREMs (Instituts de Recherche sur l'Enseignement des Mathématiques) of France, on integrating history into the teaching of mathematics.

Janet Heine Barnett, Associate Professor of Mathematics at the University of Southern Colorado, holds a doctorate in set theory from the University of Colorado. Her scholarly interests include the use of history as a mechanism for promoting mathematical understanding, and as a vehicle for promoting teacher reflection on pedagogical issues. Current projects include the mathematical history of Paris (jointly with G. Heine), and a historical study of the role of proof and intuition in the development of mathematical concepts and the implication of these developments for mathematics teaching. Her article "A Brief History of Algorithms in Mathematics" appears in the NCTM 1998 Yearbook.

Maxim Bruckheimer co-authored many books for students and teachers in the sixties. After teaching at the City University in London, he helped found the Open University in 1969. As Dean of Mathematics, he was responsible for the first multimedia courses which established the University's reputation. In 1974 he moved to the Weizmann Institute of Science in Israel, where he became head of the Department of Science Teaching. A major interest is the history of mathematics in mathematics education. He and Abraham Arcavi have produced historical worksheet materials for teachers and, with others, illustrated historical activities for use in the junior high school classroom.

Jaime Carvalho e Silva is Associate Professor of Mathematics in the Mathematics Department of the University of Coimbra (Portugal). His main area of research is partial differential equations and he has strong interest in the history of mathematics and mathematics education. He has written several papers on different aspects of the history of mathematics in Portugal and is the author or co-author of eight textbooks for secondary school and university. He has coordinated the mathematics programs for secondary education in Portugal since 1995 and is the author of the most visited mathematics internet site in Portuguese.

Ubiratan D'Ambrosio is Professor Emeritus of Mathematics at the State University of Campinas, UNICAMP, in São Paulo, Brazil. He has been involved for many years in the history of mathematics and its use in teaching. In particular, he was instrumental in founding the discipline of ethnomathematics and has been one of its leading expositors. He served his university as the Director of the Institute of Mathematics, Statistics and Computer Science and as Pro-Rector for University Development. He is currently president of the Brazilian Society of the History of Mathematics and also president of the International Study Group on Ethnomathematics.

Jean-Luc Dorier is currently Professor in a University Teacher Training Institute (I.U.F.M.) in Lyon and head of a research team in the field of didactics of mathematics, in Grenoble. His work focuses on the teaching and learning of linear algebra at university level. His interest in this didactical question led him to an extensive study of the history of linear algebra, and he developed a research program which includes theoretical issues on the connection between research in history and in didactics of mathematics. His other fields of interest include the teaching of vectors (in mathematics and in physics, as well as their use in technological fields), the modelling issue in mathematics, especially in economics oriented curricula, and the teaching of arithmetic at college level.

António Leal Duarte is Assistant Professor of Mathematics in the Mathematics Department of the University of Coimbra (Portugal). His main area of research is matrix theory, namely inverse problems for matrices with a given graph. He has strong interest in the history of mathematics and graph theory and has written articles on the history of mathematics in Portugal. Reading and book collecting, namely rare mathematical books, are his favorite hobbies

Fulvia Furinghetti is Associate Professor of "Elementary mathematics from an advanced standpoint" in the Department of Mathematics of the University of Genoa (Italy). She is the co-ordinator of a group of mathematics teachers and researchers working in the field of mathematics education. Her educational research concerns the integration of history in mathematics teaching, approaches to proof, mathematical beliefs, teacher education and training. In the field of the history of mathematics she studies mathematics journals of the past, in particular, journals for teachers or students which treat problems of teaching and learning mathematics.

Uwe Gellert has been lecturing since 1993 at the Free University of Berlin (Germany). He has worked on ethnomathematics and the history of mathematics, focusing on social aspects of mathematics and mathematics education. His PhD thesis is a socio-cultural analysis of the beliefs on which elementary teachers' professional conceptions are based. He currently is Assistant Professor at the Department of Education and Psychology at the Free University of Berlin.

Livia Giacardi teaches mathematics at the University of Turin and does research in the history of mathematics. She has published books and papers in Italy and abroad dealing with the mathematics of ancient civilizations, the history of Leibnizian infinitesimal calculus, and nineteenth century Italian geometrical studies (research and teaching). She is preparing, as general editor, a CD-ROM (conceived as a hypertext) reproducing, with critical notes, the forty notebooks written for the university courses by Corrado Segre (1863–1924), the founder of the Italian school of Algebraic Geometry. She is a member of the boards of various Italian scientific associations.

Lucia Grugnetti is Associate Professor of Foundations of Mathematics at the Mathematics Department in the University of Parma, researcher in mathematics education, history of mathematics, history of mathematics in mathematics education, editor of the bilingual *Journal L'Educazione Matematica*, Scientific Counsellor of one of the research groups in mathematics education in Parma, one of the two International Organizers of RMT (Rally Mathématique Transalpin), and member of the Committee on Mathematics Education of EMS (the European Mathematical Society). She has been the President of CIEAEM (Commission Internationale pour l'étude et l'amélioration de l'enseignement mathématique) in the period 1993–1997.

Georges Guérette was born in a small town in rural Quebec, Canada. He has taught mathematics and computer science at the high school level in the region of Sudbury, Ontario, Canada, for over 30 years. He is especially interested in finding new ways to get students motivated in learning mathematics and in promoting students' participation in mathematics competitions. He is also actively involved in the implementation of the new curriculum in mathematics in the province of Ontario.

Torkil Heiede was born in 1931. He was educated at Copenhagen University and was from 1963 until he retired in 1997 a professor of mathematics at the Royal Danish School of Educational Sutdies. He is still active speaking and writing on diverse mathematical subjects and on their history, and also on the place of history in mathematics education at all levels. Since 1966 he has been a co-editor of the Scandinavian journal *Normat*; he is a founder member of the Con Amore Problem Group, and he has now for some years been secretary of the Danish Society for the History of the Exact Sciences. He contributed to the MAA Notes No 40 *Vita Mathematica: Historical Research and Integration With Teaching* (1996) with a paper called "History of Mathematics and the Teacher."

George W. Heine is the proprietor of Math and Maps, a mathematical and statistical consulting enterprise. He holds a doctorate in Applied Probability from the University of Colorado (Denver), where his main research interests were simulated annealing and Monte Carlo methods. Current interests include a study of the mathematical history of Paris (jointly with J. Barnett), and the relationship between philosophy, mathematics, and religion in the Middle Ages, especially in Islam.

Gavin Hitchcock is a senior lecturer in mathematics at the University of Zimbabwe. His research interests lie in general topology and history of mathematics; he is also concerned with the enlivening of mathematics teaching by means of posters, models, games, drama, and history. Recent publications include: "Dramatizing the Birth and Development of Mathematical Concepts: Two Dialogues," in *Vita Mathematica: Historical Research and Integration in Teaching*, ed. Ronald Calinger (MAA, 1996): 27–41; and "Entertaining Strangers: A Dialogue Between Galileo and Descartes", *Comparative Criticism* 20 (Cambridge University Press, 1998): 63–85.

Wann-Sheng Horng is Professor of Mathematics at the National Taiwan Normal University, Taipei, Taiwan. He studied at the City University of New York in the late 1980s to become a professional historian of mathematics. Since then he has also been paying attention to the issues of history and pedagogy of mathematics. His publications cover basically two fields: (1) history of Chinese mathematics, especially its socio-cultural aspects in the nineteenth century; (2) integration of history of mathematics, especially a contrast of different mathematical cultures, into mathematical teaching.

Ian Isaacs commenced his teaching career in the education system in Jamaica, West Indies. Later he taught for several years in the graduate and undergraduate programs at the School of Education, The University of the West Indies, Jamaica campus. He was for several years the chief mathematics examiner for the Caribbean Examination Council where he assisted in developing the basic and advanced level mathematics programs for secondary schools. He is one of the co-authors of the text *Joint School Mathematics for the Caribbean.* Ian has served as lecturer and later senior lecturer at the Northern Territory University, Australia for over a decade. His main interest is in problem solving, and he has contributed to several national and international conferences. He has made significant contributions to the Teachers' Mathematics Education programs in the West Indies, Australia and more recently in Western Samoa. He has retired and is now serving in the Australian Volunteers Abroad Program, currently in Tanzania, in Africa.

Anne Michel-Pajus is Professeur de Chaire Supérieure in mathematics in "Classes Préparatoires" (tertiary level) at Lycée Claude Bernard in Paris. She is involved in in-service education at the Université Denis Diderot in Paris, where she works in the IREM (Institut de Recherche sur l'Enseignement des Mathématiques) on the history and epistemology of mathematics, including their use in teaching mathematics. She is one of the authors of *A History of Algorithms* (1999), edited by J.L. Chabert, and one of the co-editors of *Mnemosyne,* a magazine for teachers, published by her IREM.

Vivekanand Mohan-Ram's teaching career spans over forty years, commencing in Guyana, South America, Trinidad and Barbados in the West Indies and now in Darwin, Australia. His experiences as a mathematics teacher/educator were gained at secondary schools, teachers' colleges in Guyana, University of the West Indies, The Caribbean Examination Council, and at Northern Territory University, Australia where he teaches Mathematics Education, Technology, and Psychology. He is one of co-authors of the text *Oxford Mathematics for the Caribbean, Book 1.* One of his main interests is constructing three-dimensional mathematics models to facilitate the teaching and learning of mathematical concepts. He has contributed to several national and international mathematics conferences by conducting mathematics workshops demonstrating the benefits derived from integrating pop-up engineering and geometrical concepts.

Luis Moreno-Armella is a senior researcher in mathematics education in the Mathematics Education Department of the Center for Research and Advanced Studies (Cinvestav) at Mexico City. He received his BSc in the Faculty of Mathematics, at the National University of Colombia, and his MSc and PhD in mathematics in the Mathematics

Department at Cinvestav. His main research interests are the epistemology and history of mathematics from the perspective of mathematics education, and the mediation role of computational tools for the learning of mathematics. His publications include papers in such journals as *Educational Studies in Mathematics*, *International Journal in Mathematics for Science and Technology* and invited chapters in such books as *Exploiting Mental Imagery with Computers in Mathematics Education*, R. Sutherland & J.Mason (eds), Springer-Verlag.

João Filipe Queiró is Associate Professor of Mathematics in the Mathematics Department of the University of Coimbra (Portugal). His main area of research is spectral inequalities for matrices and operators. He is interested in the history in mathematics, in particular in Portugal. He has also dealt with university policies in Portugal and published a book on the subject in 1995.

Luis Radford obtained his PhD from Université Louis Pasteur, Strasbourg, France. He spent several years working in a national training program for teachers of mathematics in Guatemala, his native country. Currently he is a Full Professor at Laurentian University, Ontario, Canada. His domain of research includes the teaching and learning of mathematics, epistemology, semiotics, and the history of mathematics. One of his current research projects deals with the study of constitutive cultural aspects of diverse historical modes of algebraic thinking.

Ann Richards taught mathematics and science in secondary schools for twenty years and spent three years as a mathematics adviser to primary and secondary schools. For the past fourteen years she has worked with primary pre-service teaching students at the (now) Northern Territory University in Darwin, Australia. This work has been primarily in the area of mathematics education but also in science and technology education. Ann's interests are in the areas of visual imagery, technology in mathematics education and ethnomathematics.

Eleanor Robson works in the Oriental Institute, University of Oxford, where she teaches the languages, history, and archaeology of the ancient Near East. She is a Fellow of All Souls College. After a degree in mathematics she began to train as a cuneiformist in order to work on Mesopotamian mathematics. A revised version of her doctoral dissertation was published as *Mesopotamian mathematics, 2100–1600 BC* (Clarendon Press, 1999). She is particularly interested in the social and intellectual environments in which mathematics was developed, recorded, and transmitted in the ancient Middle East, from earliest times to late Antiquity.

Man-Keung Siu obtained his BSc from University of Hong Kong and his PhD from Columbia University writing a thesis on algebraic K-theory under the supervision of Hyman Bass. He is now a professor of mathematics at his undergraduate alma mater. He has published in the fields of algebra, combinatorics, applied probability, mathematics education, and history of mathematics. The Chinese Mathematical Society selected his book *Mathematical Proofs* (1990, in Chinese) as one of the seven outstanding books in mathematical exposition in 1991. He received a CASME Award from the Commonwealth Association of Science and Mathematics Educators in 1981 for the making of a slide show to popularize mathematics.

Frank Swetz is Professor Emeritus of Mathematics and Education at the Pennsylvania State University. His research interests have focused on societal impact on the development of mathematics, its learning and teaching. He is an advocate of using the history of mathematics in its teaching and has given many workshops for teachers on this topic. Among his recent publications are *From Five Fingers to Infinity: A Journey Through the History of Mathematics* (1994) and *Learning Activities from the History of Mathematics* (1994). At present, he is completing a study of the magic square of order three, its cultural and historical significance.

Constantinos Tzanakis (born in 1956) is Associate Professor of Mathematics at the Department of Education of the University of Crete, Greece. He has studied mathematics (Athens University, Greece), and astronomy (Sussex University, UK) and obtained his PhD in theoretical physics (Université Libre de Bruxelles, Belgium). His area of research is mathematical physics (statistical mechanics, relativity theory and geometrical methods in physics) and mathematics and physics education (the relation between history and epistemology of mathematics and physics and their teaching).

Guillermina Waldegg is a Senior Researcher in Science Education at the Educational Research Department of the Centre for Research and Advanced Studies in Mexico City. She received her BSc degree in physics at the National University of Mexico and her MSc and PhD in mathematics education at the Centre for Research and Advanced Studies. Her main research interests are the epistemology and the history of science in relation to science education. She has particularly worked on the concepts of number, infinity, and mathematical and physical continuity. Her publications include research articles and text books for both science teachers and science students. She is the editor of *Educacion matematica*, a leading journal on mathematics education published in Spanish.

Robin Wilson is a Senior Lecturer in Mathematics and a Fellow of Keble College, Oxford University. He is widely known as a popular expositor in mathematics, and has written and edited over twenty books, ranging from combinatorics and graph theory, via Gilbert & Sullivan, to the history of mathematics. In the last of these fields he has coedited *Let Newton be, Möbius and his band,* and *Oxford figures,* and is currently working on a history of combinatorics. His research interests lie in the history of combinatorics and in British mathematics from 1840 to 1950.

Greisy Winicki Landman teaches at Oranim, the School of Education of the Kibbutz Movement in Israel. She works with pre-service primary school teachers and with secondary school teachers. She teaches courses concerning didactics of mathematics teaching, and she is interested in the use of the history of mathematics to improve teaching techniques. Nowadays she is also working with in-service teachers on the introduction of original Hebrew mathematics texts dating from the Middle Ages.